# MANAGEMENT OF DRIP/TRICKLE OR MICRO IRRIGATION

# MANAGEMENT OF DRIP/TRICKLE OR MICRO IRRIGATION

Megh R. Goyal, PhD., P.E., Senior Acquisitions Editor

Apple Academic Press Inc. and

Professor in Agricultural and Biomedical Engineering,
University of Puerto Rico—Mayagüez Campus

Apple Academic Press

TORONTO    NEW JERSEY

© 2013 by
Apple Academic Press Inc.
3333 Mistwell Crescent
Oakville, ON L6L 0A2
Canada

Apple Academic Press Inc.
1613 Beaver Dam Road, Suite # 104
Point Pleasant, NJ 08742
USA

First issued in paperback 2021

*Exclusive worldwide distribution by CRC Press, a Taylor & Francis Group*

ISBN 13: 978-1-77463-200-0 (pbk)
ISBN 13: 978-1-926895-12-3 (hbk)

**Library of Congress Control Number: 2012935649**

**Library and Archives Canada Cataloguing in Publication**

Goyal, Megh Raj

Management of drip/trickle or micro irrigation/Megh R. Goyal.

Includes bibliographical references and index.
ISBN 978-1-926895-12-3
1. Microirrigation. I. Title.

S619.T74G69 2012          631.5›87          C2011-908700-6

# Contents

# List of Abbreviations

| | |
|---|---|
| ASABE | American society of agricultural and biological engineers |
| CU | Coefficient of uniformity |
| DU | Distribution uniformity |
| ET | Evapotranspiration |
| $ET_c$ | Crop evapotranspiration |
| gph | Gallons per hour |
| gpm | Gallons per minute |
| $k_c$ | Crop coefficients |
| $K_p$ | Pan coefficient |
| LEPA | Low energy pressure system |
| lps | Liters per second |
| PE | Polyethylene |
| PET | Potential evapotranspiration |
| psi | Pounds per square inch |
| RA | Extraterrestrial radiation |
| RH | Relative humidity |
| RS | Solar radiation |
| SAR | Sodium absorption rate |
| SDI | Subsurface drip irrigation |
| WUE | Water use efficiency |

# Preface

The mission of this compendium is to serve as a text book or a reference manual for graduate and under graduate students of agricultural, biological and civil engineering; horticulture, soil science, and agronomy. I hope that it will be a valuable reference for professionals that work with micro irrigation and water management; for professional training institutes, technical agricultural centers, irrigation centers, Agricultural Extension Service, and other agencies that work with micro irrigation programs.

"MANAGEMENT OF DRIP/TRICKLE OR MICRO IRRIGATION" includes information on principles of micro irrigation, filtration systems, automation, installation, chemigation, chloration, service and maintenance, evaluation of uniformity coefficients, design, ET, and economic viability. It also contains a glossary of terms and a bibliography. The majority of the chapters in this book are based on my research/teaching/extension materials and publications on micro irrigation, at the University of Puerto Rico—Mayagüez Campus. English edition is a translation, revision, and amplified version of "the Spanish electronic-version by Goyal, Megh R., 2005. Manejo de Riego por Goteo. Recinto Universitario de Mayagüez."

This book could not have been written without the valuable cooperation of a group of engineers, agronomists, and students worldwide. At the University of Puerto Rico—Mayagüez Campus, I am grateful to: Álvaro Acosta, Carmen I. Álamo, Elvin Caraballo, Octavio Colberg, Manuel Crespo, Guillermo Fornaris, Eladio A. González, Milton Martínez, José V. Pagan, Allan L. Phillips, Antonio Poventud, Nelson I. Rojas, Carmen L. Santiago, Víctor A. Snyder, Luis E. Rivera Martínez, Víctor Hugo Ramírez Builes, Eric W. Harmsen, and Miguel A. Lugo López [QEPD]. The author also thanks executive officers at University of Puerto Rico—Mayagüez Campus, for the opportunity to initiate micro irrigation program in 1979 under my supervision. I also thank my students at University of Puerto Rico—Mayagüez and at Haryana Agricultural University—Hisar (India), who have enriched my knowledge in micro irrigation and water management.

My special appreciations are due to: Vincent F. Bralts, Michael Boswell, I.Pai Wu, Kenneth H. Solomon, and <toromicroirrigation.com>/<toro.com> [Formerly James Hardie Irrigation]. I request the readers to offer me their constructive suggestions that may help to improve the next edition of this book.

I would like to thank editorial staff, Sandy Jones Sickles and Ashish Kumar—Director at Apple Academic Press, Inc. for making every effort to publish the book when the world community should be aware of the limited water resources not only for irrigation use but also for human consumption.

Finally, my whole hearted thanks to my wife Subhadra and our children Vijay, Neena, and Vinay for the understanding and collaboration of sharing the responsibility, time, and devotion necessary to prepare this manual. With my whole heart and best affection, I dedicate this book to my grandson Jeremaih Kumar and my grand daughter Naraah Nicole. Both of them have motivated me to live longer and to live happier to serve the world community.

— **Megh R. Goyal, PhD., P.E.**
**June 1, 2012**

# Forward

In the world, water resources are abundant. The available fresh water is sufficient even if the world population is increased by 4 times the present population i.e., about 25 billion. The total water present in the earth is about 1.41 billion $Km^3$ of which 97.5% is brackish and only about 2.5% is fresh water. Out of 2.5% of fresh water, 87% is in ice caps or glaciers, in the ground or deep inside the earth. According to Dr. Serageldin, 22 of the world's countries have renewable water supply of less than 1000 cubic meter per person per year. The World Bank estimates that by the year 2025, one person in three in other words 3.25 billion people in 52 countries will live in conditions of water shortage.In the last two centuries (1800–2000) the irrigated area in the world has increased from 8 m ha to 260 m ha to produce the required food for the growing population. At the same time the demand of water for drinking and industries have increased tremendously. The amount of water used for agriculture, drinking and industries in developed countries are 50% in each and in developing countries it is 90% and 10% respectively. The average quantity of water used for agriculture and other purposes in the world are about 69% and 31% respectively.Water scarcity is now the single threat to global food production. To overcome the problem, there is a compulsion to use the water efficiently and at the same time increase the productivity from unit area. It will involve spreading the whole spectrum of water thrifty technologies that enable farmers to get more crops per drop of water. This can be achieved only by introducing Drip / trickle /Micro irrigation in large scale throughout the world.

Drip irrigation is a method of irrigation with high frequency application of water in and around the root zone of plants (crop) and consists of a network of pipes with suitable emitting devices. It is suitable for all crops except rice especially for widely spaced horticultural crops. It can be extended to wastelands, hilly areas, coastal sandy belts, water scarcity areas, semi arid zones and well irrigated lands.By using drip irrigation, the water saving compared to conventional surface irrigation is about 40 – 60% and the yield can be increased up to 100%. The overall irrigation efficiency in surface irrigation, sprinkler irrigation and drip irrigation are 30 – 40%, 60 – 70% and 85-95% respectively. Apart from this, saving of labor and fertilizer used and less weed growth are other advantages.The studies conducted and information gathered from various farmers in India has revealed that drip irrigation is technically feasible, economically viable and socially acceptable. Since the allotment of water is going to be reduced for agriculture, there is a compulsion to change the irrigation method to provide more area under irrigation and to increase the required food for the growing population.

I personally reviewed this manual. Professor Megh R. Goyal is a reputed agricultural engineer in the world and has wide knowledge and experience in Soil and Water Conservation Engineering particularly drip irrigation. He has contacted / consulted many experts who are involved in the subject matter to bring the experience and knowledge about drip irrigation in this book. He has also given many figures illustrations and tables to understand the subject. I congratulate the author for writing this valuable book and the information provided in this book will go a long way in bringing large area under drip irrigation in the world especially in water scarcity countries. On behalf of Indian scientists on micro irrigation, I am indebted to Apple Academic Press for undertaking this project.

**Professor (Dr.) R. K. Sivanappan**

Former Dean- cum- Professor of College of Agricultural Engineering and Founding Director of Water Technology Centre at Tamil Nadu Agricultural University, Coimbatore, India. Worldwide consultant on Micro Irrigation. Author of about 750 scientific papers, 25 books, 50 reports on water management and drip irrigation. Father of Drip Irrigation in India as mentioned by Mrs. Sandra Postel in her book "Pillar of sand – can the irrigation miracle last by W.W.Norton and company – New York". Recipient of Honorary Ph..D. degree by  Linkoping University, Sweden.

August 27, 2010.
Coimbatore - India

# Forward

With only a small portion of cultivated area under irrigation and with the scope to the additional area which can be brought under irrigation, it is clear that the most critical input for agriculture today is water. It is accordingly a matter of highest importance that all available supplies of water should be used intelligently to the best possible advantage. Recent research around the world has shown that the yields per unit quantity of water can be increased if the fields are properly leveled, the water requirements of the crops as well as the characteristics of the soil are known, and the correct methods of irrigation are followed. Very significant gains can also be made if the cropping patterns are changed so as to minimize storage during the hot summer months when evaporation losses are highest, if seepage losses during conveyance are reduced, and if water is applied at the critical times when it is most useful for plant growth.

Irrigation is mentioned in the Holy Bible and in the old documents of Syria, Persia, India, China, Java and Italy. The importance of irrigation in our times has been defined appropriately by N.D Gulati: "In many countries irrigation is an old art, as much as the civilization, but for humanity it is a science, the one to survive". The need for additional food for the world's population has spurred rapid development of irrigated land throughout the world. Vitally important in arid regions, irrigation is also an important improvement in many circumstances in humid regions. Unfortunately, often less than half the water applied is used by the crop – irrigation water may be lost through runoff, which may also cause damaging soil erosion, deep percolation beyond that required for leaching to maintain a favorable salt balance. New irrigation systems, design and selection techniques are continually being developed and examined in an effort to obtain the highest practically attainable efficiency of water application.

The main objective of irrigation is to provide plants with sufficient water to prevent stress that may reduce the yield. The frequency and quantity of water depends upon local climatic conditions, crop and stage of growth and soil-moisture- plant characteristics. Need for irrigation can be determined in several ways that do not require knowledge of evapotranspiration [ET] rates. One way is to observe crop indicators such as change of color or leaf angle, but this information may appear too late to avoid reduction in the crop yield or quality. Other similar methods of scheduling include determination of the plant water stress, soil moisture status or soil water potential. Methods of estimating crop water requirements using ET and combined with soil characteristics have the advantage of not only being useful in determining when to irrigate, but also enables us to know the quantity of water needed. ET estimates have not been made for the developing countries though basic information on weather data is available. This has contributed to one of the existing problems that the vegetable crops are over irrigated and tree crops are under irrigated.

Water supply in the world is dwindling because of luxury use of under ground sources; competition for domestic, municipal and industrial demands; declining water quality; and losses through seepage, runoff, and evaporation. Water rather than land is one of the limiting factors in our goal for self-sufficiency in agriculture. Intelligent use of water will avoid problem of sea water entering into aquifers. Introduction of new irrigation methods has encouraged marginal farmers to adopt these methods without taking into consideration economic benefits of conventional, overhead and drip irrigation systems. What is important is "net in the pocket" under limited available resources. Irrigation of crops in tropics requires appropriately tailored working principles for the effective use of all resources peculiar to the local conditions. Irrigation methods include border-, furrow-, subsurface-, sprinkler-, sprinkler, micro, and drip/trickle and xylem irrigation.

Drip irrigation is an application of water in combination with chemigation within the vicinity of plant root in predetermined quantities at a specified time interval. The application of water is by means of drippers which are located at desired spacing on a lateral line. The emitted water moves due to an unsaturated soil. Thus, favorable conditions of soil moisture in the root zone are maintained. This causes an optimum development of the crop. Drip / micro or trickle irrigation is convenient for vineyards, tree orchards and row crops. The principal limitation is the high initial cost of the system that can be very high for crops with very narrow planting distances. Forage crops cannot be irrigated economically with drip irrigation. Drip irrigation is adaptable for almost all soils. In very fine textured soils, the intensity of water application can cause problems of aeration. In heavy soils, the lateral movement of the water is limited, thus more emitters per plant are needed to wet the desired area. With adequate design, use of pressure compensating drippers and pressure regulating valves, drip irrigation can be adapted to almost any topography. In some areas, drip irrigation is used successfully on steep slopes. In subsurface drip irrigation, laterals with drippers are buried at about 45 cm depth, with an objective to avoid the costs of transportation, installation and dismantling of the system at the end of a crop. When it is located permanently, it does not harm the crop and solve the problem of installation and annual or periodic movement of the laterals. A carefully installed system can last for about 10 years.

The publication of this book is an indication that things are beginning to change, that we are beginning to realize the importance of water conservation to minimize the hunger. It is hoped that the publisher will produce similar materials in other languages.

In providing this resource in micro irrigation, Megh Raj Goyal, as well as the Apple Academic Press, is rendering an important service to the entire world, and above all to the poor. Dr. Goyal, Father of Irrigation Engineering in Puerto Rico has done an unselfish job in the presentation of this manual that is simpler, thorough, complete and useful during the world economical crisis.

**Gajendra Singh, Ph.D.**
President 2010-2012, Indian Society of Agricultural Engineers
Former Vice Chancellor, Doon University, Dehradun, India
Former Deputy Director General (Engineering), Indian Council of Agricultural Research, New Delhi
Former Vice – President/ Dean/ Professor and Chairman, Asian Institute of Technology, Thailand

# Warning/Desclaimer

## USER MUST READ IT CAREFULLY

The goal of this text book is to guide the world community on how to manage the "DRIP/TRICKLE or MICRO IRRIGATION" system efficiently for economical crop production. The reader must be aware that the dedication, commitment, honesty, and sincerity are most important factors in a dynamic manner for a complete success. It is not a one time reading of this manual. Read and follow every time, it is needed. To err is human. However, we must do our best. Always, there is a space for learning new experiences.

The editor, the contributing authors, the publisher and the printer have made every effort to make this book as complete and as accurate as possible. However, there still may be grammatical errors or mistakes in the content or typography. Therefore, the contents in this book should be considered as a general guide and not a complete solution to address any specific situation in irrigation. For example, one size of irrigation pump does not fit all sizes of agricultural land and to all crops.

The editor, the contributing authors, the publisher and the printer shall have neither liability nor responsibility to any person, any organization or entity with respect to any loss or damage caused, or alleged to have caused, directly or indirectly, by information or advice contained in this book. Therefore, the purchaser/reader must assume full responsibility for the use of the book or the information therein.

The mentioning of commercial brands and trade names are only for technical purposes. It does not mean that a particular product is endorsed over to another product or equipment not mentioned. Author, cooperating authors, educational institutions, and the publisher Apple Academic Press Inc. do not have any preference for a particular product.

# Biodata

Megh R. Goyal – Aggarwal was born in India. He received his B.Sc. degree in Agricultural Engineering in 1971 from Punjab Agricultural University, Ludhiana - India; his M.Sc. degree in 1977 and Ph.D. degree in 1979 from the Ohio State University, Columbus; his Master of Divinity degree in 2001 from Puerto Rico Evangelical Seminary, Hato Rey – Puerto Rico.

Since 1971, he has worked as Soil Conservation Inspector; Research Assistant at Haryana Agricultural University and the Ohio State University; and Research Agricultural Engineer at Agricultural Experiment Station of UPRM. At present, he is a Retired Professor in Agricultural and Biomedical Engineering in the College of Engineering of University of Puerto Rico – Mayaguez Campus; and Senior Acquisitions Editor in Agriculture and Biomedical Engineering for Apple Academic Press Inc.. He was first agricultural engineer to receive the professional license in Agricultural Engineering in 1986 from College of Engineers & Surveyors of Puerto Rico. On September 16, 2005, he was proclaimed as "**Father of Irrigation Engineering in Puerto Rico for the 20ᵗʰ Century**" by the ASABE – Puerto Rico Section, for his pioneer work on micro irrigation, evapotranspiration, agroclimatology, and soil & water engineering. During his professional career of 42 years, he has received awards such as: Scientist of the Year, Blue Ribbon Extension Award, Research Paper Award, Nolan Mitchell Young Extension Worker Award, Agricultural Engineer of the Year, Citations by Mayors of Juana Diaz and Ponce, Membership Grand Prize for ASAE Campaign, Felix Castro Rodriguez Academic Excellence, Rashtrya Ratan Award and Bharat Excellence Award and Gold Medal, Domingo Marrero Navarro Prize, Adopted son of Moca, Irrigation Protagonist of UPRM, Man of Drip Irrigation by Mayor of Municipalities of Mayaguez/ Caguas/ Ponce and Senate/ Secretary of Agric. of ELA – Puerto Rico. He has authored more than 200 journal articles and text books on: "Elements to Agroclimatology (Spanish)", "Management of Drip Irrigation (Spanish)", "Biofluid Mechanics of Human Body".

Readers may contact him at: goyalmegh@gmail.com

# Chapter 1

## Methods to Measure Soil Moisture

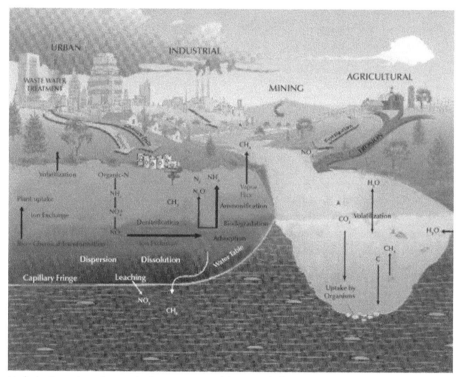

Hydrologic cycle and watershed components of the water balance.

## INTRODUCTION

The soil moisture is one of the factors that affect the crop production. The plants require an adequate amount of soil moisture that may vary according to the crop species and stage of growth or development of a plant [1]. The soil can only store a limited amount of water, and only a part of this storage is available to the plant. For this reason, it is essential to know the soil moisture content per unit mass or per unit soil volume, and its water potential or availability of the soil moisture. This provides valuable information to understand many of the chemical, mechanical, and hydraulic properties of the soil. This information helps to design an efficient irrigation system for supplying water to the soil for the plant use. Different methods have been developed to determine the soil moisture. The use of each of these methods depends mainly on the economical resources of the operator, his knowledge and a desirable degree of precision. This chapter discusses basic principles of soil, water, and plant relations; and the use, operation, advantages, and disadvantages of various methods to determine soil moisture. We hope that this information can enrich the knowledge of the farmers, scientists, and agricultural technicians.

This chapter includes basic relations among soil, water, and plants for an irrigated agriculture; and the balance and distribution of water in the soil horizons. The absorption of water into soil is determined by: Interceptions, runoff, infiltration, hydraulic conductivity, deep percolation, available soil moisture to the plant, and evaporation [4, 5 and 7].

## SOIL, WATER, AND PLANT RELATIONS

The understanding of the relations among soil, water, and plants is essential for irrigated agriculture. In the case of drip irrigation system, it is particularly important because of high initial cost of the installation. Even in areas of high amounts of rainfall, the scarcity of water can limit the development of crops. This may be attributed to an uneven distribution of rainfall, a high runoff or a deep infiltration in soils with low capacity of water retention. Therefore, the importance of irrigation is not limited only to arid or semiarid regions. Amount of water available to the plant is affected by crop water requirements and soil characteristics. The soil moisture at any given time is equal to the amount of rainfall and irrigation received by the soil minus the water loss from soil evaporation, plant transpiration, and deep infiltration. The availability of water to plants also depends on root characteristics and soil properties such as: Soil structure, soil texture, soil porosity, soil hydraulic conductivity, soil field capacity, and permanent wilting percentage. The absorption of water into the soil is determined by: Rainfall characteristics, irrigation, soil cover, the process of interception, runoff, infiltration, redistribution of water and deep percolation, retention, evaporation, and transpiration.

**Interception** is an amount of water that is intercepted by a plant canopy and soil cover. The water loss by interception is expressed as percentage of total rainfall and fluctuates between 15 and 20%. The high values are for abundant vegetation and for low applications of irrigation depth. The intercepted water never reaches the soil because it evaporates directly from the plant surface. In case of drip irrigation, the water is applied directly into the soil near the plant. Therefore the water loss due to interception does not occur.

**Runoff:** Rainfall on fallow (uncultivated) land increases runoff and the probability of high soil erosion. In many soils, with differentiation of horizons, the water infiltrates and soon flows across the contact surface between plowed and unplowed subsoil and eventually flows downward. If the fertilizer has been applied to a soil under these conditions, the runoff water will probably have a high concentration of minerals. This leads to fertilizer losses and increases the risk of environmental contamination.

**Infiltration:** The infiltration rate is an amount of water that penetrates the soil surface in a given interval of time. The infiltration rate is affected by soil properties such as: Apparent soil density, the pore distribution according to soil texture and structure and the stability of the soil aggregates. The duration of infiltration is extremely important. In the beginning of infiltration process through a relatively dry soil, the infiltration rate will be initially high and later would gradually diminish to a constant value that may be close to the soil hydraulic conductivity.

**Hydraulic Conductivity** is a vertical speed of water movement in the soil when the water is subjected to an equal net force due to gravity [This definition requires that the hydraulic potential should be expressed in units of lengths: hydraulic or pressure head]. It is a soil property that can be easily measured in the field or in the laboratory. Gravity is a dominant cause of water movement in two very important situations:

1. Infiltration occurs over a long period of time when surface has been wetted to a significant depth.
2. There is a deep percolation (redistribution) of water from wetted superficial horizons to inferior horizons, after the infiltration through the surface of the soil has ceased. This situation determines the time and soil moisture tension at field capacity of a soil.

**Moisture Redistribution and Deep Percolation:** After the infiltration has ceased, the drainage of water begins through the wetted superior horizons. The water loss is retained and redistributed by the dry inferior horizons or passes through the profile and becomes part of the subterranean water (deep percolation). The velocity of water redistribution or percolation is basically a function of the hydraulic conductivity. In the beginning, the hydraulic conductivity is high because of high soil moisture and high percolation rate. With elapsed time, the water drains from the soil and the hydraulic conductivity and the percolation rate are lowered. The process continues until the hydraulic conductivity is so low to cause almost zero drainage of water.

**Retention of Available Soil Moisture:** The water available for plants is a quantity of soil moisture retained between the field capacity (at a tension of 0.33 bars) and the permanent wilting percentage (at a tension of 15 bars). Traditionally, the available fraction is determined assuming that the field capacity corresponds to soil moisture retention at 0.33 bars of the tension. The retention capacity of available moisture to the plants varies greatly among soils [8]: Being higher in the Vertisols followed by the Inceptisols, Millisols, Ultisols, Alfisols, and finally the Oxisols groups. Sandy soils tend to have a low moisture retention capacity independent of the soil order. The soil capacity to supply water to the plants can be modified through adequate agricultural practices.

**Evaporation:** After a period of rainfall or irrigation, a percentage of irrigation depth is lost due to direct evaporation from the soil surface. The evaporation loss depends on duration, rate and frequency of the irrigation depth or rainfall, and the fraction of soil surface exposure. Light and frequent applications of water generally result in high evaporation losses, including the case of a drip irrigation system. For a fully exposed soil, uniform irrigation (the entire surface is wetted) and potential evapotranspiration of 5 mm/day: the water loss due to direct evaporation was between 25 and 90% of the potential evapotranspiration for irrigation intervals of 20 days and two days, respectively. The amount and type of soil cover can drastically modify evaporation loss under a given regime of applied water. As the crop foliage develops, the resulting shade reduces the water loss by evaporation, allowing a portion of the applied water to be lost through transpiration from the vascular system of the plant. This increases efficiency of water use. The application of organic mulch on the soil surface can be very effective in the control of evaporation. The mulching probably can reduce the losses by evaporation and promoted the infiltration rate.

**Transpiration** is an evaporation of soil moisture from the vascular system of the plant. The volume of transpired water depends on the evapotranspirative demand (potential evapotranspiration), the stage of crop growth, and the amount of available moisture in the root zone. For many crops, it has been found that the transpiration starts to reduce and the plants begin to show water stress once approximately half of the available soil moisture in the root zone has been extracted by the plant. Therefore, the moisture retention capacity of the soil plays an important role in the determination of frequency, duration, and depth of irrigation to satisfy the water needs of the plant.

## PRINCIPLES OF SOIL AND WATER RELATIONS

### Soil Composition

The soil is a complex mass of minerals and organic matter (Figure 1.1), arranged in a structure containing air, water, and solutes. The mineral portion of the soil is formed by the fragmentation and decomposition (interperization) of rocks by physical and chemical processes. It consists mainly of silica and silicates with other minerals such as potassium, calcium, and phosphorus. The organic matter is formed by the activity and accumulation of residues of various species of macroscopic and microscopic organisms. Following are principal benefits of the organic matter:

1. To provide source of essential nutrients to the plants, particularly nitrogen.
2. To improve and to stabilize the soil structure to form stable aggregates that facilitates plowing.
3. To improve aeration and drainage in clayey and silty soils.
4. To improve the field capacity in sandy soil.
5. To improve the retention of available water to the plants, in sandy soil.
6. To act as a cushioning agent that reduces the chances of abrupt changes in soil pH.
7. To affect the formation of organic-metallic compounds. This way, soil nutrients are stabilized.

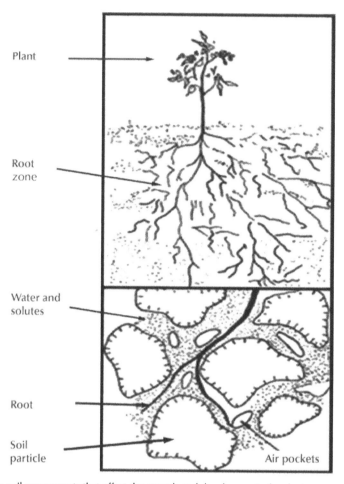

Plant

Root
zone

Water and
solutes

Root

Soil
particle

Air pockets

**Figure 1.1.** The soil components that affect the growth and development of a plant.

The water constitutes liquid phase of the soil and is required by the plants for the metabolism and transportation of soil nutrients. Soil water is needed for the physiological process of transpiration. The soil contains dissolved substances and is called as soil solutes. The soluble salts are always present in the soil water. Some are essential nutrients for the plants, while others in excessive amounts are detrimental. The gaseous phase constitutes the atmosphere of the soil and is indispensable for the respiration of the microorganisms and for providing a favorable atmosphere for the development and absorption of nutrients by roots. Therefore, the soil consists of three main phases: solid, liquid, and gas. The relative portions, of these three phases, vary continuously depending on the climate, vegetation, and soil management. Figures 1.1 and 1.2 show soil composition that is ideal for plant growth. The irrigation practices must be adequate so that the moisture, the air, and the nutrients are available in the correct proportions when needed.

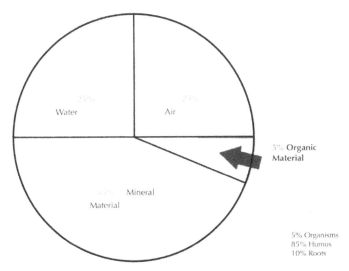

**Figure 1.2.** Effect of soil texture on the available water.

## Soil Texture

The soil is composed of infinite variety of sizes and forms of soil particles. The individual mineral particles are divided in three categories: Sand, silt, and clay (Figures 1.3 and 1.4). This classification is significant to the plant growth. Many of the reactions and important chemical and physical properties of soil are associated with the surface area of the soil particles. The surface area increases significantly as the particle size is reduced.

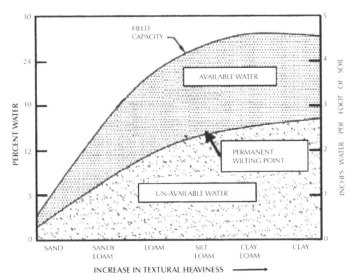

**Figure 1.3.** Volumetric content of the four principal soil components that is adequate for ideal growth of plants.

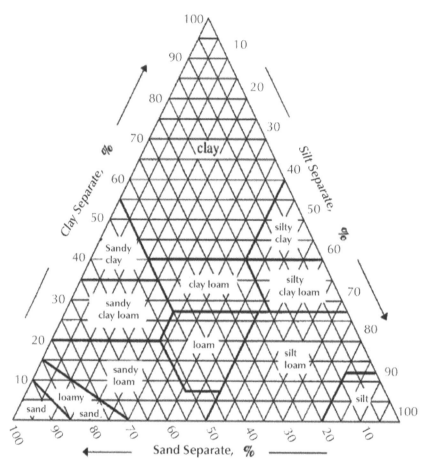

**Figure 1.4.** Soil texture classification [USDA Soil Conservation Service, Washington D.C., USA].
Legend: Fi = Fine, Co. = Coarse, v*fi = very fine, med. = medium, v.co. = very coarse

A description of soil texture can give us an idea about the interactions between soil and plant. In the mineral soil, the interexchange capacity (ability of retention of essential elements by the plants) is closely related to the clay percentage in the soil and the soil class. The capacity water retention of a soil is determined by the size distribution of particles (Figures 1.2 and 1.3). The fine textured soil (with high percentage of clay and silt) retains more water than sandy soil. The fine textured soil is generally more compact, movement of water and air is slow, and is more difficult to plow [12, 13]. Twelve classes of soil texture are recognized based on the percentage composition of sand, silt, and clay (Figure 1.4). Medium textured soils such as silty, sandy silt, and sticky silt are probably best for plant growth. Despite of this, relationship between soil texture and crop yield cannot be generalized to all soils (Figure 1.4).

## Soil Structure

The individual soil particles (sand, silt, and clay) can be united to form soil aggregates. The soil structure is an arrangement (direction, shape, and arrangement) of individual particles and soil aggregates with respect to one another. There are generally four principal types of soil structure such as: Laminar, prismatic, cuboide, and spherical as shown in Figure 1.5. When the soil units (particles and/or aggregates) are arranged around horizontal plane with much more longer horizontal axis than vertical axis, the soil structures are classified as laminar such as: Plates, leaves, or lenses. When the soil units are fixed around a vertical line forming pillars and united by relatively flat surfaces, the structure is known as prismatic or columnar. The third type of structure is called cuboide (in form of angular or sub-angular block) and it is characterized by approximately equal length in all three directions. The fourth arrangement is known as the spheroidal (granulate) and includes all the round and loose aggregates and that can be separated easily. The soil structure influences the plant growth. This is mainly due to its effect on the movement and retention of moisture, aeration, drainage, and erosive properties of soil. These can be maintained and improved with cultural practices of crop and irrigation. However, these can also be destroyed by inadequate soil management.

## Moisture (or Soil Water)

Some soils are very wet and may lack sufficient moisture available at a desired time, to obtain a good crop yield [2]. Therefore, classification, retention, and movement of soil water have drawn attention of many investigators during the last century. In 1897, Briggs explained the mechanism of retention of soil moisture on the basis of the hypothesis of capillary pores [8]. He classified soil water as gravitational, capillary, and hygroscopic based on the fact that there existed a continuous and tense film around the soil particles and the retention of soil moisture was dependent on the pore spaces. The water moved from coarse to fine particles. The speed of the water movement was related with specific curvature of particles, the surface tension and the viscosity of the liquid. Ten years later, Buckingham proposed another hypothesis on the basis of energy concepts. He suggested "Capillary Potential" to indicate the attraction between the soil particles and the water. In 1935, Schofield proposed the following equation to express the energy or tension with which the water was retained to the soil:

$$pF = Log_{10} \text{ [Height of water column]} \tag{1}$$

The movement and relation of soil water is now interpreted based on energy concept. Richard, Russel, Veihmeyer, Bouyoucos, and many other investigators have used this concept to develop devices to measure the tension with which the water is retained by the soil [8]. Soil water can be classified as: Gravitational water, capillary water, and hygroscopic water. This classification is merely physical and can be adapted to a concept of free energy on a tension scale. Figure 1.6 shows biological and physical classification of the soil water [2].

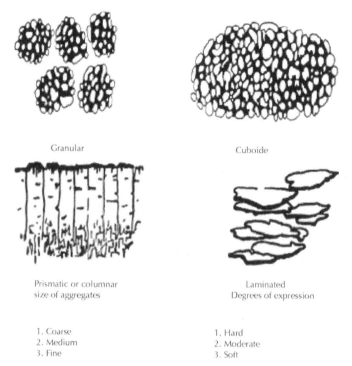

Granular

Cuboide

Prismatic or columnar
size of aggregates

Laminated
Degrees of expression

1. Coarse
2. Medium
3. Fine

1. Hard
2. Moderate
3. Soft

**Figure 1.5.** Classes of soil structure.

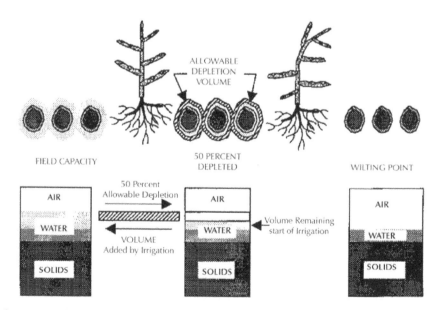

**Figure 1.6.** Soil composition affects available water to the plant.

## Classification of soil water

When the soil is wetted by rainfall or abundant irrigation, the water will fill all the pore spaces creating a thick water film around soil particles. Under these conditions, a saturation state is established. For this reason, the water is not strongly adhered or retained to soil particles. If appropriate conditions of water-drainage exist, capillary pores begin to drain due to the gravitational force. When all the macro pores have been drained but capillary pores continue to be full, this limit is called field capacity. Gravitational water is a soil water between its point of saturation (tension of zero atm.) and the soil field capacity (tension of 0.33 atm.). The gravitational water is undesirable. From the agricultural point of view, this fraction of water occupied by the pore spaces under optimal conditions of plowing must be occupied by the soil air. Because of low soil moisture tensions, this can be readily available unless prevented by some undesirable soil characteristics [8, 13].

As soon as the soil reaches field capacity, the gravitational component is no longer a principal factor for the water movement. Now absorption of water by plant roots and the evaporation are the main factors. As the soil moisture is extracted, thickness of the water film around soil particles is diminished and the water tension increases. At high soil moisture tension, the plants can not absorb sufficient water fast enough to compensate for the loss by transpiration. And the plants show signs of wilting. If the plants are able to recover of the wilting when these are placed in a saturated humid atmosphere, then the state of wilting has started. When soil moisture reaches a tension after which, the plant leaves do not recover of the wilting state even though these are placed in a saturated humid atmosphere, then this soil moisture content (at a tension of 15 atm.) is called a permanent wilting percentage.

This value varies very little with the ability of the plant to absorb water. The average values averages are about 1% for sandy soil, 3–6% for silty and greater than 10% for clayey soils, at 15 atmosphere of tension [6].

The water that remains in the soil at the permanent wilting state is not available to the plant. The plant will die if it remains longer under these conditions. The interval between the field capacity (tension 0.33 atm.) and the point of permanent wilting (tension 15 atm.) is called available water to the plant (Figures 1.7 and 1.8). Beyond the wilting state, the water is not available to the plants. The hygroscopic coefficient is soil moisture retained at a tension of 31 atmospheres. The soil moisture in the interval between field capacity and hygroscopic coefficient is called capillary water. The capillary water moves easily in the soil system, but it does not drain freely from the soil profile.

Also, the capillary water is for superior plants and the microorganisms. The hygroscopic water is a soil water above the hygroscopic coefficient (at tension > 31 atm.). The soil water in this range is not essentially available to the plant. The hygroscopic water moves at extremely slow rates in the vapor state.

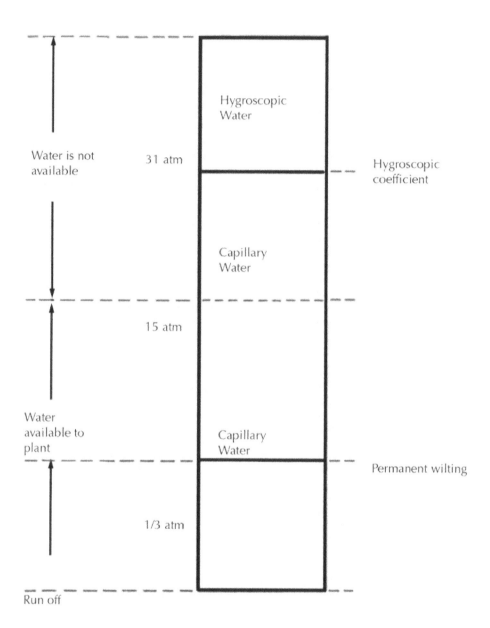

**Figure 1.7.** Physical and biological classification of soil water.

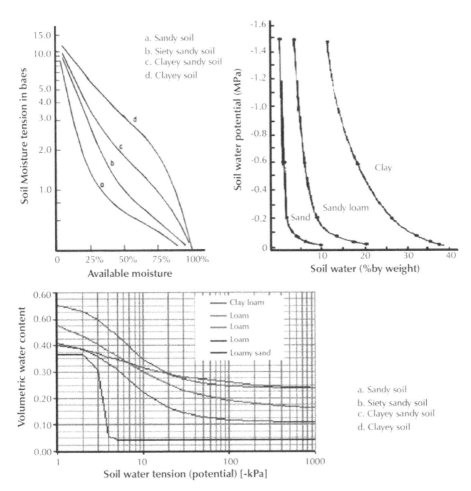

**Figure 1.8.** Soil moisture retention curves for different types of soil.

## Soil and water potential

As mentioned above, the movement and the retention of the soil water have been visualized on the basis of a concept of energy potential. The fact is that movement of all the soil water is affected by gravitational force of the earth. The laws of capillarity movement of the soil do not begin or finish at a given value of soil moisture tension or at specific pore size. The moisture tension is different from one location to another and through an elapsed time. The soil water is present in several forms: colloidal water, free water (frequently in capillary pores of the soil), and water vapor. In physical terms, the soil solution contains different amounts and forms of energy: kinetic or dynamic energy, the potential energy, and static energy.

Since the movement of the soil water is quite slow, its kinetic energy (that is proportional to the square of its speed) is generally considered insignificant. Therefore, the potential energy (that depends on the elevation or internal condition of the water) is very important. The most effective form to express the soil water content, the retention, the movement and water availability to the plants is a free energy per unit mass, which is called a potential. The free energy is an available energy (without change in temperature). The potential energy is increased when the soil water is extracted by processes such as: evaporation, infiltration, and deep percolation. As this process occurs, the plant must do an extra work to extract the next available moisture. This implies that the ability of the roots to absorb the soil water is directly related to the total water potential.

Under normal conditions, the soil water potential varies extensively. This energy difference between two points causes the movement of water from a site of greater energy (greater potential) to a site of smaller energy (smaller potential). The water does not move against the energy gradient, but moves due to an energy gradient. In general terms, it is difficult to know the amount of absolute free energy of a given substance as it is for the soil water. We can only know the difference between the free energy of soil water at a given state and the free energy of the water at a reference state. For the liquid phase of water, our state of reference will be a soil saturated with pure water at a given temperature, ambient pressure, and a height from a datum line. The energy of the soil water at any other state and elevation is a difference between the energy of the water at the given and the energy of the water at a reference state. This difference is called water potential.

### Components of soil water potential
The total potential of the soil water consists of a series of individual components that can alter the free or potential energy of the soil water. These components are presented in the following sections:

#### Gravitational potential
The gravitational potential of the soil water at a given state is determined by the elevation of this point from a datum line.

#### Pressure potential
The pressure potential of soil water is due to an increase or decrease of pressure of the free energy of the soil water. The pressure of the soil water (liquid phase) can be affected by the following factors:
1. Capillary suction (Capillary potential): The capillary potential is an energy that is required to move a unit or mass of water against the capillary forces from the water surface to a desired point. This way, it describes the effects that have the capillary forces on the free energy of the soil water.
2. Hydrostatic pressure in static water under an aquifer level: The hydrostatic pressure is a potential change in the free energy.
3. Water pressure induced by flow: Pressure potential is also affected by the amount and rate of flow of the soil water.

4.  Pressure induced potential: Pressure induced potential is a change in free energy of the soil water due to any source that has not been mentioned so far. For example: Local compressed air in the soil, mechanical forces on the soil or the suction (negative pressure).

*Osmatic potential*
This includes the effects on the soluble salts in the free energy of the water to the soil and the effects on the differences in the ion disassociations absorbed on the surface of colloidal particles of clay and organic matter.

*Matrix potential*
The potential matrix expresses the physical-chemical attractions between the water and soil particles. It includes the capillary attraction and the molecular forces that retain the water of hydratation in the soil colloids. Since it is very difficult to evaluate the hydrostatic pressure, osmotic, or adhesion potential separately, it is a general practice to include these potentials in the capillary potential or matrix potential, because these three are due to pressure deficiency [2]. The total potential of the soil water can be expressed in units of force or pressure by means of the sum of individual components.

$$\text{Total} = \underset{\text{potential}}{\text{gravitational}} + \underset{\text{potential}}{\text{pressure}} + \underset{\text{potential}}{\text{osmatic}} + \underset{\text{potential}}{\text{etc.}} \qquad (2)$$

In practice, the water potential can be measured placing the soil sample on a porous membrane plate and to determine the tension (by centrifugal or air pressure) required to extract water from the soil. If we know potential energy soil water, then we have valuable information on the availability of soil water to the plant.

**Soil Moisture Tension (or Suction)**
The soil water is in a form of a water film that surrounds the soil particles. The film is thick when there is enough soil moisture. The effects of external forces of absorption (absorption by the plant roots and evaporation) reduce the thickness of the film. The moisture tension is a measurement of a force with which the moisture is retained by the soil. When the tension increases, the thickness of the water film decreases. It is easier to extract water from wet thick films while high tension is necessary to extract water from thin films. The soil moisture tension is a negative pressure or vacuum or suction. The moisture tension is measured in bars, centibars, atmospheres, cm of water, mm of mercury, psi, kPa, and so forth. Soil moisture tension is generally expressed in centibars (one bar is equivalent to 0.987 atmospheres). One atmosphere is equivalent to 14.7 psi or a mercury column of 760 mm or a water column of 103 cm. It is a general practice to indicate tension of 100 cm of water instead of a tension of oil. In the past, units of "pF" were used to the express the energy of a water retained in the soil.

**Soil Moisture Tension Curves**
The tension and soil moisture percentage are inversely related. At low tensions, the soil can retain more moisture. The farmer should never allow that soil moisture is at a permanent wilting percentage [12, 14]. For this purpose, one should know soil moisture

content of a given volume of soil. The soil moisture tension curve for a particular soil can be used as a guide to know the condition of a soil.

Figures 1.7 and 1.8 reveal curves for soil moisture tension for different types of soils. The curve characteristics depend on the soil porosity, the specific surface of soil particles, the soil texture, the soil structure, soil depth, rainfall or irrigation depth, and soil cover. The soil moisture at a given tension can be determined by using a pressure membrane apparatus. This apparatus (Figure 1.9) includes a porous membrane on which wet soil samples are placed. The suction is applied by means of a compressed air. The water is extracted from the soil sample below the membrane plate. The soil retains only the moisture whose hydrostatic potential is identical to the pressure applied in the chamber.

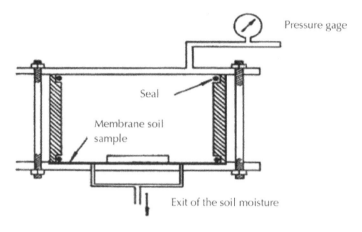

A. Pressure membrane apparatus: Operates by using a compressed air

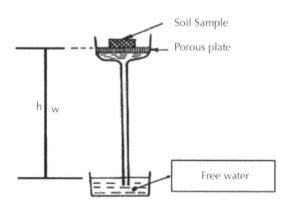

B. Pressure membrane apparautus aperated by hanging water column.

**Figure 1.9.** Pressure membrane apparatus (commonly employed) to find the soil moisture tension.

## Availability of Soil Moisture to the Plants

The available water to the plant is a difference in the soil moisture at field capacity (tension 0.33 atm.) and at permanent witting percentage (tension 15 atm.). One should not allow the soil moisture to reduce to a permanent wilting percentage. The root system of the plants is not homogenous. Generally, the roots are branched and thicker in the top soil; and are finger narrowed and branched into secondary and tertiary roots at greater soil depths. Soil moisture, at different root zones, is unequally distributed, as shown in Figure 1.10. The plant has taken advantage of all the moisture in the 30 cm of soil layer.

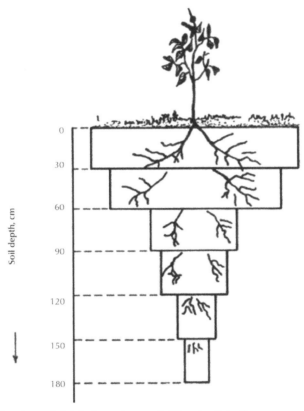

**Figure 1.10.** Soil moisture deficit in the root zone at different soil depths.

After this layer, the plant will continue absorbing water from the deeper layers. The surface area of absorption by roots reduces with depth, become there are lesser quantity of roots in contact with the available water. The water absorption by the roots compensates the water loss by transpiration through the leaves. On a warm and dry day, the plant has a faster absorption rate of water to compensate for the water loss. If the available water in the soil is not enough or the root surface for absorption has reduced, then there exists a temporary wilting of the plant during the hot and drought periods. This condition disappears in the evening, because the absorption rate is sufficient to supply the loss by the transportation rate. Therefore the root zone must be irrigated before using all the available water, with the objective of avoiding reduction in the crop yield.

## METHODS TO MEASURE THE SOIL MOISTURE

Several methods and instruments have been developed to determine the soil moisture. Many of these methods involve measuring soil properties that may change. The measurement of soil moisture helps us to determine the changes in the moisture content. Therefore, we can have information in the determination of water available to the plants.

Such information on soil moisture condition serves as a guide to the farmers or agricultural technicians for irrigation scheduling. It is also important for the irrigation management to provide a suitable irrigation depth. In the short and long term, it implies saving in time and money, since the crop yield is reduced due to excess or insufficient irrigation.

### Visual and Tactile Appearance of the Soil
#### *Use*
This method is an oldest method to estimate the soil moisture. It consists of a visual inspection and tactile appearance of a soil sample. Generally, it is used when equipment is not available or we cannot wait to know the soil moisture condition. However, the experienced farmer can estimate the soil moisture with a good precision.

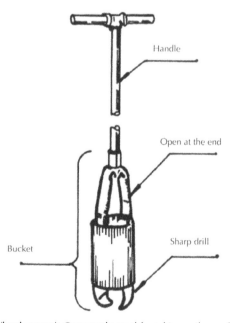

Handle

Open at the end

Bucket

Sharp drill

**Figure 1.11.** Soil auger (bucket type): Commonly used for taking soil samples at different depths.

## Procedure

By means an auger (Figure 1.11), a soil sample at a known depth is extracted. A visual and tactile inspection of the sample is conducted. Table 1.1 helps to estimate the soil moisture.

## Advantages

1. This method is simple to use.
2. It does not require use of expensive tools and equipments.
3. It provides a quick estimation of the soil moisture.

**Table 1.1.** Guide for the estimation of a soil moisture by using an extracted soil sample.

| Soil moisture deficit, % | Feel and criteria for a deficit of moisture, cm of water per meter of soil | | | |
|---|---|---|---|---|
| | **Coarse Texture** | **Moderate to coarse texture** | **Medium texture** | **Fine to extra fine texture** |
| Field Capacity | When it is compressed, no water comes out of soil. But the palm of hand becomes dirty. | When it is compressed, no water comes out of soil. But the palm of hand becomes dirty. | When it is compressed, no water comes out of soil. But the palm of hand becomes dirty. | When it is compressed, no water comes out of soil. But the palm of hand becomes dirty. |
| 25 | Tendency to form a mass quickly; sometimes with precision. A small ball can be formed but disintegrates easily. | A small ball with difficulty can be formed that is broken easily and that is not sticky. | A small ball can be formed that is molded easily. Sticky if there is relatively high clay content. | Cylinder is formed easily, when it is kneaded between fingers. Has a sticky contact. |
| 25–50 | Dry in appearance. A small ball cannot be formed by kneading it. | It is possible to form a small ball with precision, but usually it does not stay compact. | A relatively small ball can be formed that is sticky when it is pressed with fingers. | A small ball or small cylinder can be formed, when it is kneaded between the thumb and the index finger. |
| 50–75 | Dry in appearance, it is not possible to form a small ball with precision. | Dry in appearance; a small ball cannot be formed solely using precision*. | It crumbs, but stays relatively compact when the pressure is applied. | Relatively moldable, a small ball can be formed when a small amount of soil* is pressed. |
| 75–100 (100% = Point of permanent wilting) | Dry, loose in grains, and disintegrates between fingers. | Dry, loose, disintegrates between fingers. | Dusty, dry, and in small scabs that are reduced to dust when breaks itself. | Hard, very parched, tightened, sometimes in scabs; and disintegrates on the surface. |

*The small ball forms when kneading the soil sample.

### Disadvantages

1. It is not a very precise method to determine the soil moisture.
2. It is a subjective method that results in different interpretations by different persons who examine the soil sample under the same conditions.
3. It is necessary to take soil sample, this disturbing the root zone.

The visual appearance of the plants is frequently used as a guide to determine the need for irrigation. Reduction in the yellowing and change in color of the leaves during the evening are symptoms of inadequate soil moisture. It is recommended to apply irrigation before these symptoms will even appear.

### Gravimetric Method

#### Use

It is a determination of the soil moisture content by drying the soil sample in an oven. The method requires: Use of certain laboratory equipment to obtain accurate results; and a skill of the operator for precision.

#### Procedure

With the use of bucket type anger, a soil sample is taken from a known depth. To have a representative sample, samples are taken at several locations. Then, we take only 100 to 200 grams of soil sample. The sample is identified and its wet weight is recorded. The weighted sample is left in an oven at a constant temperature of 105°C for a period of 24 hours. After this period, weight of dry sample is recorded. The total moisture content in the soil is determined from the following equation:

$$PW = \frac{(SW - Sd)}{Sd} \times 100 \tag{3}$$

where: PW = Percentage of water by weight on dry basis;

SW = Weight of the wet soil sample; and   Sd = Weight of the dry sample.

The percentage of soil moisture is calculated based on the weight of a dry soil. Once we have the percentage of moisture by weight, we can express the percentage of water by volume. This provides us information on the volume of water in a given soil. The following equation is used to calculate the percentage of moisture by volume:

$$PV = PW \times \frac{Da}{D(H_2O)} \tag{4}$$

where: PV = Percentage of moisture in the soil by volume.
PW = Percentage of moisture by weight.
Da = Apparent density = [Mass of soil dried in an oven furnace]/[total Volume that occupies the soil]

$D(H_2O)$ = Density of water = 1 g/cm$^3$ or 1000 Kg/cm$^3$.

Following equation is used to calculate the total volume of the soil sample:

$$V = \frac{L\pi d^2}{4} \tag{5}$$

where: $\pi = 3.14$

d = Inner diameter of the cylinder that was used to take the sample.

L = Length of the cylinder.

### Advantages

1. It is a precise method to find the soil moisture if the samples are taken carefully

### Disadvantages

1. One requires laboratory equipment and certain degree of precision to obtain the reliable data.
2. One requires 24 hours to carry out the procedure.
3. The determination of the moisture for soils rich in organic matter can introduce an error due to an oxidation of organic matter.
4. It is a destructive method, because the soil is disturbed and samples are lost. Also, the root system of the plant is disturbed.
5. Several soil simples should be taken to have a representative sample.

## Tensiometer

### Use

Tensiometer is an instrument that indicates the tension at which the water is adhered to the soil particles (Figures 1.12 and 1.13).

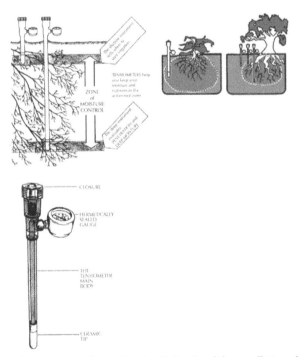

**Figure 1.12.** Principal components of a tensiometer (Bottom) and the installation of a tensiometer in the root zone of a crop (top).

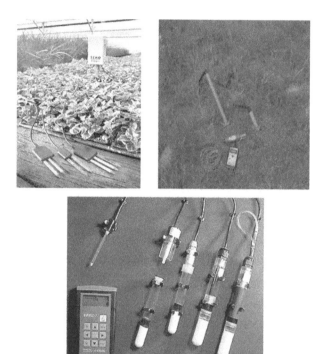

**Figure 1.13.** Electronic tensiometer.

## *Operation*

The instrument is placed in the soil taking into consideration the following factors: 1) Root depth; 2) Soil type and its variability; 3) Land; and 4) Type of irrigation system. Once the tensiometer is installed, the water within the stem of an instrument makes contact with the water retained in soil, flowing in both directions through the porous ceramic tip until the equilibrium is established. The soil water is lost through transpiration, evaporation, and absorption by the plants. This causes a tension or suction in the system and this tension increases as the soil moisture is lost. This tension is measured by a vacuum gage of a tensiometer. When the soil is wetted again by rainfall or by irrigation, the soil tension reduces due to the flow of water through the porous ceramic tip. Therefore, the tensiometer readings can be related to the available water to the plants. However, it is not a direct method of measurement of soil moisture. It is advisable to calibrate the tensiometer during the crop growth by finding soil moisture content with a gravimetric method. This calibration curve can be used for relating tensiometer readings with actual moisture contents [3, 6, 13, 14].

## *Advantages*

1. This is a good guide to decide when to apply the irrigation.
2. The tensiometer can be used to determine vertical and horizontal movement of the moisture. This is necessary when there are problems of salt accumulation.
3. The instrument provides a direct measurement of soil moisture suction.

4.  Tensiometer is especially appropriate for light soils, within limitations of a tensiometer 10–80 bars of tension.

### Disadvantages

1.  Tensiometer can only operate up to 80 cbars at sea level. Generally, after 80 cbars of tension, air enters the porous ceramic tip and breaks the water column. When this has happened, tensiometer readings are not correct.
2.  Tensiometer is a delicate instrument that must be protected from mechanical damages due to agricultural implements and operations.
3.  Tensiometer is placed generally in a fixed location of the field. It can not be moved from one place to another during the period of crop growth.

## Measurement of Electrical Resistance (Porous Ceramic Blocks)

### Use

This method estimates soil moisture content by using resistance or conductance properties of soil. It is achieved by installing electrical resistance ceramic blocks at desired soil depth. Nylon, fiber, and the combination of these materials with plaster have been used for the manufacture of electrical resistance blocks.

### Procedure

A representative area of the field is selected. With the use of proper size drill, a hole is made in the soil up to a desired depth. Then a porous plaster block with two or three electrodes is placed inside this hole. There must be a good contact between the soil and the block to allow a perfect seal. For this, a soil paste is prepared and is pored into the hole. The cables or terminals of the electrodes must be taken out of the soil surface (Figure 1.13).

Once the sensors have been installed, the moisture balance is established between the porous tip and the soil. The modifications in soil moisture conditions may change electrical properties of the soil. For a wet soil, electrical resistance is low. As the soil moisture is lost, the electrical resistance increases. This resistance is read by a portable counter. It is advisable to calibrate the equipment by determining moisture of soil samples with a gravimetric method. This way, we can establish a relationship between resistance readings and actual soil water content.

### Advantages

1. This method estimates soil the moisture.
2. This instrument is especially appropriate to measure changes in the soil moisture for tensions between 1 and 15 atmospheres.

### Disadvantages

1.  The useful life of the ceramic blocks is limited.
2.  The original calibration of the porous block changes with time, because pores can be clogged by salts.
3.  The plaster blocks are usually ineffective for soil tensions of less than one atmosphere.

4. The soluble salts in the soil solution can reduce the electrical resistance and may give high values of soil moisture content than actual values.

5. The porous blocks may not be homogenous and this results in inaccurate readings.

6. The precision of this method is reduced due to temperature, concentration of salts in the soil solution, physical characteristics of plaster to produce the block and the flight of current towards the soil.

## Neutron Scattering Method
### Procedure

This method consists of emission of neutron radiation of high energy from an emitter or a radiation source towards the soil. These fast neutrons travel through the soil material and gradually hit nuclei of different atoms thus reducing kinetic energy. The higher loss of energy occurs when these neutrons hit neutrons of mass similar to these (Figure 1.14).

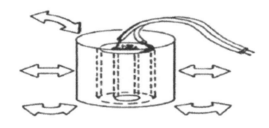

A. Gypsum block

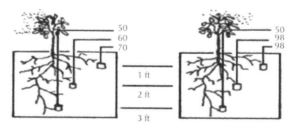

B. Depth of irrigation applied

**Figure 1.14.** Gypsum blocks: Commonly used to determine the depth of irrigation.

The hydrogen, a component of the water, is dominant factor to reduce the speed of fast neutrons. Because of these characteristics, these can change fast neutrons to slow moving neutrons in a faster way than the other elements. Because most of atoms of hydrogen in the soil comprise part of the water molecule, the portion of neutrons that are slowed down can be related to the soil moisture content.

The use of neutron emission source requires installation of access tubes in the soil to lower the slow neutron detector. These devices are installed at the beginning of the sowing season and are removed at the end of the last harvest. The neutron detector is connected to a portable recorder to facilitate the readings (Figures 1.15–1.20).

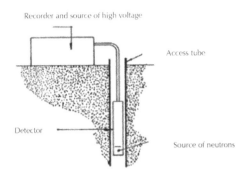

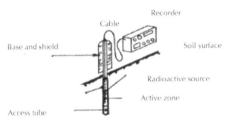

**Figure 1.15.** Determination of soil moisture by neutron scattering method.

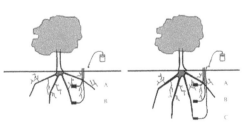

**Figure 1.15a.** Soil moisture meter.

**Figure 1.15b.** Location of soil moisture sensors.

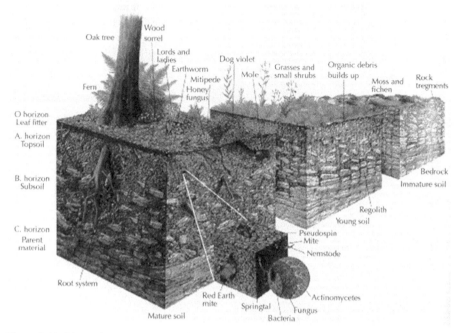

**Figure 1.16.** Principal components of soil.

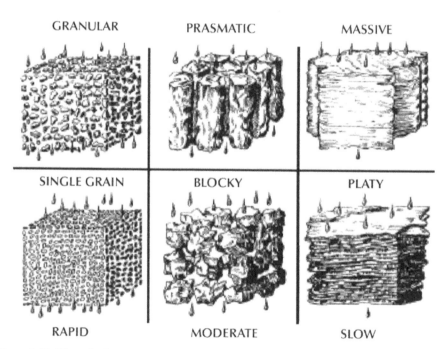

**Figure 1.17.** Effect of soil structure on water movement.

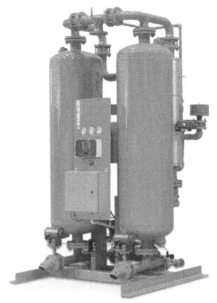

**Figure 1.18.** Pressure membrane apparatus that uses compressed air.

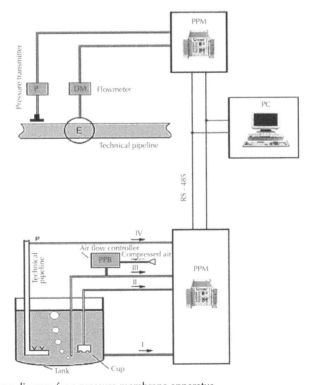

**Figure 1.19.** Flow diagram for a pressure membrane apparatus.

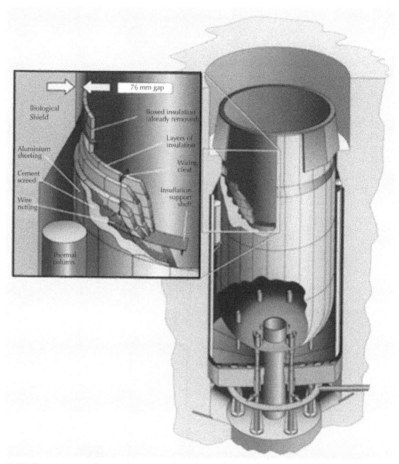

**Figure 1.20.** Pressure membrane that uses a hanging column of water.

The calibration of this instrument should be done for a desired location by knowing the soil moisture with a gravimetric method. After calibration, the readings are taken at a desired depth. It is recommended to install a sensor for each 30 cm of soil depth.

### Advantages

1. This system can cover a larger volume of soil and is relatively independent of the soil type.
2. It can be used for longest periods without any change in the radiation source.
3. The method does not involve taking of soil samples.
4. Any range of soil moisture content can be analyzed. This avoids limitations of tensiometer or electrical resistance methods that can only measure the soil moisture within a certain range.

*Disadvantages*
1. The equipment uses a radiation source. The technician must have basic skills and knowledge of the operation. It may cause heath risks.
2. The system is expensive and solely used for research purpose.
3. The moisture measurement, in soils with organic matter, is not precise and reliable because of presence of excessive hydrogen atoms. The readings for the surface soil layer are not precise because of escape of neutrons towards the surface.

## Alternate Methods

Additional approaches to measure the soil moisture are absorption of gamma rays, the dependency of the thermodynamic properties of the soil on the moisture content, the use of ultrasonic waves and the dielectric properties of the soil. Some of these methods have been tried with the global positioning system and geographic information system. However, most of these are under development and these are not in common use at the farm. Because of the high cost and complicated procedures.

## SUMMARY

Plants need a specific amount of soil moisture to ensure an adequate growth and development. This amount varies with the crop species. A limited amount of water can be retained by the soil, and a fraction of this water is available to the plant. This chapter discusses methods to measure soil moisture: visual and tactile appearance of the soil, gravimetric method, tensiometer, electrical resistance, and the neutron scattering method. Advantages and disadvantages of each method are presented. This chapter also discusses soil structure, soil texture, soil water and soil moisture available to the plant.

## KEYWORDS

- Aggregate
- Atmosphere
- Avaiable water,
- Bar
- Capillary water
- Centibar (cbar)
- Clay
- Dissolve
- Drainage
- Evaporation
- Field capacity
- Gaseous phase

- **Gravitational water**
- **Gypsum blocks**
- **Hydraulic conductivity**
- **Hygroscopic coefficient**
- **Hygroscopic water**
- **Infiltration**
- **Irrigation depth**
- **Low tension**
- **Neutron scattering**
- **Nitrogen**
- **Nutrient**
- **Organic matter**
- **Percentage, moisture**
- **Permanent wilting percentage**
- **pF**
- **pH**
- **Phosphorus**
- **Potassium**
- **Principal**
- **Respiration**
- **Root zone**
- **Sand**
- **Soil moisture**
- **Soil structure**
- **Soil texture**
- **Soil water content (Soil moisture)**
- **Soluble salts**
- **Tensiometer**
- **Transpiration**
- **Water content**
- **Water potential**
- **Water redistribution or percolation**

## BIBLIOBGRAPHY

Refer to bibliography, section 1.1 for the literature cited.

# Chapter 2

## Evapotranspiration

### INTRODUCTION

Due to increased agricultural production, irrigated land has increased in the arid and sub-humid zones around the world. Agriculture has started to compete for water use

---

[1]This chapter is modified and translated from "Goyal, Megh R. y Eladio A. Gonzalez—Fuentes. 1990. Evapotranspiración. Capítulo III en: *Manejo de Riego Por Goteo* editado por Megh R. Goyal. Páginas 71–100. Rio Piedras, PR: Servicio Extension Agricola, UPRM." For additional details, one may contact by E-mail at: goyalmegh@gmail.com.

with industries, municipalities, and other sectors. This increasing demand along with increments in water and energy costs have absolutely made necessary to develop new technologies for the adequate management of water. The intelligent use of water for crops requires understanding of evapotranspiration (ET) processes.

ET is a combination of two processes: evaporation and transpiration. Evaporation is a physical process that involves conversion of liquid water into water vapor into the atmosphere. Evaporation of water into the atmosphere occurs on the surface of rivers, lakes, soils, and vegetation. Transpiration is basically a process of evaporation. The transpiration is a physical process that involves flow of liquid water from the soil (root zone) to the surface of leaves/branches and trunk; and conversion of liquid water from the plant tissue into water vapors into the atmosphere. The water evaporates from the leaves and plant tissue, and the resultant water vapor diffuses into atmosphere through the stomates. An energy gradient is created during the evaporation of water, which causes the water movement into and out of the plant stomates. In the majority of green plants, stomates remain open during the day and stay closed during the night. If the soil is too dry, the stomates will remain closed during the day in order to slow down the transpiration.

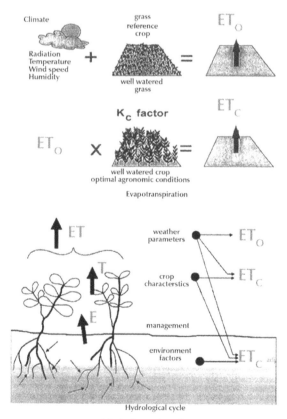

Hydrological cycle

Evaporation, transpiration, and ET processes are important for estimating crop irrigation requirements and for irrigation scheduling [4]. To determine crop irrigation requirements, it is necessary to estimate ET by on site measurements or by using meteorological data. On site measurements are very costly and are mostly employed to calibrate ET methods utilizing climatological data. There are number of proposed equations that require meteorological data and some are used to estimate the ET for periods of one day or more. All of these equations are empirical in nature. The simplest methods require only data about average temperature of air, length of the day, and the crop. Other equations require daily radiation data, temperature, vapor pressure, and wind velocity. Figure 2.1 shows the instruments for a weather station. None of the equations should be rejected, because the data is not available. Not all methods are equally precise and reliable for different regions of the world. There is no unique meteorological method that can be universally adequate under all climatic conditions.

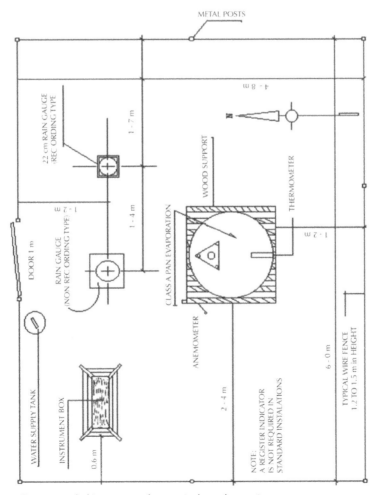

**Figure 2.1.** Recommended instruments for a typical weather station.

## POTENTIAL EVAPOTRANSPIRATION (PET)

Potential evapotranspiration is a water loss from the soil surface completely covered by vegetation. Meteorological processes determine the ET of a crop [7]. Closing of stomates and reduction in transpiration are usually important only under drought or under stress conditions of a plant. The ET depends on three factors: (1) Vegetation, (2) Water availability in the soil, and (3) Behavior of stomates. Vegetation affects the ET in various forms. It affects ability of soil surface to reflect light. The vegetation changes amount of absorbed energy by the soil surface. Soil properties, including soil moisture, also affect the amount of energy that flows through the soil. The height and density of vegetation influence efficiency of the turbulent heat interchange and the water vapor of the foliage [15].

Changes in the soil moisture affects direct evaporation from the soil surface and available water to the plants. As the plants are water stressed, stomates close resulting in the reduction of a water loss and $CO_2$ absorption. This is a factor that is not considered in the PET equation. Under normal conditions (with enough water), there is a variation among stomates of different plant species. Besides, these variations are usually small and the concept of PET results useful for the majority of crop species with complete foliage [10, 13].

## MATHEMATICAL MODELS FOR PET

There are different methods to estimate or measure the ET and the PET. The precision and reliability vary from one method to another, some provide only an approximation. Each technique has been developed with the available climatological data to estimate the ET. The direct measurements of PET are expensive and are only used for local calibration of a given method using climatological data. The most frequently used techniques are: hydrological method or water balance method, climatic methods, and micro meteorological methods. Many investigators have modified the equations that are already established. For example, one may find modification of Blaney–Criddle formula, Hargreaves–Samani, Class A pan evaporation and so forth. Allen [1] investigated 13 variations of the Penman equation. He found that the Penman–Monteith formula was most precise. Modified equations are actually recommended by the FAO and the USDA—Soil Conservation Service. Most of investigators agree that Penman, Class A pan evaporation, Blaney–Criddle, and Hargreaves–Samani equations, can be trusted. High precision can be obtained with local calibration of a given method. Every researcher has its preferred formula that may give good results. Hargreaves and Samani [7] presented their formulae as to be simplest and practical. I can add that, "There is no evidence of a superior method." Allen and Pruitt [2] presented the FAO modified Blaney–Criddle method, which involves relatively easy calculations and give precise estimates of PET (when it is calibrated for local conditions). Every researcher has preference. However each formula, depending where it was evaluated, may or may not result in the first or the last place.

### Hydrologic Method or Water Balance

This technique employs periodic determination of rainfall, irrigation, drainage, and soil moisture data. The hydrologic method uses water balance equation:

$$PI + SW - RO - D - ET = 0 \tag{1}$$

where: PI = Precipitation and/or irrigation.

RO = Runoff.

D = Deep percolation.

SW = Change in the soil moisture, and ET = Evapotranspiration.

In equation (1), every variable can be measured with precision with the lysimeters (Figure 2.2 a, b, c). The ET can be calculated as residual, knowing values of all other parameters (Figure 2.3).

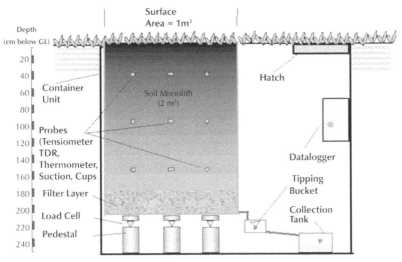

**Figure 2.2a.** A typical lysimeter with its components in Australia.

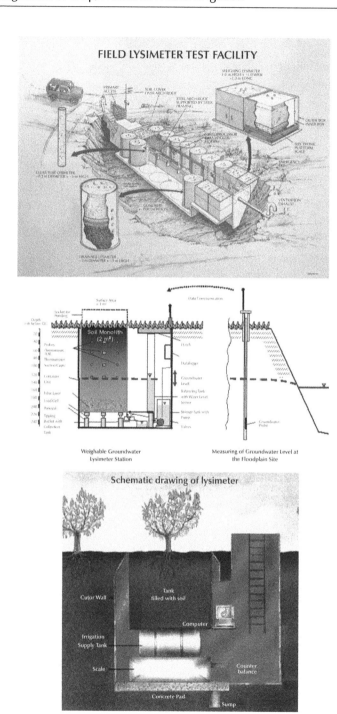

At Universiy of California, Davis.

**Figure 2.2b.** A field lysimeter test facility.

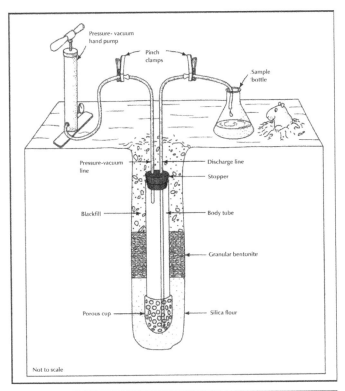

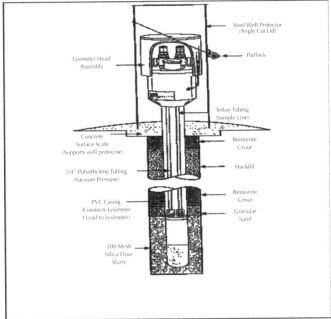

**Figure 2.2c.** Plastic tube lysimeter.

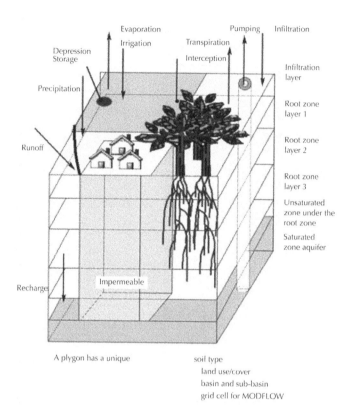

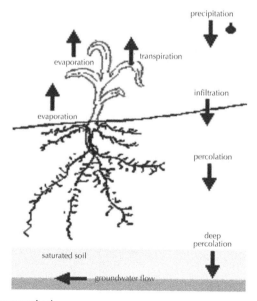

**Figure 2.3.** Water balance method.

## Climatic Methods

Using weather data [3], numerous equations have been proposed. Also, numerous modifications have been made to the available formulae for application to a particular region.

### Penman

The Penman formula was presented in 1948. It employs net radiation, air temperature, wind velocity, and deficit in the vapor pressure. He gave the following equation:

$$PET = \frac{R_n/a + b\,E_a}{c + b}$$

(2)

where: PET = Daily potential evapotranspiration, mm/day.

C = Slope of saturated air vapor pressure curve, mb/°C.

$R_n$ = Net radiation, cal/cm² day.

a = Latent heat of vaporization of water = [59.59 – 0.055 T], cal/cm² – mm

= 58 cal/cm² – mm at 29°C.

$E_a = 0.263 [(e_a – ed) * (0.5 + 0.0062 * u_2)]$     $(2_a)$

$E_a$ = Average vapor pressure of air, mb = $(e_{max} – e_{min})/2$.

$e_d$ = Vapor pressure of air at minimum air temperature, mb.

$u_2$ = Wind velocity a height of 2 meters, km/day. b = Psychrometric constant = 0.66, mb/°C.

T = $[(T_{max} – T_{min})/2]$, in degrees °C.

$(e_{max} – e_{min})$ = Difference between maximum and minimum vapor pressure of air vapor, mb.

$(T_{max} – T_{min})$ = Difference between maximum and minimum daily temperature, °C.

### Penman modified by Monteith [14]

After modification, the resultant equation is as follows:

$$LE = \frac{-s\,(R_n – S) + Pa * Cp\,(e_s – e_a)/r}{[(s + b) * (r_a + r_c)]/r_a}$$

(3)

where: LE = Latent heat flow.

$R_n$ = Net radiation.

S = Soil heat flow.

Cp = Air specific energy at constant pressure.

s = Slope of saturated vapor pressure of air, at air average temperature of wet bulb thermometer.

Pa = Density of humid air.

$e_s$ = Saturated vapor pressure of water.

$e_a$ = Partial water vapor pressure of air.

$r_a$ = Air resistance.

$r_c$ = Leaf resistance.

b = Psychrometric constant.

This method has been successfully used to estimate the ET of a crop. The Penman-Monteith equation is limited to research work (experimentation) since the $r_a$ and $r_c$ data are not always available.

### Penman modified by Doorenbos and Pruitt

$$PET = c * [W*R_{n + (1 - W)} * F(u) * (e_a - e_d)] \tag{4}$$

where: PET = Potential evapotranspiration, mm/day.

W = A factor related to temperature and elevation.

$R_n$ = Net radiation, mm/day.

F(u) = Wind related function.

$(e_a - e_d)$ = Difference between the saturated vapor pressure of air at average temperature and vapor pressure of air, mb.

c = A correction factor.

The Penman formula is not popular, because it needs data that is not available at majority of the weather stations. Estimations of PET using Penman formula cannot be complex. The equation contains too many components, which should be measured or estimated, when data is not available.

### Thornthwaite

This method uses monthly average temperature and the length of the day.

$$PET = 16 \, L_d \, [10 \, T/I]^a \tag{5}$$

where: PET = Estimated evapotranspiration for 30 days, mm.

$L_d$ = Hours of the day divided by 12.

$A = (6.75 \times 10^{-7} \, I^3) - (7.71 \times 10^5 \, I^2) + 0.01792 \, I + 0.49239 \tag{5_a}$

T = Average monthly temperature, °C.

$I = i_1 + i_2 + \ldots + i_{12}$, where, i = $[T_m/5] \times 1.514 \tag{5_b}$

The Thornthwaite method underestimates PET during the summer when maximum radiation of the year occurs. Besides, the application of equation to short periods of time can lead to an error. During short periods, the average temperature is not an adequate measure of the received radiation [14]. During long terms, the temperature and the ET are similar functions of the net radiation. These are related, when the long periods are considered.

### Blaney–Criddle

The original Blaney–Criddle equation was developed to predict the consumptive use of PET in arid climates. This formula uses percentage of monthly sunshine hours and monthly average temperature.

$$PET = K_m \, F \tag{6}$$

where: PET = Monthly potential evapotranspiration, mm.

$K_m$ = Empirically derived coefficient for the Blaney–Criddle method.

F = Monthly ET factor = [25.4 * PD * (1.8 T + 32)]/100     (6$_a$)

T = Monthly average temperature, °C.

PD = Monthly percentage daily sunshine hour.

This method is easy to use and the necessary data are available. It has been widely used in the Western United States with accurate results, but not the same in Florida, where ET is underestimated during summer months.

### Blaney–Criddle modified by FAO [4]

$$PET = C * P * [0.46 * T + 8]  \tag{7}$$

where: PET = Potential evapotranspiration, mm/day.

T = Monthly average temperature.

P = Percentage daily sunshine hours, Table 2.1.

C = Correction factor, which depends on the relative humidity, light hours, and wind.

Doorenbos and Pruitt [4, 5] recommended individual calculation for each month. They indicated that it may be necessary to increase its value for high elevations.

Table 2.1. Percentage average daily sunshine hours (P) based on annual day light hour for different latitudes.

| Latitude, degrees North | January | February | March | April | May | June |
|---|---|---|---|---|---|---|
| South* | July | August | Sept. | October | Nov. | Dec. |
| 60 | 0.15 | 0.20 | 0.26 | 0.32 | 0.38 | 0.41 |
| 58 | 0.16 | 0.21 | 0.26 | 0.32 | 0.37 | 0.40 |
| 56 | 0.17 | 0.21 | 0.26 | 0.32 | 0.36 | 0.39 |
| 54 | 0.18 | 0.22 | 0.26 | 0.31 | 0.36 | 0.38 |
| 52 | 0.19 | 0.22 | 0.27 | 0.31 | 0.35 | 0.37 |
| 50 | 0.19 | 0.23 | 0.27 | 0.31 | 0.34 | 0.36 |
| 48 | 0.20 | 0.23 | 0.27 | 0.31 | 0.34 | 0.36 |
| 46 | 0.20 | 0.23 | 0.27 | 0.30 | 0.34 | 0.35 |
| 44 | 0.21 | 0.24 | 0.27 | 0.30 | 0.33 | 0.35 |
| 42 | 0.21 | 0.24 | 0.27 | 0.30 | 0.33 | 0.34 |
| 40 | 0.22 | 0.24 | 0.27 | 0.30 | 0.32 | 0.34 |
| 35 | 0.23 | 0.25 | 0.27 | 0.29 | 0.31 | 0.32 |
| 30 | 0.24 | 0.25 | 0.27 | 0.29 | 0.31 | 0.32 |
| 25 | 0.24 | 0.26 | 0.27 | 0.29 | 0.30 | 0.31 |
| 20 | 0.25 | 0.26 | 0.27 | 0.28 | 0.29 | 0.30 |
| 15 | 0.26 | 0.27 | 0.27 | 0.28 | 0.29 | 0.29 |
| 10 | 0.26 | 0.27 | 0.27 | 0.28 | 0.28 | 0.29 |
| 5 | 0.27 | 0.27 | 0.27 | 0.28 | 0.28 | 0.28 |
| 0 | 0.27 | 0.27 | 0.27 | 0.27 | 0.27 | 0.27 |

**Table 2.1.** (Continued)

| Latitude, degrees North | July | August | Sept. | October | Nov. | Dec. |
|---|---|---|---|---|---|---|
| South* | January | February | March | April | May | June |
| 60 | 0.40 | 0.34 | 0.28 | 0.22 | 0.17 | 0.13 |
| 58 | 0.39 | 0.34 | 0.28 | 0.23 | 0.18 | 0.15 |
| 56 | 0.38 | 0.33 | 0.28 | 0.23 | 0.18 | 0.16 |
| 54 | 0.37 | 0.33 | 0.28 | 0.23 | 0.19 | 0.17 |
| 52 | 0.36 | 0.33 | 0.28 | 0.24 | 0.20 | 0.17 |
| 50 | 0.35 | 0.32 | 0.28 | 0.24 | 0.20 | 0.18 |
| 48 | 0.35 | 0.32 | 0.28 | 0.24 | 0.21 | 0.19 |
| 46 | 0.34 | 0.32 | 0.28 | 0.24 | 0.21 | 0.20 |
| 44 | 0.34 | 0.31 | 0.28 | 0.25 | 0.22 | 0.20 |
| 42 | 0.33 | 0.31 | 0.28 | 0.25 | 0.22 | 0.21 |
| 40 | 0.33 | 0.31 | 0.28 | 0.25 | 0.22 | 0.21 |
| 35 | 0.32 | 0.30 | 0.28 | 0.25 | 0.23 | 0.22 |
| 30 | 0.31 | 0.30 | 0.28 | 0.26 | 0.24 | 0.23 |
| 25 | 0.31 | 0.29 | 0.28 | 0.26 | 0.25 | 0.24 |
| 20 | 0.30 | 0.29 | 0.28 | 0.26 | 0.25 | 0.25 |
| 15 | 0.29 | 0.28 | 0.28 | 0.27 | 0.26 | 0.25 |
| 10 | 0.29 | 0.28 | 0.28 | 0.27 | 0.26 | 0.26 |
| 5 | 0.28 | 0.28 | 0.28 | 0.27 | 0.27 | 0.27 |
| 0 | 0.27 | 0.27 | 0.27 | 0.27 | 0.27 | 0.27 |

*Southern latitudes have six months of difference as shown in Table 2.1.

### Blaney and Criddle modified by Shih

$$PET = 25.4 * K * [MR_s * (1.8\ T + 32)]/[TMR_s] \qquad (8)$$

where: PET = Monthly potential evapotranspiration, mm.

K = Coefficient for this modified method.

$MR_s$ = Monthly solar radiation, cal/cm².

T = Monthly average temperature, °C.

$TMR_s$ = Sum of monthly solar radiation during the year, cal/cm².

### Jensen–Haise

The Jensen–Haise equation [9] resulted from about 3000 measurements of the ET taken in the Western Regions of the United States for a 35 years period. It is an empirical equation.

$$PET = R_s (0.025 * T + 0.08) \qquad (9)$$

where: PET = Potential evapotranspiration, mm/day.

$R_s$ = Daily total solar radiation, mm of water.

T = Air average temperature, °C.

This method seriously underestimates ET under conditions of high movements of atmospheric air masses. However, it gives reliable results for calm atmospheres.

## Stephens–Stewart

Stephens–Stewart utilized solar radiation data. It is similar to the original Jensen–Haise method [9]. The equation is as follow:

$$PET = 0.01476 * [(T + 4.905) * MR_s]/b \qquad (10)$$

where: PET = Monthly potential evapotranspiration, mm.

T = Monthly average temperature, °C.

$MR_s$ = Monthly solar radiation, cal/cm².

b = Latent vaporization energy of water = [59.59 – 0.055 $T_m$], cal/cm²-mm.

= 58 cal/cm²-mm at 29°C.

## Pan evaporation

Class A pan is commonly used instrument to measure evaporation. The evaporation pan (Figure 2.4) integrates the climate factors and has proven to give accurate estimations of PET. It requires a good service, maintenance, and management. Table 2.2 gives class A pan coefficients under different conditions [Doorenbos and Pruitt, 4 and 5]. The relationship between PET and pan evaporation can be expressed as:

$$PET = K_p * PE \qquad (11)$$

where: PET = Potential evapotranspiration, mm/day.

$$K_p = \text{Pan coefficient.}$$
$$PE = \text{Class A pan evaporation.}$$

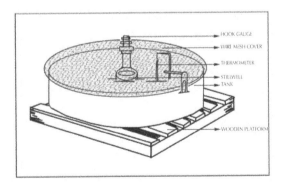

View of climatological station

**Figure 2.4.** A typical class A pan.

**Table 2.2.** Pan coefficient ($K_p$) for the class A pan evaporation under different conditions.

| Class A Pan | | Condition A | | | Condition B* | | | |
|---|---|---|---|---|---|---|---|---|
| Average of HR% | | Pan surrounded by grass | | | Pan surrounded by dry uncovered soil | | | |
| | | Low | Medium | High | | Low | Medium | High |
| | | 40 | 40–70 | 70 | | 40 | 40–70 | 70 |
| Wind** km/day | Distance from the green crop, m | | | | Distance from the dry fallow, m | | | |
| Light | 0 | 0.55 | 0.55 | 0.75 | 0 | 0.70 | 0.80 | 0.85 |
| 175 | 10 | 0.65 | 0.75 | 0.85 | 10 | 0.60 | 0.70 | 0.80 |
| | 100 | 9.70 | 0.80 | 0.85 | 100 | 0.55 | 0.65 | 0.75 |
| | 1000 | 0.75 | 0.85 | 0.85 | 1000 | 0.50 | 0.60 | 0.70 |
| Moderate | 0 | 0.50 | 0.60 | 0.65 | 0 | 0.65 | 0.75 | 0.80 |
| 175–425 | 10 | 0.60 | 0.70 | 0.75 | 10 | 0.55 | 0.65* | 0.70 |
| | 100 | 0.65 | 0.75 | 0.80 | 100 | 0.50 | 0.60 | 0.65 |
| | 1000 | 0.70 | 0.80 | 0.80 | 1000 | 0.45 | 0.55 | 0.60 |
| Strong | 0 | 0.45 | 0.50 | 0.60 | 0 | 0.60 | 0.65 | 0.70 |
| 425–700 | 10 | 0.55 | 0.60 | 0.65 | 10 | 0.50 | 0.55 | 0.65 |
| | 100 | 0.60 | 0.65 | 0.70 | 100 | 0.45 | 0.45 | 0.60 |
| | 1000 | 0.65 | 0.70 | 0.75 | 1000 | 0.40 | 0.45 | 0.55 |
| Very Strong | 0 | 0.40 | 0.45 | 0.50 | 0 | 0.50 | 0.60 | 0.65 |
| | 10 | 0.45 | 0.55 | 0.60 | 10 | 0.45 | 0.50 | 0.55 |
| | 100 | 0.50 | 0.60 | 0.65 | 100 | 0.40 | 0.45 | 0.50 |
| | 1000 | 0.55 | 0.60 | 0.65 | 1000 | 0.35 | 0.40 | 0.45 |

*For areas of extensive uncovered and not developed agricultural soils.
Reduce values of $K_p$ by 20% under hot wind conditions and by 5–10% for moderate wind conditions, temperature and humidity.
**Total wind movement in km/day.

## Hargreaves method

Hargreaves method uses a minimum of climatic data. The formula is as bellow:

$$PET = MF * (1.8\,T + 32) * CH \tag{12}$$

where: PET = Potential evapotranspiration, mm/month.

MF = Monthly factor depending on the latitude.

T = Monthly average temperature, °C.

CH = Correction factor for the relative humidity (RH) = To be used for RH > 64%

$$= 0.166\,[(100 - HR)]^{1/2} \tag{12a}$$

The Hargreaves original formula for the PET was based on radiation and temperature as given below:

$$PET = [(0.0135 * RS)] * [T + 17.8] \qquad (13)$$

where: RS = Solar radiation, mm/day.

T = Average temperature, °C.

To estimate solar radiation (RS) using extraterrestrial radiation (RA), Hargreaves and Samani [7, 8] formulated the following equation:

$$RS = K_{rs} * RA * TD^{0.50} \qquad (13_a)$$

where: T = Average temperature, °C.

RS = Solar radiation.

RA = Extraterrestrial radiation.

$K_{rs}$ = Calibration coefficient.

TD = Difference between maximum and minimum temperatures.

## Hargreaves and Samani modified method

Finally after several years of calibration, equation (13) was modified as follows:

$$PET = 0.0023 R_a * [T + 17.8] * (TD)^{0.50} \qquad (14)$$

where: PET = Potential evapotranspiration, mm/day.

$R_a$ = Extraterrestrial radiation, mm/day.

T = Average temperature, °C.

TD = Difference between maximum and minimum temperatures, °C.

This equation requires only maximum and minimum temperature data. This data is normally available. This formula is precise and reliable.

## Linacre method

The Linacre equation is as follow:

$$PET = \frac{700\ T_m / [100 - L_a] + 15\ [T - T_d]}{[80 - T]} \qquad (15)$$

where: PET = Potential evapotranspiration, mm.

$$T_m = (T_a + 0.0062 * Z) \qquad (15a)$$

Z = Elevation, m.

T = Average temperature, °C.

$L_a$ = Latitude, degrees.

$T_d$ = Daily average temperature, °C.

The variations in PET values by this formula are 0.3 mm/day based annual data and 1.7 mm/day based on daily data.

## Makkink method

This formula provides good results in humid and cold climates, and in arid regions. Makkink used radiation measurements to develop a following regression equation:

$$PET = R_s * [s/(a + b)] + 0.12 \tag{16}$$

where: PET = Potential evapotranspiration, mm/day.

$R_s$ = Total daily solar radiation, mm/day.

b = Psychrometric constant.

s = Slope of saturated vapor pressure curve at average air temperature.

### Radiation method

Doorenbos and Pruitt [4] presented following radiation equation, which is a modified Makkink formula [16]:

$$PET = c * (W * R_s) \tag{17}$$

where: PET = Potential evapotranspiration for the considered period, mm/day.

$R_s$ = Solar radiation, mm/day.

W = Correction factor related to temperature and elevation.

C = Correction factor, which depends on the average humidity and average wind speed.

This method was employed in the Equator zone, in small islands and in high latitudes. Solar radiation maps provide the necessary data for the formula.

### Regression method

The simple lineal regression equation is given as follow:

$$PET = [a * R_s] + b \tag{18}$$

where: PET = Potential evapotranspiration, mm/day.

a and b = Empirical constants (regression coefficients), which depend on the location and season.

$R_s$ = Solar radiation, mm/day.

This regression method is simple and easy to use. However, it is not frequently used because of highly empirical nature.

### Priestly–Taylor method

In the absence of atmospheric air mass movement, Priestly and Taylor showed that the PET is directly related to evaporation equilibrium:

$$PET = A [s/(S + B)] [(R_n + S)] \tag{19}$$

where: PET = Potential evapotranspiration, mm/day.

A = Empirically derived constant.

s = Slope of saturated vapor pressure curve, at average air temperature.

B = Psychrometric constant.

$R_n$ = Net radiation, mm/day.

This method is of semi-empirical in nature. It is reliable in humid zones, and is not adequate in arid regions. Table 2.3 shows advantages and disadvantages of different methods of estimation of potential evapotranspiration.

**Table 2.3.** Advantages and disadvantages of different methods to estimate PET.

| | Method | Advantages | Disadvantages |
|---|---|---|---|
| 1. | Penman | Easy to apply. | Underestimates ET under high movement conditions of atmospheric air masses. |
| 2. | Penman (FAO) | Provide satisfactory results. | The formula contains many components, which may result complex calculations. |
| 3. | Water balance | Easy to process the data and integrate with the observations. | Low precision on the daily measures and difficult to obtain the ET when it is raining. |
| 4. | Thornthwaite | Is reliable for long terms. | Underestimate the ET during the summer. Is not precise for short terms. |
| 5. | Blaney–Criddle | Easy to use, the data is usually available. | The crop coefficient depends a lot on the climate. |
| 6. | Blaney–Criddle (FAO) | The given crop coefficient depend less on the climate. | In high elevations, coasts, and small islands there is no relation between temperature and solar radiation. |
| 7. | Stephens–Stewart | Is reliable on the western side of the United States (where it was developed). | Need to be evaluated in other locations. |
| 8. | Jensen–Haise | Is reliable under calm atmospheric conditions. | Underestimates ET under conditions of high movement of atmospheric air masses. |
| 9. | Evaporation pan | Integrate all climatological factors. | Evaporation continues during the night in the pan, which affects the PET estimates. |
| 10. | Hargreaves | Requires a minimum of climatological data. | Underestimates PET on the coasts and under high movements of air masses. |
| 11. | Hargreaves and Samani | Requires only maximum and minimum temperature data. | Needs to be evaluated in many locations for its acceptance. |
| 12. | Radiation | Is reliable in Equator, small islands and high altitudes. | Monthly estimates are often necessary outside Equator. |
| 13. | Makkink | Good for humid and cold climates. | It is not reliable in arid regions. |
| 14. | Linacre | Is precise on annual basis. | Precision decreases on daily base. |
| 15. | Priestly–Taylor | Reliable on humid areas. | Not adequate for arid zones. |

## CALIBRATION OF PET METHOD FOR LOCAL CONDITIONS

The methods that use weather data are not adequate for all the locations, especially in tropical areas and at high elevations. Local calibration is always necessary to obtain reliable and good estimates of the crop water requirements. Table 2.4 shows the data used in different equations in this chapter. For the Blaney–Criddle method, ET can be estimated using measurements of the soil moisture in lysimeters, and that can measure water entering and going out. Only ambient temperatures and rainfall data are necessary for complete calibration when determining appropriate monthly crop coefficient. The Jensen–Haise method [9] is recommended for periods of 5–30 days. To make a calibration, local field data or lysimeters can be used for periods of 5 days. For a monthly calibration, the ET can be estimated by soil moisture measurements, inlet and outlet flows, in lysimeters, and so forth. The Penman equation can provide precise estimations from a month to an hour depending on the calibration method. For short periods, lysimeters can provide the necessary data for the ET. Usually local calibration is completed through a calibration correction factor.

## CROP EVAPOTRANSPIRATION (ET$_c$)

To obtain the ET$_c$ (consumptive use), it is necessary to know crop and ambient conditions. This includes climate, soil moisture, crop type, growth stage, and the amount soil coverage by the crop.

**Table 2.4.** Parameters used in different formulas to estimate the PET.

| | Method | Temp. | HR | Wind | Sun | Radiation | Evaporation Class A | Ambient |
|---|---|---|---|---|---|---|---|---|
| 1 | Blaney–Criddle | * | 0 | 0 | 0 | | 0 | |
| 2 | Radiation | * | 0 | 0 | * | | 0 | (*) |
| 3 | Penman | * | * | * | * | | 0 | (*) |
| 4 | Class A Pan | | 0 | 0 | | * | * | |
| 5 | Thornthwaite | * | | | * | | | |
| 6 | Hargreaves | * | 0 | | | | 0 | |
| 7 | Linacre | * | | | | | * | |
| 8 | Jensen–Haise | * | | | | | | * |
| 9 | Makkink | * | | | | | | * |
| 10 | Priestly–Taylor | * | | | | | | * |
| 11 | Regression | | | | | | | * |

* = Measured data essential, 0 = Estimation required, (*) = Available data, but not essential.

The ET$_c$ indicates amount of water consumed for a given crop stage and the irrigation requirements can be determined. The procedure involves use of PET estimations and the experimental crop coefficients for the ET. This method is extensively used to schedule irrigation. The Blaney–Criddle method does not need a crop coefficient. The estimate of the ET$_c$ is only made in one step. Doorenbos and Pruitt [4] provided an appropriate crop coefficient to estimate the ET for specific crops. These procedures resulted in precise estimates for periods of 10 days to a month.

## CROP COEFICIENTS

The crop coefficients (K$_c$) are related to crop species, crop physiology, crop growth stage, days after planting, degree of ground coverage, and the PET. When using the coefficients, it is important to know, how these were obtained. The following is an empirical relation between ET$_c$ and PET:

$$K_c = [ET_c/PET] \qquad (20)$$

The combined $K_c$ includes evaporation from the soil surface and the plant surface. The evaporation from the soil surface depends on the soil moisture and soil characteristics. The transpiration depends on the amount and nature of leaf area index of a plant and the available soil moisture to the root zone. The $K_c$ can be adjusted to the available soil moisture and evaporation on the surface. Crop coefficient curve shows variation of $K_c$ with days after planting (see Figure 2.5).

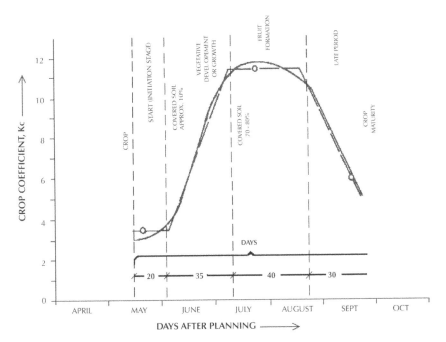

**Figure 2.5.** A typical crop coefficient curve.

## Reference Crop (Alfalfa)

The alfalfa is frequently selected as a reference crop, because it has high ET rates in arid regions [9]. Under these conditions, the PET is equal to the daily ET. When the crop is actively growing, alfalfa has a height of about 20 cm and there is sufficient available soil moisture. The PET obtained with alfalfa is usually higher for the Bermuda—grass, particularly in windy arid regions. The daily ET rates can be measured with the sensitive lysimeters.

## Crop Coefficient

Crop coefficients are given in Table 2.5. It is possible to estimate the consumptive water use (ET) using the crop coefficient and the calculated PET relation:

$$ET_c = Kc * (PET) \tag{21}$$

where: $ET_c$ = Crop evapotranspiration (consumptive water use), mm/day.

$K_c$ = Crop coefficient.

PET = Potential evapotranspiration, mm/day.

**Table 2.5.** Crop coefficients (Kc) at different growth stages of a given crop.

| Crop | Initial 1/ 2/ | Vegetative Development | Fruit Formation | Late Period | Crop Maturity | Total Growth Period |
|---|---|---|---|---|---|---|
| Banana | -------- | 0.70–0.85 | 1.00–1.10 | 1.90–1.00 | 0.75–0.85 | 0.70–0.80 |
| Beans: Green | 0.30–0.40 | 0.65–0.75 | 0.95–1.05 | 0.90–0.95 | 0.85–0.95 | 0.85–0.90 |
| Dry | 0.30–0.40 | 0.70–0.80 | 1.05–1.12 | 0.65–0.75 | 0.25–0.30 | 0.70–0.80 |
| Cabbage | 0.40–0.50 | 0.70–0.80 | 0.95–1.11 | 0.90–1.00 | 0.80–0.95 | 0.70–0.80 |
| Grape | 0.35–0.55 | 0.60–0.80 | 0.70–0.90 | 0.60–0.80 | 0.55–0.70 | 0.55–0.75 |
| Corn: Sweet | 0.30–0.50 | 0.70–0.90 | 1.05–1.20 | 1.00–1.15 | 0.95–1.10 | 0.80–0.90 |
| Corn: Field | 0.30–0.50 | 0.70–0.85 | 1.05–1.20 | 0.8–0.95 | 0.55–0.60 | 0.75–0.90 |
| Onion: Dry | 0.40–0.60 | 0.70–0.80 | 0.95–1.10 | 0.85–0.90 | 0.75–0.85 | 0.80–0.90 |
| Green | 0.40–0.60 | 0.60–0.75 | 0.95–1.05 | 0.95–1.05 | 0.95–1.05 | 0.65–0.80 |
| Pepper | 0.30–0.40 | 0.60–0.75 | 0.95–1.10 | 0.85–1.00 | 0.80–0.90 | 0.70–0.80 |
| Potato | 0.40–0.50 | 0.70–0.80 | 1.05–1.20 | 0.85–0.95 | 0.70–0.75 | 0.75–0.90 |
| Rice | 1.10–1.15 | 1.10–1.50 | 1.10–1.30 | 0.95–1.05 | 0.95–1.05 | 1.05–1.02 |
| Sorghum | 0.30–0.40 | 0.70–0.75 | 1.00–1.15 | 0.75–0.80 | 0.50–0.55 | 0.75–0.85 |
| Soy | 0.30–0.40 | 0.70–0.80 | 1.00–1.05 | 0.70–0.80 | 0.40–0.30 | 0.75–0.90 |
| Sugarcane, (stalk) | 0.40–0.50 | 0.70–1.00 | 1.00–1.30 | 0.75–0.80 | 0.50–0.60 | 0.85–1.05 |
| Tobacco | 0.30–0.40 | 0.70–0.80 | 1.00–1.20 | 0.90–1.00 | 0.75–0.85 | 0.85–0.95 |
| Tomato | 0.40–0.50 | 0.70–0.80 | 1.05–1.25 | 0.80–0.95 | 0.60–0.65 | 0.75–0.90 |
| Watermelon | 0.40–0.50 | 0.70–0.80 | 0.95–1.05 | 0.80–0.90 | 0.65–0.75 | 0.75–0.85 |

1/ Value for high humidity of HR 70% and low wind velocity, U = 5 meters per second.
2/ Value for low humidity of HR 20% and high wind velocity, U = 5 meters per second.

## SUMMARY

ET is a combination of two processes: Evaporation and transpiration. Evaporation is a conversion of liquid water into water vapor. The transpiration is the flow of liquid water from the soil to the surface of leaves/branches and trunk; and conversion of liquid water from the plant tissue into water vapor. ET processes are important for estimating crop irrigation requirements and for irrigation scheduling. PET is a water loss from the soil surface completely covered by vegetation. This chapter includes various Methods to estimate the ET and PET: Hydrologic method or water balance, climatic method, Penman equation, the Penman–Monteith equation, the Penman equation modified by Doorebos and Pruitt, Thornthwaite, Blaney–Criddle, the Blaney–Criddle equation modified by FAO, the Blaney–Criddle equation modified by Shih, Jensen–Haise, Stephens-Stewart, the Pan evaporation, Hargreaves method, Hargreaves and Samani modified method, Linacre method, Makkink method, Radiation method, Regression method, and Priestly–Taylor method. The methods that use weather data like Blaney–Criddle, Jensen–Haise, and Penman are not adequate for all the locations, especially in tropical areas and at high elevation. To obtain the crop water use $ET_o$, it is necessary to know crop and ambient conditions. The crop coefficients are related to the crop type, the crop physiology, crop stage, days after planting, the degree of coverage, and the PET.

## KEYWORDS

- **Atmosphere**
- **Available water**
- **Blaney–Criddle method**
- **Class A pan**
- **Class A pan coefficient**
- **Consumptive use**
- **Crop coefficient**
- **Crop evapotranspiration**
- **Crop water requirement Consumptive water use**
- **Drainage**
- **Evaporation**
- **Evapotranspiration**
- **Ground cover**
- **Irrigation requirement**
- **Percolation**
- **Potential evapotranspiration**
- **Precipitation**
- **Relative humidity**
- **Root zone**
- **Soil moisture**
- **Solar radiation**
- **Thornthwaite method**
- **Transpiration**

## BIBLIOGRAPHY

Refer to bibliography Section 2.1 literature cited.

# Chapter 3

## The Tensiometer: Use, Installation, and Maintenance

---

### INTRODUCTION

The tensiometer is an indirect method of measuring soil moisture in irrigated agriculture [1]. For an efficient irrigation system, it is necessary that water quantities are available for the plants during the crop season and precise measurements of soil moisture

---

[1]This Chapter is modified and translated from "Goyal, Megh R., José A. Santaella Pons y Luis E. Rivera Martínez. 1990. El Tensiómetro: Su Uso, Instalación y Mantenimiento. Capítulo V en: *Manejo de Riego Por Goteo* editado por Megh R. Goyal, páginas 143–176. Río Piedras, PR: Servicio de Extension Agrícola, UPRM." For more details, one may contact by E-mail: goyalmegh@gmail.com

are conducted. Plants require a specific amount of moisture in the root zone. The soil moisture needs may depend on the crop type and growth stage. During the irrigation process, irrigation application must provide the amount of moisture needed by the soil. The soil field capacity is a maximum amount of retention of water between soil particles. The difference between the permanent wilting percentage and the field capacity is an amount of water available for the plant [4]. When a quantity of water is applied, a percentage of this water is lost due to deep percolation, evaporation from the soil deep surface and surface runoff. It is of paramount importance that the available soil moisture must be measured during the crop season. Currently, there are various methods that can be used as a guide to determine when it is safe to apply the irrigation. In this chapter, we will describe selection, installation, service and maintenance of a tensiometer. *Methods to measure soil moisture are:* Manual palpitations, tensiometer, neutron scattering, evaporation pan, and gypsum blocks.

### WHAT IS A TENSIOMETER?

A tensiometer is an instrument that indicates the tension with which the water is adhered to the soil particles [11]. It is an indirect method to indicate if the soil has adequate available moisture for the plant growth. The changes occur on the film of water surrounding soil particles and these changes are expressed as fluctuations in the soil moisture tension. In fact, the tensiometer measures indirectly the soil moisture, in the root zone.

### HOW DOES THE TENSIOMETER WORK?

The principal components of the tensiometer are shown in Figure 3.1. The porous ceramic tip is a key component of the instrument. This should be correctly inserted in the soil. After the tensiometer has been properly installed, the water in the plastic tube comes in contact with the moisture in the soil pores, flowing in both directions until the equilibrium is reached. It is assumed that the soil moisture tension is same as indicated by the vacuum gage of the tensiometer [5]

Use of a tensiometer in different situations.

As the soil continues to loose moisture due to water use by the roots, the tension (or the suction) of soil is increased. This tension is measured with a vacuum gage of the tensiometer (Figure 3.1). When the soil is again wetted by the rainfall or irrigation, the tension falls since the soil water flows through the pores of the ceramic tip. The tensiometer readings can be related to water available to plants but cannot be used to determine the amount of water in the soil. However, a soil moisture curve can help to identify the status of soil moisture if one knows the soil tension.

## PARTS OF A TENSIOMETER

Agricultural workers and technicians must have the knowledge about the parts, use, installation and maintenance of the tensiometer [5]. This knowledge will help them to obtain correct readings while using the tensiometer.

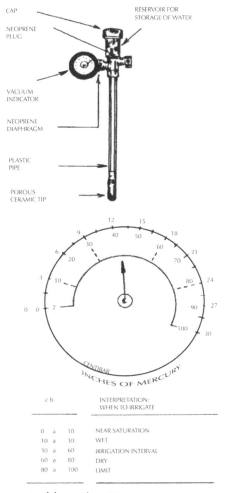

**Figure 3.1.** Principal components of the tensiometer.

## Plastic Tube

It provides rigidity to the instrument and allows storage of water between the indicator and the ceramic tip. This rigidity is a key to obtain exact readings of soil tension. The instrument must be built as rigid as possible to minimize shocks. The plastic tube of the tensiometer has a diameter of 2.2 cm (7/8 inches). This transparent tube is resistant to environmental conditions and chemical substances. The support inside the wall of the pipe provides a maximum resistance to protect connections of the indicator. The superior part of the pipe above the indicator allows immediate observation of the presence of air in the tensiometer.

## Porous Ceramic Tip

It is fabricated of porous ceramic with an excess bubbling pressure of 20 pounds per square inch. The tip is connected directly to the plastic pipe without the use of any adhesive agent. The tension constantly changes due to small quantities of water that move inside and out of the instrument through the ceramic tip. This creates equilibrium with the soil moisture suction. The movement of water in the soil is slow when the values of suction are high [7, 10].

## Reservoir

It provides space for the storage to substitute the water lost through the porous tip. It also helps in the    maintenance of the instrument. Through the reservoir, water is added and air bubbles are eliminated. This part is also made of transparent plastic. The Neoprene rubber plug is resistant to climatic changes and can be substituted easily if necessary.

## Vacuum Gage

It indicates the vacuum or suction that occurs inside the tensiometer due to water movement across the porous ceramic tip. The analog suction indicator (or vacuum gauge) is sealed shut to resist outside conditions and can be completely submerged in water without suffering any harm. It is built in such a way to keep water and air out of the indicator. The body is covered with neoprene diaphragm for protection against shocks, dirt and moisture. It also offers compensation for variations in temperature and barometric pressure. The indicator is graduated to read from 1 to 100. Each division is equal to one centibar. At sea level, one centibar of suction is equal to a water column of 10 cm. One bar = 0.987 atmosphere = 100 centibars. The reading of a "two" indicates that the soil is saturated of water. In practice only the graduations of 10–80 are useful. When soil looses moisture due to water loss or crop water use, the readings on vacuum gage may go over 80. When this occurs, air enters the ceramic tip and the readings are incorrect.

Due to mechanical problems associated with the construction of the vacuum gage, the position of zero on the gage is an error. If we closely observe the gage, one may notice that there are only four divisions between the "0 and 10" marks while there are five divisions between 10 and 20 and so on. Each division represents two centibars. Therefore zero position represents a reading of two centibars instead of a zero. The first division after the zero centibar represents a value of four centibars. Though we are only interested in soil suction values from 10 to 80, yet it is very important that the reader knows this information. Other compatible indicators also exist for the tensiometer. Digital indicators can be connected to electrical sensors that allow the reader to view readings of several tensiometers simultaneously. The response of these transducers is fast and is more reliable than the conventional gages. These sensors function on the basis of the relationship between the suction and voltage when synchronizing the tensiometer with the digital indicator.

Tensiometer with electrical sensor is employed in an automatic irrigation system. Unfortunately, the digital indicator is very expensive.

## PREPARATION AND CALIBRATION

The tensiometer should be prepared and calibrated before any installation in the field. Distilled water is used to fill the plastic tube and reservoir cup. If the distilled water is not available, then boiled water is also an option. Water contains dissolved air. During the distillation or boiling process, this air is lost to the atmosphere leaving the liquid free of air. Gases dissolved in crude water form bubbles inside the plastic tube. These bubbles interfere with the readings of an indicator or transducer. To reduce or eliminate the growth of algae and bacteria and to facilitate water movement inside the instrument, a special conditioning blue dye is added. This special solution is prepared by adding one cubic centimeter of the blue dye to one liter of water liquid, following the instructions on the label of the bottle. Remove the cap of the tensiometer and the plastic bag on the porous ceramic tip. Then fill the instrument with the water solution without covering it yet. Keep the ceramic tip submerged in clean water for few days before the actual installation. Again, fill the tensiometer with the solution (including the reservoir cap). After the system is completely filled up, apply suction with the hand pump to remove the air bubbles and eliminate any partial vacuum. In such a way, the air is removed from the ceramic tip and the plastic cup of the tensiometer. Surprisingly, big bubbles may appear coming out from the base of the plastic tube. This indicates that the connection of the ceramic tip to the tube is not tight enough. Any appearance of the air bubbles above the vacuum gage implies that the connection between the gage and the tube is not completely sealed. Any possible leakage within the instrument must be corrected. Small bubbles may be due to the air dissolved in water. To test the transducer (vacuum gage), allow the tensiometer to gradually dry until the gage reads a high suction value. After this, place the ceramic tip in water. The reading must lower within seconds and reach zero within three to five minutes. This test proves that the conductivity of the porous ceramic tip is satisfactory. A thin plastic polyethylene tube (spaghetti) can easily be inserted in the wider plastic cup of the tensiometer to eliminate any air bubbles from the tensiometer. When all the air bubbles disappear, then the compartments are filled with the conditioning solution. The rubber cap of the tensiometer is placed by twisting it until the neoprene seal makes contact with the lower part of the reservoir. Finally, a quarter of a turn is allowed to seal the neoprene rubber plug without causing any damage. Remember, not to tight too much [8].

## SELECTING SOIL DEPTH AND LOCATION FOR THE INSTALLATION

The following factors should be considered for the selection of a tensiometer station:

1. Crop species and the depth of a root system.
2. Soil type and variation.
3. Natural surroundings and land topography.
4. Type of an irrigation system.
5. Desired level of accuracy and control.

The main objective of the tensiometer is to determine the status of soil moisture in the rhizosphere. Therefore, it is very important that the ceramic tip is correctly placed in root zone of a developing plant. It is of equal importance that the instrument is placed at a specific location where irrigation and rainfall can actually reach the ceramic tip. Plants (such as ornamental ones) with shallow root system of <45 cm depth only need one tensiometer per station. The instrument should be installed at a depth equivalent to ¾ of actual root depth. The tensiometer should be located deeper into the soil as the root depth increases with the crop stage (see Figure 3.2). Plants with deep root system must have at least two tensiometers per station. Shallow tensiometer is placed at one fourth of the root depth. The second tensiometer is placed at ¾ of the root depth (see Figure 3.3). In a fruit orchard, it is recommended to place shallow tensiometer at 30–45 cm depth and second tensiometer at 60–90 cm of root depth (see Table 3.1). Utilizing two tensiometers per station helps technicians recognize the condition of a soil moisture in the root zone. Irrigation begins when the shallow tensiometer indicates high suction value. Irrigation continues until the readings on the deeper tensiometer begin to fall. This is very important to know that the irrigation water has penetrated to the desired depth and that all the roots above this point have been watered adequately. Tensiometers must be installed in strategic locations in the field because different types of soil have different levels of infiltration and water retention capacity. If the land has distinctive soils at several locations of the field, we must install a tensiometer at each location. If the land is flat and uniform and the irrigation system is permanent, then a pair of tensiometers for every 20 acres is enough. In case of land with non uniform slopes, a pair of tensiometers must be installed at both highest and lowest points. In most irrigation systems, we should consider soil variations to provide special needs of crops at high or low areas of the field [1, 2, 3, 7].

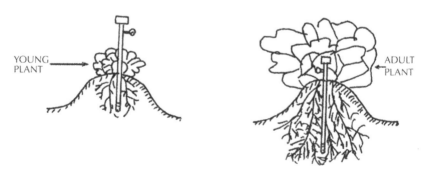

**Figure 3.2.** Selection of a location for the installation of a tensiometer.

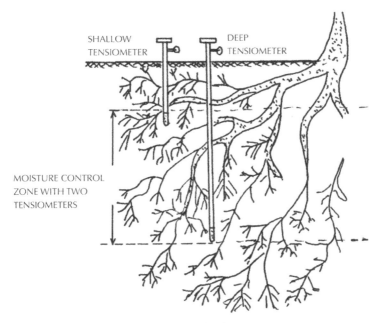

**Figure 3.3.** The use of two tensiometers in a fruit orchard.

**Table 3.1.** Depth for location of a tensiometer.

| Crop | | Depth of a tensiometer, cm. | | |
|---|---|---|---|---|
| **English** | **Spanish** | **Shallow** | **Deep** | **Very Deep** |
| Alfalfa | Alfalfa | 45–60 | 90–120 | 150 |
| Almond | Almendra | 60 | 120 | 180 |
| Apple | Manzana | 50 | 100 | 150 |
| Apricot | Albaricoque | 60 | 120 | 180 |
| Artichokes | Alcachofas | 45 | 45 | – |
| Asparagus | Espárragos | 45–60 | 90 | – |
| Avocado | Aguacate | 30 | 60 | 90 |
| Banana | Guineo | 30 | 60 | – |
| Barley | Cebada | 45 | 90 | – |
| Beans (bush) | Habichuelas | 25 | – | 45 |
| Beans (lima) | Habas | 45 | 90 | – |
| Beans (Pole) | Hab. Polacas | 45 | 90 | – |
| Beets (sugar) | Remolacha dulce | 45 | 90 | – |
| Beets (Table) | Remolacha de mesa | 30–45 | 90–120 | – |
| Blueberry | Vaccinio | 30 | 60 | – |
| Broccoli | Brecól | 30 | 50 | – |
| Cabbage | Repollo | 30 | 50 | – |

**Table 3.1.** *(Continued)*

| Crop | | Depth of a tensiometer, cm. | | |
|---|---|---|---|---|
| **English** | **Spanish** | **Shallow** | **Deep** | **Very Deep** |
| Cantaloupe/Honey dew | Melón | 45 | 90 | – |
| Carnation | Claveles | 30 | 90 | – |
| Carrot | Zanahoria | 30 | 60 | – |
| Cauliflower | Coliflor | 30 | 60 | – |
| Celery | Apio | 25 | 50 | – |
| Cherry | Cerezas | 60 | 120 | 150 |
| Christmas tree | Arbol de Navidad | 30 | 60 | – |
| Citrus,    orange, lemon | Cítricos, china, limón | 45 | 90 | – |
| Grapefruit | Toronja | 45 | 90 | – |
| Coffee | Café | 45–60 | 90–120 | – |
| Corn (Sweet) | Maíz Dulce | 30 | 75 | – |
| Corn (Field) | Maíz de campo | 45 | 90 | – |
| Cotton | Algodón | 45 | 90 | 120 |
| Cranberry | Arándanos agrios | 45 | 90 | – |
| Cucumber | Pepinillo | 45 | 90 | – |
| Date Palm | Palma de dátiles | 60 | 120 | 150 |
| Egg plant | Berenjena | 30 | 60 | – |
| Fig tree | Higos | 45 | 90 | – |
| Garlic | Ajo | 30 | 60 | – |
| Grain & Flax | Granos | 45 | 90 | – |
| Grape vines | Uvas | 60 | 120 | 150 |
| Hops | Lúpulo | 60 | 120 | 150 |
| Joroba | Joroba | 60 | 120 | 150 |
| Kivi | Kivi | 45 | 90 | 120 |
| Lettuce | Lechuga | 30 | – | – |
| Macadamias | Macadamias | 30 | 30 | 90 |
| Maize | Maíz | 45 | 90 | – |
| Melon | Melones | 45 | 90 | – |
| Milo | Millo | 60 | 120 | – |
| Mint | Menta | 30 | 60 | – |
| Monterrey Pine | Pino | 30 | 60 | – |
| Mustard | Mostaza | 45 | 90 | – |
| Nectarine | Nectarina | 45 | 90 | – |
| Okra | Quimbombó | 45 | 90 | – |
| Olive | Oliva | 60 | 120 | 150 |

**Table 3.1.** (Continued)

| Crop | | Depth of a tensiometer, cm. | | |
|------|------|---------|------|-----------|
| English | Spanish | Shallow | Deep | Very Deep |
| Onion | Cebolla | 30 | 120 | – |
| Papaya | Papaya | 30 | 60 | – |
| Parsnip | Chirivia | 45 | 90 | – |
| Pasture Permanent | Pasto Permanente | 20–37 | – | 60–75 |
| Peach | Melocotones | 45 | – | 150 |
| Peanut | Maní | 30 | 60 | – |
| Pear | Pera | 45 | 90 | 120 |
| Pea | Guisantes | 45 | 90 | – |
| Beet (sugar) | Remolacha dulce | 45 | 90 | – |
| Pecan | Nuez encarcelasa | 45 | 90 | 120 |
| Pepper | Pimiento | 37 | 75 | – |
| Persimmons | Placamineros | 45 | 90 | – |
| Pineapple | Piña | 37 | 75 | – |
| Pistachio Nuts | Nueces | 60 | 120 | 150 |
| Pomegranate | Granada | 45 | 90 | – |
| Potato (Irish) | Papa (Irlanda) | 25 | 45 | – |
| Potato (Sweet) | Batatas | 45 | 90 | – |
| Plum | Ciruela | 60 | 120 | 180 |
| Prune | Ciruela Pasa | 60 | 120 | 150 |
| Pumpkin | Calabaza | 45 | 90 | 120 |
| Radish | Rábano | 30 | – | – |
| Raspberry | Frambuesa | 45 | 90 | – |
| Sorghum | Sorgo | 45 | 90 | – |
| Soy bean | Habichuela Soya | 45 | 90 | 120 |
| Spinach | Espinaca | 30 | 60 | – |
| Squash (Summer) | Calabazín | 37 | 75 | – |
| Strawberry | Fresa | 15 | 30 | – |
| Sudan Grass | Yerba Sudán | 45–60 | 90–120 | – |
| Sugar Cane | Caña | 45 | 90 | – |
| Sunflower | Girasol | 60 | 120 | 150 |
| Tobacco | Tabaco | 20–37 | 75 | – |
| Tomato | Tomate | 45 | 90 | – |
| Turnip | Nabo | 45 | 90 | – |
| Walnut | Nueces | 60 | 120 | 180 |
| Watermelon | Melón de Agua | 45 | 90 | 120 |
| Wheat—Hay | Trigo—Heno | 45 | 90 | – |

## LOCATION OF A TENSIOMETER IN AN IRRIGATION SYSTEM

### Furrow Irrigation

Tensiometer is installed on the inner side of the raised bed of a furrow (see Figure 3.4). Crops are planted in rows on the side of furrows. When the bed spacing is from 1.5 to 1.8 meters, tensiometers are located in the direction of the crop row. For fruit trees irrigated by ditches, two tensiometers are installed: One at the side the tree facing the noon sun and the second tensiometer at a distance of 30–45 cm from the furrow.

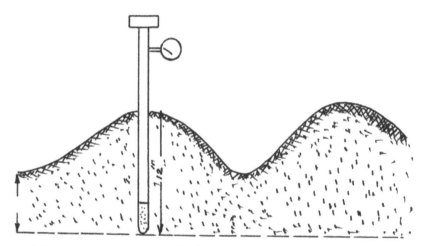

**Figure 3.4.** The tensiometer at the inner side of a furrow.

### Overhead (Sprinkler) Irrigation

Tensiometers should be installed at a location where the plant receives afternoon sunlight. The location must not be covered with branches of a tree and should not be located at the lowest point with accumulation of runoff. When portable and permanent irrigation systems are used, the tensiometer should not be installed too close to the irrigation line or a sprinkler. For fruit trees, two tensiometers should be installed between two trees down the row. The space between the sprinkler and tensiometer should be free of any obstructions such as tree trunk, branches, etc. For row crops, the tensiometer should be installed at the bed of a planted crop. In case of forage crops or narrow row crops, specific locations are not important.

### Drip Irrigation

In fruit trees, tensiometer should be placed on the sunny side of the tree at 30 or 45 cm (1–1½ feet) from the emitter. In newly planted fruit trees, the shallow tensiometer should be located near the root zone without giving importance to the location of an emitter. The row crops require tensiometers down the row. When measuring water movement from the emitter, one can install an extra tensiometer at a distance of 30–45 cm (1–1½ feet) from the emitter and down the lateral [11].

## INSTALLATION

It is important that there exists a good contact between the ceramic tip and the soil. The size of the hole needed to insert the tensiometer into the soil should be of diameter equal to that of a tube of the instrument. The external diameter of the plastic tube and ceramic tip is 2.2 cm ($^7/_8$ in). Generally, a galvanized iron pipe of 1.25 cm (½ in) is used to make the hole. In a hole with a larger diameter, contact is lacking and air penetrates the ceramic tip causing error in observations. Also the runoff water can enter into a space between the hole and wall of the tube, thus causing incorrect readings. The hole is dug to a desired depth to install the tensiometer. The metal pipe must be removed carefully to keep the hole clean and round. The tensiometer is inserted slowly and gently from top to bottom. Do not apply pressure on the gage to push it into the hole. Apply a gentle pressure directly downward on the cap of the reservoir. It is desirable to leave a space of 2.5 cm between lowest part of a gage and the soil surface. This allows free expansion and contraction of the diaphragm of the gage. At ground level, some loose soil is placed around the plastic tube and is pressed lightly so that runoff water does not enter into a space between the tube and the hole.

The tensiometer is identified with the station number and installation depth. After installation, fill up the reservoir compartment with a solution made up of distilled water and a blue dye. Remove air bubbles from the reservoir compartment and plastic tube. A good extraction of air increases the sensitivity of tensiometer and the accuracy of reading. After installation, it is recommended to wait for at least two days until the water from the instrument reaches equilibrium with the surrounding soil.

## READINGS OF TENSIOMETER

The best time of the day for reading the tensiometer is early in the morning. At this time, the water movement between the soil and the plants is at a minimum level because of a state of equilibrium. Readings should be taken at the same hour on a regular basis. Frequency of readings depends on the crop, soil and climatic conditions and the irrigation method. Larger is the interval of irrigation, lesser is the frequency of readings. Tensiometer should be read daily for plants with shallow roots and for drip irrigation. In sandy soils and hot climates, and for other irrigation methods, the readings are taken three times a week.

In moderate climates, one reading a week is enough. Before taking the readings, lightly tap the vacuum gage. A quick movement of the needle will tell if the soil is drying or receiving moisture. Readings must be taken always before servicing the instrument. Maintenance is given to the tensiometer in the same order in which it is read. The use of the tensiometer is most efficient when the data is copied on graph paper (see Figure 3.5) and the curve is drawn. The curve describes the tendency of crop water use. This will answer questions about the factors that affect the water consumption. These factors include climatic changes, soil variations, fruit development, etc. The graph helps to indicate when and how much to irrigate. The graph should be updated on a daily basis.

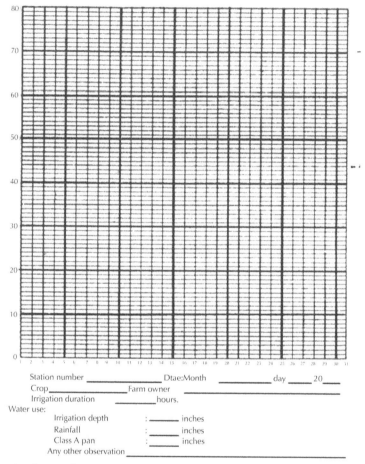

**Figure 3.5.** Graph paper for recording tensiometer readings.

The record of tensiometer readings can help us to give information about rainfall, diseases, fertilizers, etc. This information can create a progressive record of the crop.

To adjust irrigation based on tensiometer readings, it is important to correctly identify the soil type. Different soil types have distinctive water absorption capacities. It implies that each soil needs a specific amount of water for a specific crop. This method requires a lot of patience. It generally takes months to determine the best irrigation interval on the basis of a tensiometer.

If a deep tensiometer keeps a low reading after four days of irrigation, then it is an indication of excessive water use. Contrary to this, if the tensiometer indicates high tension after four days, then it indicates that the irrigation was not enough and the depth of irrigation should be increased. Finally, if a surface tensiometer indicates a high tension and the deep shows a one low tension, then it is necessary to irrigate more than once daily, (See Table 3.2).

**Table 3.2.** Intervals of optimum soil tension (in centibars) before the irrigation.

| Crop | Tension |
|------|---------|
| Avocado | 50 |
| Broccoli | 45–55 |
| Cabbage | 60–70 |
| Carrot | 55–65 |
| Cauliflower | 60–70 |
| Grape vines | 40–70 |
| Lawn grass | 20–30 |
| Lemon | 30–40 |
| Lettuce | 40–60 |
| Maize (corn), sweet | 50–70 |
| Melon | 35–40 |
| Onion | 55–65 |
| Orange (Citrus) | 20–40 |
| Potato | 30–50 |

## INTERPRETATION OF THE READINGS OF A TENSIOMETER

1. Near saturation (2–10 cb): Soil stays close to saturation on an irrigation day and a few days later. If a low reading persists, the soil might be flooded or the tension column is broken.
2. Field capacity (11–20 cb): Irrigation must stop in this interval to prevent water loss by deep percolation and loss of nutrients. Sandy soils will be at their field capacities at the lower range while the higher range will be for clayey soils.
3. Irrigation interval (30–60 cb): Usual interval to initiate the irrigation. Generally speaking, in sandy soils irrigation is applied with readings of 30–40 cbars. Usually in loamy soils, irrigation starts with readings of 40–50. In clayey soils irrigation will be applied in range from 50 to 60. Commencing the irrigation in these intervals assures that soil water will be available to the plant.
4. Dry (60–70 cb): The stress interval. Crops are not necessarily under water stress. In some soils, water is available for the plant but is not enough.
5. Breakage of tension (80 cb): This is the limit of precision of the tensiometer. Readings over 80 are possible but the tension breaks the water column between 80 and 85 cbars.

## AUTOMATION OF AN IRRIGATION SYSTEM WITH TENSIOMETERS

Tensiometers can be used with an automatic irrigation system (Figure 3.6). These instruments are inserted in the soil at a specific depth depending on crop type, soil and irrigation method.

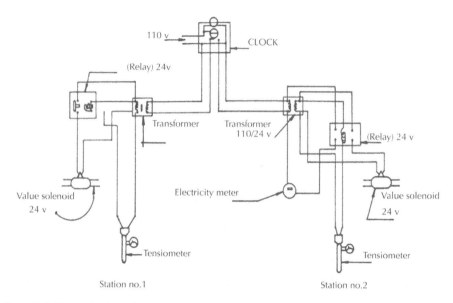

**Figure 3.6.** Electronic circuit for an irrigation system.

Each tensiometer includes an electrical conductor that converts measurement of the soil tension into an electrical signal. This signal is received by a control panel, which operates a solenoid valve to regulate the flow of water from the system to the field.

This instrument is adjusted to readings of vacuum gage during the irrigation cycle. When soil moisture is lowered and the tensiometer marks the predetermined maximum tension and an electronic signal is emitted. This signal activates the irrigation system by opening the solenoid valve. When the soil begins to saturate and reaches maximum moisture, the electronic tensiometer sends a message to the control unit to close the solenoid valve and the irrigation is stopped.

Automatic irrigation system has following advantages:

1.  Better control of quantity and frequency of irrigation.
2.  Money is saved since no labor is needed to open and close irrigation lines in the field.
3.  One does not have to visit the field during odd hours to allow the irrigation.
4.  It registers tendencies of water consumption with a system of automatic recording units.

## MAINTENANCE AND SERVICE OF A TENSIOMETER

In a moist and wet soil, the tension readings are low; air will not accumulate under the plug of the reservoir cup. If the soil is relatively dry and the readings reach 40–60 cbars, air can rapidly accumulate in few days. Air bubbles increase the response time to changes in tension, thus lowering the precision of the instruments.

After installation, the instrument must be revised frequently and daily. Excess air is removed and then water is added to the reservoir cup. With periodic service and maintenance, less air will enter into the system. Then the maintenance can be done once a month. If the tensiometer is not serviced and maintained for a long period of time, then the water level in the reservoir is lowered significantly. It must be filled up again with water. Air bubbles must also be removed with a hand vacuum pump. Under saline conditions, maintenance is essential to keep the ceramic tip free of salts [2, 3].

## Vacuum Hand Pump

This is an instrument to remove air from the tensiometer [9]. It has a universal suction tip that adjusts itself to all models of tensiometer. The cap of the reservoir compartment is removed and the suction to the instrument is applied with the pump. Four or five strokes should produce a reading no greater than 80 cbars. The pump should stay attached to the top of a tensiometer without holding it. If the manual vacuum pump is not available, one can use a small flexible polyethylene tube (spaghetti) to remove the air bubbles from the tensiometer (See Figure 3.7).

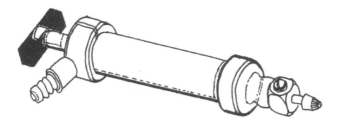

**Figure 3.7.** Vacuum hand pump.

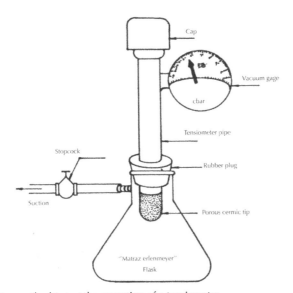

**Figure 3.8.** Suction method to test the operation of a tensiometer.

## Suction Method

This method consists of placing the tensiometer inside a "Matraz Erlenmeyer flask" through a rubber plug (see Figure 3.8). A known amount of suction is applied around the ceramic tip by opening a stopcock. The vacuum gage will show a reading corresponding to the applied suction. Then the suction valve is closed. If the reading of the instrument remains unchanged, then it implies that there is no loss of suction from the seal of the ceramic tip and the gage. If the reading changes, then there is a loss of suction somewhere in the instrument. To find this spot, we add water through the reservoir cup and apply the suction again. The instrument is allowed to flood and the top of the reservoir is covered. If the reading remains unchanged, then there is no loss. If it is changed, then there is an escape and the spot is easy to locate. This method reduces the testing time to approximately 10 minutes.

## STORAGE

After crop season is over, the tensiometer must be removed, emptied and stored. For removing the tensiometer from the soil, rotate to loosen it from the soil. Then grasp the main tube and pull it upwards. The ceramic tip should never be allowed to dry up. Immediately after removing it from the soil, cover the ceramic tip with a wet towel to keep it wet, while it is being packed. Even a few minutes of exposure to the air can permanently seal the pores. Wet packing is recommended for temporary packing. Fill up and cover the tensiometer. Clean the external ceramic tip and submerge it in water. This method keeps the instrument in operating status and ready for immediate installation. Dry packing is only recommended for long periods of storage. Carefully, clean and wash the ceramic tip with a brush. Clean the surface of the plastic with a soap solution. Place the instrument in an inverted position. Hang it in a clean and dry place free of dust. Cover the ceramic tip with a transparent plastic bag.

## LIMITATIONS

The tensiometer is filled with water and is limited by the physical properties of water. The gage indicates the difference in suction inside and outside the plastic tube. This suction depends on the atmospheric pressure. The operational limit of the tensiometer is 80 cbars at sea level. For each 300 meters of height above sea level, pressure lowers by 3.5 cbars. A reading of 80 cbars in a tensiometer at sea level equals 62.5 cbars at 1500 meters of elevation:

$$[(3.5/300 \times 1500 = 17.5), \text{ and } (-17.5 + 80 = 62.5)].$$

Tensiometer with a common vacuum gage does not respond quickly. A good precision is possible for changes in tension with tensiometer equipped with electric digital sensors.

## SUMMARY

A tensiometer is an instrument to measure the tension with which the water is adhered to the soil particles. The tensiometer shows the changes and fluctuations that occur on the water film surrounding the soil particles. Readings of the tensiometer can be related to the water available to plants, but can not be used to determine the amount of soil moisture. The parts of the tensiometer are plastic tube, porous ceramic tip, reservoir and vacuum gage. This chapter discusses the procedure to prepare and calibrate a tensiometer; to select and locate the site for installation of a tensiometer; to install a tensiometer; to provide adequate service and maintenance; and to store the tensiometer for short and long terms.

## TROUBLE SHOOTING

| CAUSE | REMEDY |
|---|---|
| **The instrument always reads 2** | |
| 1. Soil is saturated of water with the irrigation, rainfall or poor drainage. | Do not irrigate. |
| 2. The tensiometer has no water. | Fill the tensiometer with water and conduct the maintenance procedure. |
| 3. Loss of suction due to a low level of water in the tensiometer. | Review the calibration with a vacuum hand pump (The gage must read 80–85 cbars). |
| **The tensiometer does not indicate the correct tension** | |
| 4. The actual soil moisture is different from the one that is indicated by a tensiometer. | Take soil samples within 15 cm of the tensiometer station and at the same depth of the ceramic end with a drill or shovel. Determine the soil moisture content with an oven dry method. |
| **The water needs to be replaced frequently** | |
| 5. High suction readings for several days indicating little irrigation. | Water in the reservoir cup is needed. |
| 6. The soil is not properly pressed around the ceramics tip. | Revise the installation. |
| 7. The seal leaks when it is closed. | Change the neoprene plug of the cap. |
| 8. There is a leakage a in the connection of the indicator. | Revise the connection and conduct the necessary maintenance. |
| **Little response to the irrigation** | |
| 9. Low infiltration rate of the soil. | Clean the ceramic tip. |
| 10. The ceramic tip is partially sealed with salts. | Give light taps to the indicator before taking the readings. Also revise and clean the ceramic tip. |
| 11. The needle of a gage moves slowly. | Replace the ceramic tip and neoprene plug. |
| **Wide variation in readings** | |
| 12. Topography and different soil types. | Install an adequate number of tensiometers for a reliable control of the irrigation. |

**KEYWORDS**

- Automation
- Bar
- Cap or plug
- Centibar (cbar)
- Drainage
- Evaporation
- Field capacity
- Frequency of irrigation
- Gypsum blocks
- Infiltration
- Low tension
- Maintenance
- Neutron scattering
- Permanent wilting percentage
- Plant nutrient
- Polyethylene (PE)
- Pump
- Rhizosphere
- Root system

- Root zone
- Sand
- Soil moisture
- Solenoid valve
- Tensiometer
- Tensiometer station
- Term
- Vacuum gage
- Vacuum pump

**BIBLIOGRAPHY**

Refer to bibliography Section 3.1 for the literature cited.

# Chapter 4

## Irrigation Systems

Kenneth H. Solomon, Ph.D., P.E.

Founding Director of Fresno Center for Irrigation Technology, California. Past Chairman and Professor of BioResource and Agricultural Engineering Department at California Polytechnic State University

## INTRODUCTION

The art of irrigation is very ancient and has been essential for the development and growth of some civilizations. The Code of Hammurabi, sixth king of the first dynasty of Babylonia, indicates the importance of irrigated land. II Kings 3: 16–17 of the Holy Bible alludes to irrigation, in 2000 B.C. In the same year, the queen of Assyria diverted the Nile River to irrigate the Egyptian desert. The channels that they constructed during her kingdom still work. Irrigation is also mentioned in the old documents of Syria, Persia, India, China, Java and Italy. The importance of irrigation in our times has been defined accurately by N.D Gulati: "In many countries irrigation is an old art, as much as the civilization, but for humanity it is a science, the one to survive [3]." Irrigation water is an important component in the hydrologic cycle (Figure 4.1). The irrigation is applied to achieve following objectives:

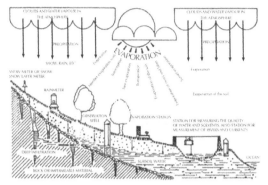

**Figure 4.1.** Irrigation as a component in the hydrological cycle.

1. To provide the necessary moisture for crop development.
2. To ensure sufficient supply of water during droughts of short duration and unpredictable climate.
3. To dissolve soil salts.
4. It is a way to apply agrochemicals.
5. To improve the ambient conditions for the vegetative growth.
6. To activate certain chemical agents.
7. To generate operational benefits.

## SELECTION OF IRRIGATION SYSTEMS

The irrigation can be accomplished by different methods: surface, subsurface, sprinkler and drip [2]. With any of these means, it is necessary to select an irrigation system, before the design, the equipment specification and installation can proceed. In order to make a suitable selection of the system, one should carefully consider the capacities and limitations of all the potential alternatives. Tables 4.1 and 4.2 indicate the aspects that must be considered in the selection of an irrigation system. These aspects are listed as follows [9]:

1. The crop and related factors: Type, root depth, water consumption, development of diseases.
2. Soil characteristics: Texture and structure, depth and uniformity, infiltration rate, potential of erosion, salinity and internal drainage, topography and degree of irregularity.
3. Water supply: Source, quantity available and reliability, quality, solids in suspension, chemical analysis.
4. Value and availability of land.
5. Limitations and obstructions of flooding.
6. Phreatic level: Level of underground aquifer.
7. Climatic conditions: Flood and furrow systems are usually associated with long irrigation cycles (many days between irrigations), so while the evaporation for the first few days after wetting may be high, the soil surface soon reaches "air dry" condition, and evaporating for the remaining days of the cycle is essentially zero—the average evaporation over the entire cycle is often low.

8. Availability and reliability of energy.
9. Economic considerations: Investment of required capital, availability of credit and interest rate, durability of the equipment and annualized cost, inflation costs, crop yield and value.
10. Available technology.
11. Social considerations: Political and legal issues, cooperation of the habitants, availability and reliability of the manual labor, level of knowledge and specialization of the manual labor, expectations of the government and inhabitants, desirable level of automation, potential damage by vandalism, and health issues.

Table 4.1. Conditions for the selection of an irrigation method.

| Conditions | Irrigation methods | | | |
|---|---|---|---|---|
| | Flood | Furrow | Sprinkler | Drip |
| Topography | Moderate to irregular | Moderate | Irregular | Irregular |
| Soil permeability | Good | Good | Excessive to good | Excessive to good |
| Potential for erosion | High | High | Low | Low |
| Crop characteristics | Sown by broadcast | Sown in rows | Crop value variable | Crop value variable |
| Water flow requirements | High | High | Moderate | Low |

Table 4.2. Comparison of irrigation methods.

| TOPIC | Flood | Furrow | Sprinkler | Micro |
|---|---|---|---|---|
| Evaporation loss | Low | Low | Medium | Minimum |
| Wetting of the foliage | High | Medium | High | Minimum |
| Water consumption by weeds | High | High | High | Minimum |
| Surface drainage | High | High | Medium | Minimum |
| Control of irrigation depth | Minimum | Minimum | Medium | High |
| Crop yield per unit of applied water | Minimum | Minimum | Medium | High |
| Uniformity in the crop yield | Little | Medium | Medium | High |
| Soil aeration | Minimum | Little | Little | High |
| Interference of other tasks by the irrigation method | Low | Low | High | Low |
| Application of fertilizers and pesticides through the irrigation water | Minimum | Minimum | Moderate | High |
| Operation and labor cost | Low | Low | Moderate | High |
| Leveling of the land is required | High | High | Low | Minimum |
| Automation of the system | Low | Low | High | High |
| Energy requirements | Low | Low | High | High |
| Quality of water | Minimum | Minimum | Moderate | High |
| Use of filters | Minimum | Minimum | Moderate | High |
| Control of diseases and pests | Minimum | Minimum | Moderate | High |

## Capacities and Limitations
### Crop, soil and topography
*Surface irrigation or irrigation by infiltration*
There are different methods of surface irrigation for almost every crop. In some regions, hay crops are irrigated by the furrow method. The flood irrigation method is adequate for forage crops after an adequate leveling. Row crops are irrigated by the furrow method. The irrigation system works better when the soil is uniform because of good infiltration of the water. The length of a furrow is limited to 100 meters in a heavy textured soil. However it can go up to 400 meters in well textured soils. Furrow irrigation is adaptable in all soil types; however soils with fast or slow infiltration rates may need an excessive manual labor. Uniform and slopes of slight incline adapt better to the infiltration method of irrigation. The irregular topography and steep slopes increase the cost of leveling and reduce the length of a furrow. The deep plowing can affect soil productivity requiring a special fertilization. Steps, benches or terraces may be required to control erosion in the case of high flow rates.

### Sprinkler irrigation
Almost every crop and soil can be irrigated with some type of sprinkler irrigation system, although the crop characteristics and crop height must be considered when the system is selected. Sometimes the sprinklers are used to germinate the seed and to cover soil. For this purpose, the light and frequent applications can be easily accomplished with some type of sprinkler system.

Soils with infiltration rates less than 5 mm/h may require special considerations. Sprinklers can be used in soils with depths too shallow to permit effective surface irrigation. In general, the sprinklers can be used in any topography that can be cultivated. Generally the leveling of the land is not required. The sprinkler system is designed (with some exceptions) so that it can apply to water at lower intensity than the infiltration rate of the soil in order that the quantity of water infiltrated at any point depends on the intensity of the application and time of application and not on the infiltration rate of the soil.

### Drip irrigation
Drip/micro or trickle irrigation is more convenient for vineyards, tree orchards and row crops. The principal limitation is the high initial cost of the system that can be very high for crops with very narrow planting distance. Forage crops cannot be irrigated economically with drip irrigation. Drip irrigation is adaptable for almost all soils. In very fine textured soils, the intensity of water application can cause problems of aeration. In heavy soils, the lateral movement to the water is limited, thus more emitters per plant are needed to wet the desired area. With adequate design, use of pressure compensating drippers and pressure regulating valves, drip irrigation can be adapted to almost any topography. In some areas, drip irrigation is used successfully on steep slopes.

### Quantity and quality of water
*Surface irrigation*
It is important that the water applied for irrigation reaches the end of the field relatively quickly to obtain a uniform irrigation. The volume flow rate required varies from 15

to 300 liters per second. The furrow method is not well suited for leaching salts from the soil, since water cannot stay in the soil for a required time. However, flood irrigation with border strips is ideal for this situation.

### Sprinkler irrigation

Sprinkler irrigation adapts particularly well to a situation with high water table, since the sprinkler equipment can make the application with a determined volume of water. Sprinkler irrigation generally requires the application of smaller quantity of water than flood irrigation.

### Drip irrigation

With drip irrigation, the application of water is less intense during a period of time longer than in other methods of irrigation. The most economic design would utilize water flowing throughout the crop area during almost all the day, every day, during peak periods of water use. If the water is not available constantly, then it may be necessary to store water. Saline water can be used successfully, with special precautions. The salts tend to concentrate in the perimeter of the wet volume of soil. For longer time intervals between the irrigations, the movement of the water in the soil can be reversed returning the salts to the root zone. When it rains after a period of accumulation of salts, the normal irrigation must be continued until approximately 50 mm of rain have fallen to prevent the damage by salts. In the arid regions, where annual rain is insufficient (<of 300 to 400 mm) to leach the salts, the artificial leaching of salts may need more time, requiring the use of a complementary sprinkler or surface irrigation method.

Sand (hydro cyclone) filters and screen filters are used to remove suspended solids in the irrigation water. A media filter can remove organic clogging agents. Chemical treatment of the water may be required to control biological activity, to adjust the pH or to avoid the chemical precipitation that can obstruct the emitters. Suitable design and periodic maintenance of the system for the treatment of the water are vital for the successful use of the drip irrigation. Drip irrigation can operate at low operating pressures and at low discharge rates.

## Crop yields

### Surface irrigation

High potential crop yield is possible with surface irrigation, especially with the border strip method. The yields must be 80–90% in the design for the storage systems, except with soils having a very high infiltration capacity. Reasonable irrigation efficiency for border strip varies from 70 to 85% compared to 65–75% for furrow irrigation. One must take into account factors such as the runoff and drainage water characteristics.

The engineer and the operator can control many of the factors that affect the irrigation efficiency. However, the potential uniformity of the water application in a surface irrigation is limited by the variation of the soil properties, mainly the infiltration capacity. Studies indicate that relatively uniform soils can have a uniformity of 80% in a single irrigation. Researchers have suggested that the calculations of uniformity of the surface irrigation based on the infiltration rate need to be reduced by 5%–10% due to variation of the soil properties.

*Sprinkler irrigation*

The sprinkler irrigation system can give performance as tabulated next page:

| | |
|---|---|
| Manual or portable | 65–75% |
| Lateral on wheels | 60–70% |
| Center pivot | 75–90% |
| Linear move | 75–90% |
| Solid or permanent set | 75–80% |

*Drip irrigation*

Properly designed and maintained drip irrigation is able to give high performance. The design efficiency can vary from the 90 to 95%. With reasonable care and maintenance, the efficiencies of systems operation can range from 80 to 90%. Where obstruction is a problem, the emitter flow efficiency can be as low as 60%.

## Considerations of Manual Labor and Energy
### Surface irrigation

The flood irrigation requires minimum use of manual labor compared to other surface methods. Furrow irrigation can be automated to a certain degree. To reduce the requirements for manual labor input, furrow irrigation requires expertise to obtain high performance. The need for this expertise can be reduced with the use of higher cost equipment. Putting siphons or tracks are useful to measure the desired flow rate. The surface irrigation method requires little or no energy, to distribute the water to the entire field. However the energy is required for taking the water to the field, and for pumping the underground water. In some cases, these energy costs can be substantial, particularly with a low efficiency of water application. Some manual labor and energy for leveling and preparing the land are always needed.

### Sprinkler irrigation

The requirement for manual labor varies depending on the degree of automation and mechanization for the equipment to be used. Manual systems require the least degree of ability, but more hours of labor. On the other end, the center pivot and linear move systems require considerable operational ability and fewer hours of labor. The energy consumption related to the requirements of operating pressure varies considerably among all methods of sprinkler irrigation. The traveling sprinkler system with progressive movement requires seven bars or more of pressure. Other systems can use of two to five bars, depending on the design of the sprinklers and nozzles.

### Drip irrigation

Due to characteristics of low flow and relatively long irrigation set times, drip irrigation can be automated totally. Therefore, manual labor is needed only for inspection and maintenance of the system and the initial installation. The manual labor requirements for maintenance are related to the sensitivity of the emitters to obstructions and the quality of the irrigation water. In a vineyard, for example, a worker can inspect and maintain approximately 20 hectares per day.

Generally, the drip irrigation method requires less energy than other pressurized irrigation systems. The operating pressure ranges from 0.5 to 1.5 bars. The pressure of the system fluctuate from approximately 2 (small flat land systems) to four bars (steep slopes and uneven lands).

## ECONOMIC CONSIDERATIONS

### Surface irrigation

An important cost for surface irrigation is to level the land. The cost is directly related to the volume of earth that must be removed, the length and size of the border strips. The typical volumes of earth to be removed are in the order of 800 cubic meters per ha. For excessive costs, the design must be economical. The typical cost of the earth to be removed is $0.65 per cubic meter. The final, finished leveling with drag-scrapers controlled by laser after removing the earth will cost approximately $110 per hectare. Finishing touches on previously leveled soil by laser will cost approximately $50.00 per hectare.

Low pressure buried plastic pipes or concrete pipes for low flows, can approximately cost the double of the concrete ditches, and up to 5–10 times more for higher flow rates. A compact land dam of medium size able to store a water volume for 24 hours costs $250/ha for small farm (<15 ha). For a larger farm, the price can be lowered up to $125/ha. A weir dam can cost two to five times more.

Table 4.3. Cost of a sprinkler irrigation system.

| Type of system | Field size | Capital cost | Energy use | Manual labor | Maintenance factor |
|---|---|---|---|---|---|
| | ha | $/ha | kvh/ha/mm | Hrs/ha/100 mm | ** |
| 1. Manual or portable | 65 | 175–250 | 0.86–2.05 | 1.65 | 0.020 |
| 2. Lateral on wheels | 65 | 740–1000 | 0.86–2.05 | 1.17 | 0.025 |
| 3. Linear move with progressive move-ment | 32 | 865–1100 | 3.42–4.79 | 0.68 | 0.060 |
| 4. Center pivot (without system for corners) | 55 | 620–1000 | 0.86–2.23 | 0.09 | 0.050 |
| 5. Center pivot (with system for corners) | 60 | 865–1100 | 0.96–2.33 | 0.09 | 0.060 |
| 6. Linear move or trench alignment | 130 | 1000–1200 | 0.86–2.23 | 0.19 | 0.060 |
| 7. Linear move or fed by hose | 130 | 1500–1850 | 1.20–2.57 | 0.19 | 0.060 |
| 8. Solid set | 65 | 2500–3000 | 0.86–2.05 | 0.97 | 0.020 |
| 9. Permanent | 65 | 2100–3000 | 0.86–2.05 | 0.09 | 0.010 |

**The maintenance cost is calculated multiplying the capital cost by the maintenance factor.

## Sprinkler irrigation

Table 4.3 summarizes the factors that affect the cost of the sprinkler irrigation system. The capital cost depends on the type of the system and size of the area to be irrigated. The typical costs of investment (Table 4.3) assume that the water is available at the ground surface and near the field. The costs are necessary for the main pipe and the pumping unit. The energy costs vary widely from place to place. The energy requirements in Table 4.3 can be used to calculate the costs of adequate unitary energy. A pump efficiency of 75% has been assumed. The units of energy are kilowatts per hour per hectare per millimeter (gross) of applied water. The operational cost (manual labor) varies according to the type of system and the local cost for manual labor. Table 4.3 gives typical values per hour of manual labor that is required per hectare per mm of applied water (gross).

It is difficult to predict the maintenance costs, but the numbers in Table 4.3 serve as a guide. The annual maintenance cost is calculated by multiplying the initial capital cost of the system by a maintenance factor.

## Drip irrigation

The cost of the drip irrigation system can vary widely, depending on the crop characteristics and the type of drip laterals (disposable or permanent tubes). The cost of drip irrigation is about $2200/ha per hectare for vegetable crops with wider row spacing. For vineyards with narrow row spacing, the costs can be go up to $3400/ha. For vegetables crops with narrow row spacing (tomato, pepper, etc.) the cost can vary from $2900 to $4900/ha. For vegetable crops with disposable drip lines, the cost of replacing the disposable lateral lines can vary from $340 to $450/ha during the year.

These estimations are for a system of high quality and include pumps, filters, controls, the network of main lines and emitters. In situations where there is more than one basic pump, sufficient equipment for filtration and control, costs vary from 20 to 25% less of the calculated expenses. The typical expenses for operation and maintenance of a drip irrigation system vary widely depending on the local conditions and the irrigation efficiency. It is advisable to calculate the cost of maintenance and operation ($ by hectare per year) as a fraction of the initial cost:

| Cost | Maintenance factor |
|---|---|
| Manual labor | 0.0015 |
| Energy | 0.03–0.07 |
| Water | 0.04–0.06 |
| Maintenance | 0.01 |
| Taxes and Insurances | 0.02 |

*Depends on the performance of the system.

## SURFACE IRRIGATION

In the case of the surface irrigation [2, 3], the water runs on the soil surface providing necessary moisture to the plants for its development. The basic components are: Water source, supply line, control mechanism, dams or dikes of control, furrows of irrigation, system of drainage and system of reusability of the water (optional). The supply lines can be PVC or flexible tubes (lay-flat); also can be open channels of earth, asphalt, concrete, plastic, iron, or brick lined. In addition to the pipes and open channels, other potentially useful structures are: Tunnels (to shorten to the channels for redelivery); hydraulic jumps (to reduce the energy of the water current without causing erosion and to reduce the speed of water, respectively), and siphons (to cross depressions of the land), etc. The control mechanisms may include: orifices, timer for mains (manual and automatic), Parshall (or Replogle or other) flumes, garbage dumps, and control gates for channels. For the control, the flood gates are used.

### Border Strip Irrigation

In border irrigation, the water is applied to a leveled area surrounded by ridges. In level basin irrigation, each irrigated area is completely at level without slopes in any direction. The most common types of "border strip" irrigation use slope in the long direction of the border. It is not necessary that the edges are not rectangular or straight nor that the ridges are permanent. This technique is called "leveled flood or irrigation by border strip." The size of each border depends on the volume of available water, the land topography, the soil characteristics and the required leveling. The size fluctuates from a few square meters to 15 hectares, approximately. An advantage of the border irrigation at leveled ground is that one needs to provide solely a specific volume of water. The border irrigation flood is more effective on uniform and accurately leveled grounds. It is possible to obtain high performance with low manual labor requirements.

Figure 2a.   Flood Irrigation

Figure 2b.   Gated pipe irrigation system

**Figure 4.2.** Surface irrigation.

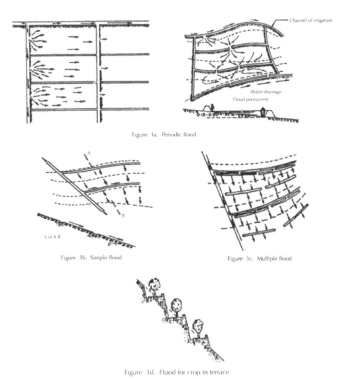

Figure 3a. Periodic flood

Figure 3b. Simple flood

Figure 3c. Multiple flood

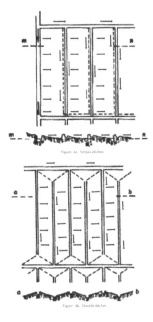

Figure 3d. Flood for crop in terrace

**Figure 4.3.** Flood irrigation.

Figure 4a. Simple ditches

Figure 4b. Double ditches

**Figure 4.4.** Flood irrigation.

**Furrow or Border Strip Irrigation**

It is a traditional system and is more commonly used in agriculture. One adapts to extensive planting methods and it is prone to diseases that are developed due to excess of soil moisture. It uses surface (rivers, lakes, pools, etc.) water resource or a deep well. This method requires that the fields are prepared with gentle slopes so that the water runs slowly by gravity and arrives at the lowest part of the farm, where it is collected by open channels for elimination or is recycled for use.

*Strip irrigation with dikes*

This type of flood irrigation is very popular. It is used mainly in narrow row crops like rice. It can be defined as the application of water between parallel strips. The strip between adjacent dikes does not have slope in the transverse direction, but these have slopes in the direction of the irrigation. The strip irrigation method with dikes uses the soil accommodated in strips (made level through narrow measurement and with slopes throughout the wide measurement). These strips are surrounded by dikes in order to avoid the lateral flow of the water. The water is transported towards the superior end of the dike and advances downwards. High irrigation performance is rarely possible due to difficulty to balance the phases of advance and retraction in the application of water. The strip irrigation with dikes is most complicated among the all surface irrigation methods. The design considerations include length and slope of each strip, volume of water per unit of width of strip, deficiency of soil moisture at the initiation of irrigation and infiltration capacity. However, due to wide variations in the field conditions, it is difficult for the irrigator to get high irrigation performance.

Advantages:

1. It is easy to use.
2. It is of low operational control.
3. The system is a traditional method of irrigation.

Disadvantages:

1. Enough land for sowings is lost.
2. It is not possible to enter the field, once it has been irrigated.
3. Fertilizer application efficiency is low.
4. Water losses by evaporation and runoff.
5. May cause erosion problems
6. Often the water cannot be reused, because it may be contaminated by fertilizers, etc.

*Furrow irrigation or irrigation by infiltration*

The furrows are channels with slopes that are formed from the soil. The infiltration occurs vertically and laterally, and through the wetted perimeter of the furrow. The systems can be designed in a diversity of ways. The optimal length of each furrow is mainly determined according to the infiltration capacity and the size of the field. The infiltration capacities in the furrows can vary much, even when the soil is uniform, due to the crop practices. The infiltration capacity of a new furrow (or just cultivated) will be greater than the one of a furrow that has been watered before. The infiltration in furrows can be reduced due to compaction of rows (between the furrows) and due to compaction by farm vehicles and equipments. Due to the many controllable parameters of design and use, the furrow irrigation can be used in several situations within the limits of the uniformity and topography of the land. The uniformity and efficiency will depend on the soil characteristics and cultivation methods.

The furrow irrigation is designed so that the water runs throughout the desired field. In this system, the water under pressure arrives at the highest elevation of the field. It is distributed by channels or tubes, towards the fields where it will enter the furrows to flood the area. From the supply lines, the water enters the furrows by means of floodgates, siphons or by opening a furrow. Water is applied when the channel is opened. One may use flood gates to control the application of water to a particular field.

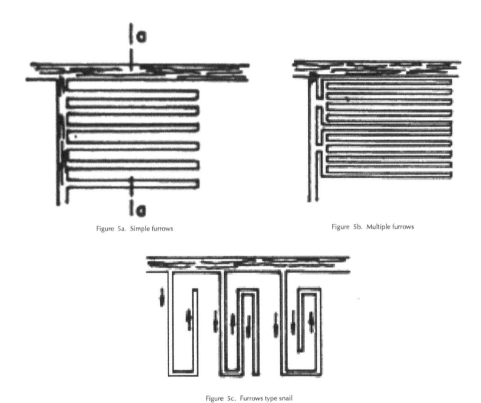

Figure 5a. Simple furrows

Figure 5b. Multiple furrows

Figure 5c. Furrows type snail

**Figure 4.5.** Irrigation by surface infiltration.

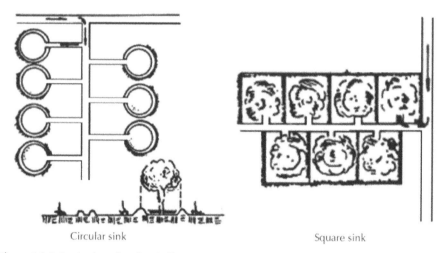

Circular sink                                    Square sink

**Figure 4.6.** Irrigation by subsurface infiltration.

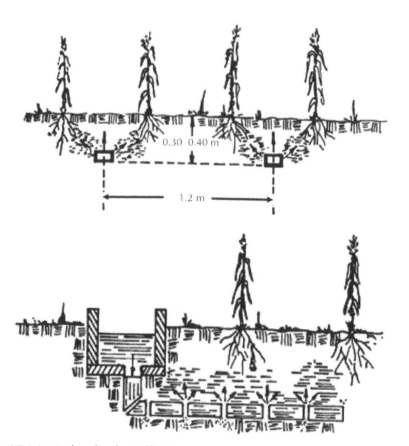

**Figure 4.7.** Irrigation by subsurface infiltration.

One can obtain high uniformity in the water application to the irrigated area by regulating the volume of water that is spilled into the furrow. For this purpose, one can use the pipes with lateral floodgates. The use of the pipes with lateral orifices has wider acceptance. The wing types of floodgates can facilitate the regulation of the water volume that arrives at each furrow. By means of these pipes, the exit of the water into the field can be automated. These floodgates can allow passage of water with flows > 4 liters per minute. The galvanized iron or aluminum pipes with lateral flood gates are easy to install, simple to adjust and can be transferred quickly from one field to other. This system can be used for row crops: sugar cane, corn, fruit orchards, etc. The furrows are oriented directly toward the slope, but the cross-sectional slope is almost uniform.

Advantages:

1. Low initial cost.
2. Less specialized manual labor.
3. Easy operation, care and maintenance.

Disadvantages:

1. The application of fertilizers is manual.
2. Water loss by evaporation.
3. There are nutrient losses.
4. Planting area is reduced.
5. Land must be leveled. This increases initial cost and more time is lost between successive crops.
6. Low efficiency of water use.
7. Incidence of pests and diseases is high.
8. It requires irrigation duration and a large quantity of water.
9. Losses by infiltration and runoff cannot be avoided.

*Furrows along the contour lines*

This system utilizes small channels with continuous slope and almost uniform by which the hilly areas are irrigated. The furrows follow the contours lines of the land. This system can be used in uneven lands, hilly areas, row crops and except for sandy soils. The furrows along the contours can be used to irrigate areas with pronounced slopes. The recommended slope must be between 1% and 2%. The system is efficient and acceptable if all the practices are followed. A proper method is necessary to avoid the overflow of the water in the furrows. The length of the furrows must be short to eliminate excess water that can destroy (dismantle) the furrows. The furrows along contours are used jointly with parallel terraces to provide protection against breakage of furrows. The superior and inferior drainage furrows must be protected.

*Leveled furrows*

The leveled furrows with no slopes are formed by a furrow opener and are used to irrigate crops seeded on the furrows or on the sides of furrows. This method requires a fast supply of water. The leveled furrows adapt better to soils with a moderate to slow index absorption and an index of retention capacity from medium to high. With the furrow irrigation, the best results are obtained in gentle and uniform slopes. The amount of water can adjusted according to the variations of the furrows. A high application efficiency can be obtained if it is designed properly and the surface runoff can be reduced. Unless high wind speed affects the water flow, the furrows can be doubled in length, since the water can be applied on both sides of the furrow. This procedure reduces cost of construction and maintenance of the distribution system.

In places where the wind speed is greater than 24 or 32 Km/h, the application of the irrigation water becomes difficult if the wind direction is opposite to the water course. The capacity of the furrows must be the sufficient to maintain the flow rate. These must be able to apply at least half of the volume of the irrigation for the crop use. For effective operation of the system, the field operations should not alter the topography, the shape and the cross-sectional area of the furrows.

### Sloping furrows

These furrows consist of small channels with a uniform continuous slope that follow the direction of the irrigation. This method can be used for row crops, including vegetables. Furrows with slopes can be used in all soils except sandy soils, with high degree of infiltration capacity and with a very little lateral distribution. This method must be used with extreme care in soils with high concentration of soluble salts. Normally, most of the energy is needed for the preparation of the furrow. The flow to each furrow must be regulated carefully so that the distribution of the water is uniform, with the minimum runoff. The lands must be leveled. Necessary facilities should be provided to collect and to dispose of the excess runoff. The method is not adaptable for shallow rooted crops or low irrigation rates for the germination of the seed.

### Corrugated furrows

This system consists of a partial wrinkling of the soil surface. The irrigation water does not cover all the land, but is distributed is small channels or undulations at regular spaces. The water applied in the undulations infiltrates into the soil and extends laterally to irrigate the intermediate spaces between the furrows. Irrigation by undulations or corrugations adapts better to the drought areas and flat lands with slopes of l to 8%. It can be used on irregular slopes, but the undulations must have a continuous slope in the direction of the irrigation and the cross-sectional slope must be so as to minimize the runoff. The undulations give best results in moderate to heavy textured soils. These are not suitable in soils with high coefficient of absorption and in saline soils. The irrigation flow rates can vary from high to low, because the numbers of undulations that are watered at the same time, and can be adjusted to suit the flow available. The manual labor requirements are high. The irrigation flow rates must be regulated carefully so that the distribution of the water is uniform and the runoff is reduced to the minimum. This method does not adapt to areas with high rainfall during the irrigation season.

## SPRINKLER IRRIGATION

In sprinkler irrigation, the water is applied on the soil surface in form of a rain [1, 2]. The spray pattern is obtained when the water at pressure is expelled through small orifices. The operative pressure is developed by an appropriate pumping unit. Use of this system expanded after the World War II with the introduction of light and movable aluminum pipes. Sprinklers and fast couplers were developed to facilitate assembly and dismantling. Manually lateral movable and rotating sprinklers were then developed. Later to save energy and water, the fixed solid-set systems came into being. In plantations and fruit orchards, these consisted of lateral plastic pipes placed between or along the tree row and sprayers of low flow rates or mini sprinklers. Giant sprinklers mounted on small carts were also developed to facilitate to coverage of extensive areas.

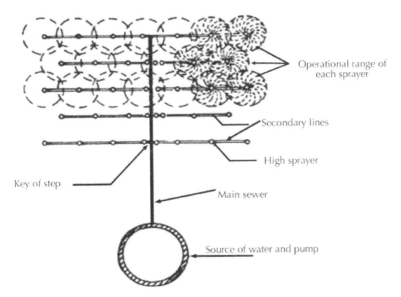

Operational range of each sprayer

Secondary lines

High sprayer

Key of step

Main sewer

Source of water and pump

**Figure 4.8.** Sprinkler irrigation.

**Figure 4.9.** Overhead irrigation.

**Figure 4.10.** Irrigation with a tank mounted on tractor.

**Figure 4.11.** Center pivot with drop sprinklers.

**Figure 4.12.** Wheel line irrigation system.

The simple automation is carried out by means of use of volumetric metering valves to provide a known volume of water, or timed valves to provide water for a known duration. The valves can work at a predetermined sequence. Then the automation systems included use of solenoid valves to close and open according to an established schedule. Several fields can be watered by connecting the units from subfield to a "Master Central Control Unit." Although the sprinklers have increased in popularity, the surface irrigation continues to predominate. The basic components of a sprinkler system are: A water source, a pump to provide a operating pressure, one or more than one main/sub mains to distribute the water to all the field, sprinklers and control valves, etc.

Advantages:

1. Preparation of land is not required. It can be used on lands of rough topography and level lands.
2. It allows total use of the land.
3. It offers great flexibility in design.
4. It can be used in any type of soil including high permeability rates.
5. It is generally more efficient than the surface irrigation.
6. More water in comparison with the surface irrigation is saved.
7. One adjusts very well to additional irrigation. It can be used for protection against frosts and the heat.
8. The use of portable pipe reduces the cost of the equipment.
9. Fertilizers can be applied efficiently through the irrigation water.
10. The root zone is developed better than with the surface irrigation.
11. The cost for manual labor is lower than for the surface irrigation.
12. The gauging of the water is easier with this system.
13. The performance efficiency is high.
14. It is possible to make frequent applications at low volumes, when it is necessary.

Disadvantages:

1. Elevated initial cost.
2. The evaporation loss is high. These losses can be reduced to a minimum by watering at night.
3. Salty water can cause damage to the crop.
4. Incidence of foliage diseases is high. Some crops can suffer by loss of flowers due to the impact of the water.
5. The freezing conditions can affect the functioning of sprinklers.
6. There is interference with the pollination.
7. The wind interferes so that distribution of water is affected.

## Types of Sprinkler Irrigation Systems
### Manually operated and portable system

These systems include lateral pipe with sprinklers installed at regular intervals. The lateral pipe is, generally, of aluminum, with Sections of 6, 9, or 12 meters in length, and fast connections for pipes. The sprinkler is installed at the top of a riser (in the orchards, the height of a riser should go under the leaf canopy). The risers are connected to the lateral tube. The length of the tube is selected to correspond with the desired spacing of the sprinklers.

The lateral pipe with sprinklers is placed on a ground and it is used until the application has been terminated. Then, the lateral tube is dismantled and is positioned in the next section. This system has a low initial cost, but requires high manual labor. It can be used in almost all crops. However it may be difficult to move the lateral pipes when the crop is mature.

### Lateral system on wheel (movable)

This system is a variation of the manual system. The lateral pipe is mounted on wheels. The height of the wheels is chosen so that the axis exceeds height of the crop for easy movement. A drive unit is commonly a motor driven with gasoline and is located near the center of the lateral pipe. The system is moved from one place to another by wheels.

### Traveling sprinkler system with progressive movements

This system uses a sprinkler (tube) gun of high pressure and high volume. The spray gun is mounted on a towable trailer. The water is supplied by means of a flexible hose or from an open ditch. The system can be used in one field for a desired time and then can be moved to the next field. The tow can be moved by means of a cable or it is possible to be pulled ahead while the hose is coiled in a spool to the border of the field. These systems can be used in almost all crops. However, due to high intensities of application, these are not suitable for clay soils.

### Center pivot system

This system consists of a lateral pipe with simple sprinklers supported by a series of towers. The towers are impelled in such way so that the lateral pipe moves around the center, about the pivot point. The speed of the complete circular motion fluctuates from twelve hours to several days. The intensity of the application of water increases with the distance from the pivot in order to give a uniform amount of application. Because the center pivot irrigates a circular area, it leaves non irrigated the corners of the field (unless additional special equipment is added to the system). Center pivots are able to water almost all field crops.

### Linear move system

The linear move system is similar to the center pivot system. The line of the pipe extends in perpendicular direction to the lateral one. The delivery of the water to the lateral is by a flexible hose or from an open ditch. The system irrigates the rectangular fields free of high obstructions.

### Low energy pressure system (LEPA)

LEPA systems are similar to the linear move systems of irrigation. The orifices in the lateral pipe and pipes can discharge at very low water pressure, exactly according to the soil moisture. Generally LEPA systems are accompanied with soil surface management regimes such as minimum tillage or most commonly micro-dams to prevent high application rates from causing runoff. With suitable soil surface management systems, high infiltration rates are not an absolute requirement.

### Solid-set systems

In this system, sufficient lateral pipes are placed in the field and are not moved during the season. Solid-set systems utilize a network of aluminum tubes for irrigation. Enough lateral lines are used to cover all the area. The system reduces to a minimum the need for manual labor during the irrigation season.

### Types of Sprinklers

1. **Rotating sprinklers, impact type:** These are commonly used over a wide range of pressure, discharge, spacing and rate of application for different crops.

2. **Sprinklers, high volume or "gum" type:** These are rotating sprinklers with a discharge up to 60 $M^3$/h at a pressure head of 60 meters. It can cover areas up to one hectare simultaneously.

3. **Sprinklers with low flow rate:** These sprinklers apply 120 to 350 liters per hour at a pressure head of 15–25 meters. These are used mainly in irrigation of fruit trees.

4. **Mini sprinklers:** Small sprinklers can apply 30–120 liters per hour at a pressure head of 15–25 meters. These are used in vegetables and nurseries.

### SUBSURFACE IRRIGATION

In many areas, soil conditions and topography are favorable to irrigation using the water below the ground surface. These favorable conditions are: The existence of impermeable subsoil at a depth > 1.8 meters; silt or silt-sandy permeable layer; uniform topography and moderate slopes. The irrigation may be applied by means of exposed drain ditches. The water table stays at a predetermined depth, normally from 30 to 40 centimeters, depending on the root characteristics of the crop.

The perforated subsurface pipes allow the infiltration through the soil. The pipes can be placed at a spacing of 45 cm and at a depth of 50 cm. These buried pipes can suffer damage by deep plowing. This method works, if the soil has a high horizontal and a low vertical permeability.

The open ditches are probably used on a greater scale. The feeding drain is excavated along the contours. The drain spacing must be sufficient to maintain and to regulate the level of the water. These are connected to a supply channel that runs downwards, following a predominant slope of the land.

### Adaptability

Subirrigation is appropriate for uniformly textured soils with a good permeability so that the water is mobilized quickly, in a horizontal and vertically direction and to a recommended depth under the root zone. The topography must be uniform or almost level or very smooth with uniform slope. Subirrigation is adapted for vegetable and root crops, forage crops and gardens.

## Important Characteristics

This method is used in soils with low capacity and when surface irrigation cannot be used and the cost of pressurized irrigation is excessive. The level of the water can be maintained at the optimal depth according to the crop requirements at different growth stages. The evaporation losses are reduced to a minimum. The irrigation does not allow the weed seeds to come to germinate.

## Limitations

Water with high concentration of salts can not be used. The selection of crops is limited. The crops with deep root system (such as some citrus) are not generally suitable for subirrigation.

## Natural subsurface irrigation

When geological and topographical conditions are favorable natural subsurface irrigation is suitable for nearly flat topography with a deep surface layer and high lateral permeability. At a depth of two to seven meters from the soil surface, there is usually an impermeable rocky substratum. A constant water table at a particular depth is maintained. A set of parallel ditches can be arranged below the soil surface under following conditions: For complementary irrigation in spring and summer in humid regions; a good drainage is needed in winter. The soils are sandy and very permeable. During excessive rainfall, water is removed by gravity or by pumping.

## DRIP IRRIGATION

The application of water is by means of drippers of that are located at desired spacing on a lateral line. The emitted water moves due to a unsaturated soil. Thus, favorable conditions of soil moisture in the root zone are maintained. This causes an optimum development of the crop.

## SUBSURFACE DRIP IRRIGATION (SDI)

In this system, laterals with drippers are buried at about 45 cm depth. The fundamental intention is to avoid the costs of transportation, installation and dismantling of the system at the end of a crop. When it is located permanently, it does not harm the crop and solve the problem of installation and annual or periodic movement of the laterals. A carefully installed system can last for about 10 years. The components of the system are basically the same as those for the drip irrigation system:

1. Manual solenoid valves.
2. Pressure controllers.
3. Pressure gauges.
4. Fertilizer tank.
5. Hydro cyclone filters, sand filters and mesh filters.

The system can be automated by means of electronic sensors on the basis of soil moisture. The laterals are installed at a depth of 45 cm depth. The system is connected to a computerized central control station. Chemical agents can be applied through the system only if these are highly soluble. It is recommended to apply water at pressure, use fumigants and herbicides to prevent the penetration of roots into the tubes. The advantages are:

1. High crop yields.
2. Conservation of water and energy.
3. Less manual labor.
4. Less damage to the lines and the system during the installation and dismantling process.
5. Life of irrigation lines and parts is increased, since the plastic pipes are not exposed to the solar light.

## XYLEM [TREE INJECTION] IRRIGATION

Xylem irrigation (Figure 4.13) is the direct application of water with necessary chemical agents into the xylem of the tree trunk using a series of injectors that depends on the age of the tree. Xylem irrigation is also called ultra micro, high frequency, tension, tree injection, or chemotherapy irrigation. There is no difference in the concept these names represent. The basic idea originated when various chemicals were injected into the internal circulatory system of tree. It is simple to inject water, fertilizers, micronutrients, growth promoters, growth inhibitors, pesticides, trace elements, gases, precursors of flavors/color and aroma, and in general any substance valuable for the improvement of fruit quality. This system is in the experimentation stage and has not been evaluated commercially.

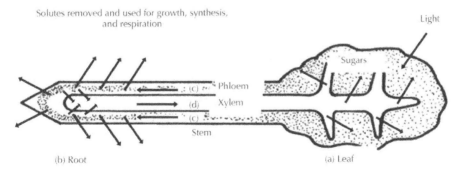

**Figure 4.13.** (a) Principle of tree injection irrigation.

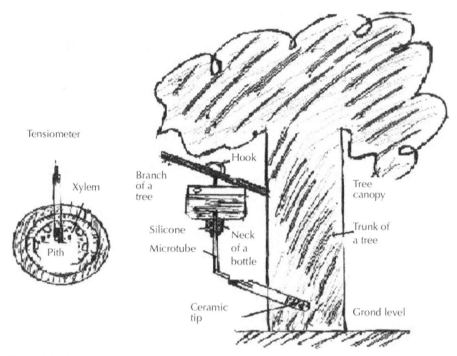

Tensiometer

Xylem

Pith

Branch
of a
tree

Hook

Silicone

Microtube

Neck
of a
bottle

Ceramic
tip

Tree
canopy

Trunk of
a tree

Grond level

**Figure 4.13.** (b) Tensiometer in a trunk tree.

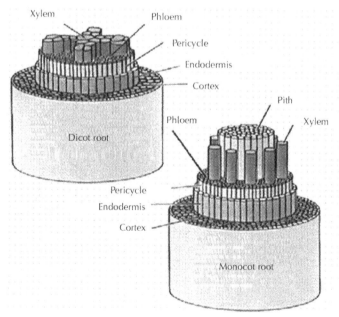

Xylem

Phloem

Pericycle

Endodermis

Cortex

Pith

Phloem

Xylem

Dicot root

Pericycle

Endodermis

Cortex

Monocot root

**Figure 4.13.** (c) Diagram illustrating the tissue layers and their organization within monocot and dicot roots.

## Advantages

### *Efficient use of water*

1.  No evaporation from the soil surface into the atmosphere.
2.  No infiltration into the subsoil where roots are incapable of absorbing moisture.
3.  No runoff.
4.  No wetting of foliage.
5.  Inhibits non beneficial consumptive use of water by weeds because the terrain is free of weeds.
6.  One can irrigate the entire field up to edges.
7.  Accurate quantity of irrigation water can be applied according to transpiration rate of the plant.
8.  Overall water application efficiency can go up to 99%.
9.  Savings up to 95% of water use can be achieved.

### *Plant response*

1.  Crop growth characteristics can be manipulated.
2.  Better fruit quality and uniformity of crop is expected.

### *Root environment*

1.  Shallow root system.
2.  Effective soil aeration.
3.  Provision of required amount of nutrients.

### *Pest and diseases*

1.  Pesticides can be injected into the plant system.
2.  Frequency of sprays can be reduced.
3.  Reduction in incidence of insects and diseases.
4.  Reduced application rates of pesticides.

### *Weed growth*

1.  It is a minimum.
2.  No weeds in dry surface between trees.

### *Agronomical benefits*

1.  Irrigation activities do not interfere with cultivation, spraying, picking and handling.
2.  Less inter-cultivation, soil crusting, and compaction problems.
3.  No surface runoff.
4.  No soil/water erosion due to irrigation.
5.  Fertigation and chemigation are possible thus savings in energy and quantity.
6.  Other necessary chemicals can be applied along with irrigation water.

### Engineering and economical benefits

1. Significant savings in energy.
2. Cost is low compared to surface sprinkler and drip irrigation systems.
3. Pipe sizes are significantly smaller compared to pipe sizes in other irrigation systems.
4. Conveyance efficiency and water use efficiency can be increased up to 99%.
5. System can be installed in uneven terrains.
6. It requires constant discharges at low pressures.
7. Water and chemical use can be programmed with the crop response.

## Disadvantages

1. May not be applicable in vegetable crops as it is more convenient to inject into tree trunks.
2. May not be used in monocot species as the xylem is not as differentiated.
3. Introduction of new substances can cause toxic effects just as a man can overdose on drugs.
4. May cause fungus growth at the injection site.
5. Holes in tree trunk must be made to install injection tips thus causing physical injury to the plant.

## Operational Problems

1. No information is available on number of injection sites, water application rates, dosages of various chemicals depending upon the age of the tree.
2. At what height and depth, should the injection points be located?
3. Injection tips can be easily clogged by gum, wax and resins of tree.
4. Effective cleaning agent needs to be found to avoid clogging of tips.
5. Algae formation in the injector lines.
6. Laterals may contain air [from the tree] and thus obstructing the flow.
7. Leakage of water at the contact point between the injector tip and tree surface.
8. Excess pressure might loosen the sealing agent [silicon] and may throw out the injection tip.
9. Expert advice is needed to locate xylem.
10. Chemigation might disturb the osmotic and electrical internal equilibrium in the plant.
11. Screening of pesticides and chemicals suitable for xylem irrigation.
12. Salts in excess of 300–500 ppm may require desalinization of water.
13. A clean, pure or soil water is necessary.

## Principle of Operation

It is based on utilizing natural negative sap pressures within a plant to suction liquids and gases directly into the inner circulatory system, analogous to a human blood transfusion. The technique is accomplished by placing an injection tip [e.g., a ceramic implant] directly in the xylem layer, the negative pressure area. Liquid or gas is then made available to the implant through *a* plastic tubing at very little or no pressure. Fluids can then traverse in the plant in any direction. The roots of the plant continue to be nourished by the natural way, with sap, water and nutrients. The roots still seek moisture and grow down using a stimulus called geotropism.

Plants give off water through a process called transpiration. The amount of water a plant "throws off" and the amount it needs are two different situations. A well known "Hill Reaction" is:

$$6CO_2 + 6H_2O + H_2 = C_2H_{12}O_6 + 6O_2 \qquad (3)$$

Opposite of transpiration is called respiration. By careful measurement of the quantities of sugar synthesized in the leaves by unitary surface and time [10–15 mg of hexose/sq dm-hr], it is readily calculable what stoichiometric quantity of water is required under the same conditions [e.g., 50–80 ml for a period of 8 hrs considering a canopy surface from 8 to 10 sq. m]. This quantity is very approximate to the quantity of water consumed by xylem irrigation during the same period under the same conditions. Primary water uptake occurs only during photosynthesis or day light [12]. System modification: It is accomplished by simply placing the ceramic piece in the root zone of house or commercial indoor plants, nursery stock or almost any plant too small to receive an implant in the trunk. The same efficient use of water and nutrients are applicable but some of the metabolic engineering techniques [Modulation of the plant metabolism with the aim of obtaining better fruits by injection of substances such as promoters of color, bouquet, flavor, aroma, metabolites, enzymes or coenzymes] may not be effective. Seeds for greenhouses can also be germinated and grown from an implant in the soil. The seed can actually be glued to the implant, then planted and grown through maturity. Tree crops can be raised with other irrigation systems and then xylem tips can be installed after first year of growth. Water usage of 40 ounces/day on older trees, 5 ounces/day for grapes has been reported. This calculates to be approximately 0.05 gpm/acre of irrigation during 12 hours photosynthesis or 36 gallons/day/acre.

### Description of Xylem Irrigation System

The system consists of a water resource, pump, chemigation system, filter, main line, sub main, laterals, and injection tips. The installation of injection tip should be done in the following manner:

1. Select the size of a ceramic tip.
2. Select the best location on the plant.
3. Bore a hole through the cambium layer approximately 1/4 diameter larger than the injector.
4. Use a sharp instrument to remove this plug of bark. It is important to bore past the phloem to cause leakage out of the plant.
5. Continue the bore into xylem [sapwood] portion to the same dimension as length of ceramic portion of the injector. The hole should allow a snug fit.

6.  Use an inert sealing agent [silicone] for sealing the injector to tree.

7.  Hook water to be injected to the tip at a pressure of 1–8 ft of head. [Necessary pressure can be allowed by gravity, low pressure pump]. Very minute quantities of chemical can be injected into the water stream using a plastic syringe [doctor's needle].

## Precautions for Xylem Irrigation System

1.  Sterilize the drill bit.
2.  Hole should allow a perfect fit.
3.  Use a good sealing agent.
4.  Water should be free from pathogens.
5.  Use a pesticide to avoid fungal growth.
6.  Injector site should be allowed to dry before starting irrigation.
7.  No leakage can be allowed between the tree and tip, as it will break the suction.
8.  High precaution is essential in determining dosage of the chemicals to avoid toxic hazards in the plant.
9.  Any injection holes, which cannot be used, should be left open. They heal with time.

## HYDROPONIC

The growth of plants without soil is known as hydroponics. From 1925 to 1935, extensive work was done to modify the nutrient culture in the nurseries. In 1930, W.F. Gericke at the University of California, defined hydroponics as a science to cultivate plants without the soil use, but using inert materials such as sand and sawdust, among others. A nutrient solution with all the essential elements is added for good development of the plants. After World War II, the Air Force Wing of the United States used hydroponics to provide food to the troops and established the first hydroponics project of 22 hectares in Chofu-Japan. During the fifties, the commercial use of hydroponics expanded quickly to United States, Italy, Spain, France, England, Germany, Sweden, USSR, Israel, India, and others. This expansion continued in Australia, New Zealand, Africa of the South, Kuwait, Brazil, Poland, Singapore, Canada, Malaysia, the East and Center of Africa and other countries, during 1950–2000. Figures 4.14 and 4.15 show a typical hydroponics system. The ventilation, darkening of the roots and support of the plant are key elements of the hydroponics. The nutrient solution is pumped to a PVC tube covered with a black polyethylene cover. The PVC tube is cut in half and the black polyethylene is placed on top to prevent the entrance of light. Holes are made at the top of the polyethylene tube through which plants grow in the nutrient solution, which flows at the bottom of the PVC tube. At the end of a crop, the system is dismantled and the tubes with accessories are washed with a chlorine solution.

Disadvantages are:

1.  The high initial capital cost.
2.  Incidence of diseases like Fusarium and Verticillium.
3.  The problems with nutritional complexes.

**Figure 4.14.** Hydroponic system.

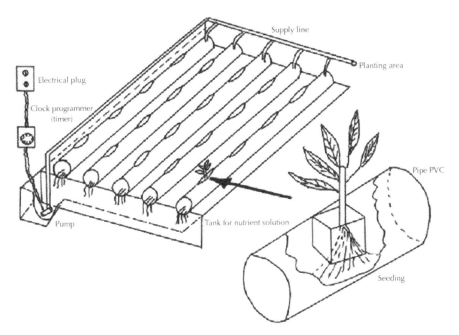

**Figure 4.15.** General components of a hydroponic system.

Advantages:
1. A greater efficiency in the regulation of nutrients.
2. Its availability in the regions of the world with no arable lands.
3. Efficient use of the water and fertilizers.
4. Easy and low cost of sterilization of growing medias.
5. A greater density of plants by area.

## SUMMARY

The need for additional food for the world's population has spurred rapid development of irrigated land throughout the world. Vitally important in arid regions, irrigation is also an important improvement in many circumstances in humid regions. Unfortunately, often less than half the water applied is beneficial to the crop – irrigation water may be lost through runoff, which may also cause damaging soil erosion, deep percolation beyond that required for leaching to maintain a favorable salt balance. New irrigation systems, design and selection techniques are continually being developed and examined in an effort to obtain the highest practically attainable efficiency of water application.

## KEYWORDS

- Automation
- Bar
- Border strip irrigation
- Center pivot irrigation
- Chemical precipitation
- Clay
- Clogging
- Contour
- Distribution system
- Drainage
- Drip irrigation
- Dripper or emitter
- Soil erosion
- Evaporation
- Fertilizer
- Fertilizer tank
- Filter
- Filter, mesh

- Filter, sand
- Filter, screen
- Furrow irrigation
- Hydroponics
- Infiltration
- Irrigation, depth
- Irrigation, duration
- Irrigation, flood
- Irrigation, sprinkler
- Irrigation, drip
- Land leveling
- Main line
- Mesh
- Micronutrients
- Plant nutrient
- Percolation
- pH
- Polyethylene (PE)
- Precipitation
- Pressure gauge
- Pump
- Root system
- Root zone
- Salinity
- Sand
- Soil moisture
- Solenoid valve
- Soluble salts
- Sprinkler irrigation
- Strip irrigation
- Avaiable water
- Saline water
- Water source
- Weed
- Xylem irrigation

## BIBLIOGRAPHY

See bibliography Section 4.1 for the literature cited.

## APPENDICES

**Appendix 1.** Drop-tube (Over head) irrigation System.

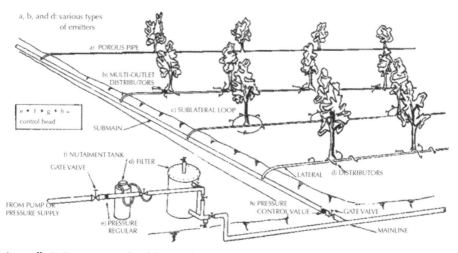

**Appendix 2.** Components of a drip irrigation system.

# Chapter 5

## Principles of Drip/Trickle or Micro Irrigation

### INTRODUCTION

The competitive demand of available water is more and more acute now in most parts of the world. The supplies of good water quality are declining every day. It is, therefore, necessary to find methods to improve water use efficiency (WUE) in agriculture [3]. The drip irrigation is one of the most efficient irrigation systems that are used in agriculture. This system has been used in arid regions of the world.

Drip irrigation also known as *trickle irrigation* or *micro irrigation* is an irrigation method that applies water slowly to the roots of plants, by depositing the water either on the soil surface or directly to the root zone, through a network of valves, pipes, tubing, and emitters [1]. The goal is to minimize water usage (see below). Drip irrigation may also use devices called micro-spray heads, which spray water in a small area, instead of emitters. These are generally used on tree and vine crops. Subsurface drip irrigation (*SDI*) uses permanently or temporarily buried dripper line or drip tape. It is becoming more widely used for row crop irrigation especially in areas where water supplies are limited. The drip irrigation is a slow and frequent application of water to the soil by means of emitters or drippers located at specific locations/interval throughout the lateral lines. The emitted water moves through soil mainly by unsaturated flow. This allows favorable conditions for soil moisture in the root zone and optimal development of plant.

## History

Drip irrigation was used in ancient times by filling buried clay pots with water and allowing the water to gradually seep into the soil. Modern drip irrigation began its development in Germany in 1860 when researchers began experimenting with subsurface irrigation using clay pipe to create combination irrigation and drainage systems. In 1913, E.B. House at Colorado State University succeeded in applying water to the root zone of plants without raising the water table. Perforated pipe was introduced in Germany in the 1920s. In 1934, O.E. Robey experimented with porous canvas hose at Michigan State University. With the advent of modern plastics during and after World War II, major improvements in drip irrigation became possible. Plastic micro-tubing and various types of emitters began to be used in the greenhouses of Europe and the United States.

A new technology of drip irrigation was then introduced in Israel by Simcha Blass and his son Yeshayahu. Instead of releasing water through tiny holes (blocked easily by tiny particles), water was released through larger and longer passageways by using friction to slow water inside a plastic emitter. The first experimental system of this type was established in 1959 in Israel by Blass, where he developed and patented the first practical surface drip irrigation emitter. In 1973, publisher Massada Limited in Israel printed a book in Hebrew by Simcha Blass named "WATER IN STRIFE AND ACTION." Blass was a well known water engineer in Israel in its first years and chief engineer in Jewish Yishuv. He planned the Israeli main water carrier. It is a common misconception that drip irrigation was invented in Israel. There is no question that much of the product innovation in this field occurred in Israel and that companies in Israel have contributed significantly to the industry, but they can not take all the credit for its development. The facts are that drip irrigation system, with plastic pipes and the new plastic drippers, was invented and first used and developed in Israel by Blass. It does not change the contributions made earlier and later.

This method subsequently spread to Australia, North America, and South America by the late 1960s. In the United States, in the early 1960s, the first drip tape, called *Dew Hose*, was developed by Richard Chapin of Chapin Watermatics (first system established at 1964, http://www.microirrigationforum.com/new/archives/drip-invention.html). In 1969, researchers under Prof. R. K. Sivanappan at Tamil Nadu Agricultural University in Coimbatore—India started research using drip irrigation using the available tubes and micro tubes in the market not only in the research station, but also in the farmer's field for banana, grapes, cotton, and vegetables. Beginning in 1987–88, Jain irrigation toiled and struggled to pioneer water-management through drip irrigation with the advice of Dr. Sivanappan. Jain irrigation (www.jains.com) also introduced some hi-tech drip irrigation concepts to Indian agriculture such as "Integrated System Approach," One-Stop-Shop for Farmers, "Infrastructure Status to Drip Irrigation & Farm as Industry." They have established "Research and Demostration Farm" for modern drip irrigation concepts at Hi-tech Agricultural Institute near Jalgaon—Maharashtra—India. Jains opened a head office in Fresno—CA (www.jains. com/Reachus/usa.htm). The area under drip irrigation is more than five lakh hectares covering about 30 crops including grapes, sugarcane, banana, cotton, horticultural crops, vegetables, and fruits.

The first drip irrigation system in Puerto Rico was installed in 1970 for fruit orchard owned by Luciano Fuentes in Coamo municipality, and in mango orchard at Fortuna Agricultural Experimental Substation in Juana Diaz municipality. In 1979, first experiment on drip irrigation was established at Fortuna Substation by Megh R. Goyal, Father of Irrigation Engineering in Puerto Rico. Today, acreage under drip irrigation for vegetables and fruits has increased to more than 60,000 acres in Puerto Rico.

It is assumed that the drip irrigation will fortify agriculture and increase efficiency of food production. With this system, the plant can efficiently use available natural resources such as: soil, water, and air. The drip irrigation is also known as "daily irrigation," "trickle irrigation," "daily flow irrigation," or "micro irrigation," The term "trickle" was originated in England, "drip" in Israel, "daily flow" in Australia and "micro irrigation" in USA. The difference is only in the name, and all these terms have the same meaning. The water in a drip irrigation system flows in three forms:

1.  It flows continuously throughout the lateral line.
2.  It flows from an emitter or dripper connected to the lateral line.
3.  It flows through orifices perforated in the lateral line.

## ADVANTAGES OF DRIP IRRIGATION

Well designed drip irrigation system can increase the crop yield due to following factors:

A.  Efficient use of the water:
    1.  It reduces the direct losses by evaporation.
    2.  It does not cause wetting of the leaves.
    3.  It does not cause movement of drops of water due to the effect of wind.
    4.  It reduces consumption of water by grass and weeds.
    5.  It eliminates surface drainage.
    6.  It allows watering the entire field until the edges.
    7.  It allows applying the irrigation to an exact root depth of crops.
    8.  It allows watering greater land area with a specific amount of water.

B.  Reaction of the plant:
    1.  It increases the yield per unit (hectare-centimeter) of applied water.
    2.  It improves the quality of crop and fruit.
    3.  It allows more uniform crop yield.

C.  Environment of the root:
   1.  It improves ventilation or aeration.
   2.  It increases quantity of available nutrients.
   3.  The conditions are favorable for retention of water at low tension.
D.  Control of pests and diseases:
   1.  It increases efficiency of sprayings of insecticides and pesticides.
   2.  It reduces development of insects and diseases.
E.  Correction of problem of soil salinity:
   1.  The increase in salts happens at a distance away from the plant.
   2.  It reduces salinity problems. A greater reduction is obtained by increasing the water flow. Salts are limited to an outer periphery of a wetted zone.
F.  Weed control:
   1.  It reduces the growth of weeds in the shaded humid space.
G.  Agronomic practices:
   1.  The activities of the irrigation do not interfere with those of the crop, the plant protection, and the harvesting.
   2.  It reduces inter cultivation, since there is less growth of weeds.
   3.  Helps to control the erosion.
   4.  It reduces soil compaction.
   5.  It allows applying fertilizers through the irrigation water—called fertigation.
   6.  It increases work efficiency in fruit orchards, because the space between the rows is maintained dry.
H.  Economic benefits:
   1.  The cost is lower compared with overhead sprinkler and other permanent irrigation systems.
   2.  The cost of operation and maintenance is low. The costs are high when the average row spacing is less than three meters.
   3.  It can be used in uneven terrains.
   4.  The water application efficiency is high. Energy use per acre is reduced due to smaller diameter pipes, and only 50–60% of water is used.
   5.  The BC ratio is favorable and the pay back period is one to two years.

## DISADVANTAGES OF DRIP IRRIGATION

A.  The drip irrigation, like other methods of irrigation, cannot adjust to all the specific crops, sites, and objectives
B.  The system has the following problems and limitations:
   1.  The drippers are obstructed (or clogged) easily with soil particles, algae or mineral salts.
   2.  The soil moisture is limited. The soil moisture volume depends on the discharge of drippers, dripper spacing, and the soil type.
   3.  The rodents or insects can damage some components of the drip irrigation system.

4. A more careful high technology management is needed compared to other irrigation systems.
5. The initial investment and annual cost are higher compared to other irrigation methods.

## DRIP IRRIGATION SYSTEM LAYOUT AND ITS PARTS

The main components of the drip irrigation system (Figure 5.1) are:

1. The water source.
2. The pump and the energy unit.
3. Main Line (larger diameter Pipe and Pipe Fittings).
4. The filtration systems: Sand separator, screen filter, media filters & manifolds, backwash controller.
5. Fertigation systems and chemigation equipment.
6. The controls in the automation of the system: Plastic Control Valves and Safety Valves.
7. The water distribution system: Smaller diameter PE tubing (often referred to as "laterals").
8. Emitting Devices at plants (ex. Drippers, micro-spray heads, irrigation mats).

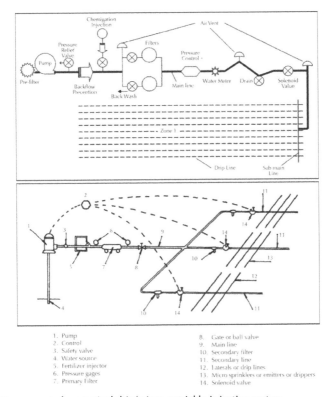

1. Pump
2. Control
3. Safety valve
4. Water source
5. Fertilizer injector
6. Pressure gages
7. Primary Filter

8. Gate or ball valve
9. Main line
10. Secondary filter
11. Secondary line
12. Laterals or drip lines
13. Micro sprinklers or emitters or drippers
14. Solenoid valve

**Figure 5.1.** Components for a typical drip/micro or trickle irrigation system.

To make connections, fittings and accessories needed for a drip irrigation system are: Gate or ball valve, safety or check or one way valve, flushing valve, union, nipple, adapters, reducers, tee, coupling, universal, elbow, double union, and cross.

## The Water Source

The water source can be a treated water, well water, open channel, rivers, and lakes. The clean water is essential for the drip irrigation. If water of poor quality is used, physical and chemical or biological polluting agents can obstruct the emitters and drip laterals. The underground well water is generally of good quality. In some cases, it can contain sand or chemical substances.

Although, superficial water from lakes, rivers, and open channels can be used, this water source has the disadvantage of possible contamination by bacteria, algae, and other organisms. Almost all water sources contain bacteria and elements that nourish it. A good filtration system is needed to remove all the polluting agents that can obstruct or clog the drippers.

## The Pump

The pumping and the power units represent a significant part of an initial installation cost of a drip irrigation system. Therefore, for the selection of appropriate equipment, it is convenient to know all the characteristics and operating conditions. It is necessary to acquire effective, reliable, and low cost pump and a power unit. A centrifugal pump (Figure 5.2) is suitable for extracting water from superficial sources or shallow wells. It can also be used to increase the pressure in a main or sub main line. The centrifugal pump is relatively cheap and efficient. It is available for wide range of flows and pressures.

While selecting a pump, one should know total pressure in the system, operating pressure, total volume of water for irrigation, and horsepower rating. A pump of a slightly greater capacity than necessary can ensure sufficient water at a desired pressure.

The pump of a required size guarantees a good uniformity of water application. On the contrary, an undersized pump and oversized drippers can cause non uniform emitter flow throughout the field. Total pressure for the system generated by a pump depends on the following factors:

1. The highest elevation: Elevation difference between a water source and an emitter located at the highest elevation of the field.

2. The highest friction loss: The pressure required for the most distant emitter. One should also consider discharge flow.

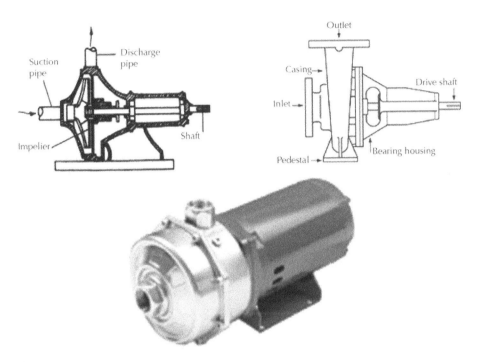

**Figure 5.2.** Pumps.

## Energy or Power Units

In drip irrigation system, the electrical motors are preferable because of high overall efficiency, ease of automation, quiet operation, and necessity of least maintenance. The gasoline or diesel engines can also be used. However, these may cause variations in the operational pressures and emitter flow. These have high maintenance cost and cause noise during operation.

## The Irrigation Controls

### The volumetric valve

The volumetric valve measures volume of irrigation application and is necessary for proper management.

### The pressure gage

The pressure gage measures water pressure in the drip irrigation system [2]. It is especially useful for non compensating pressure drippers. The pressure gages on the upstream and downstream side of the filtering system indicates pressure loss that helps in programming the flushing process of filters. The intervals and range of pressure gage must match with the operating pressure of the system for good management.

**Figure 5.3.** Pressure gage.

Pressure controller

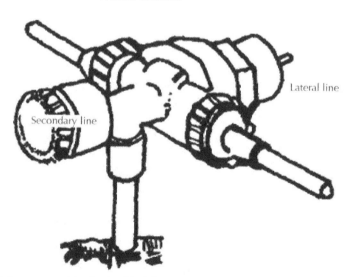

Lateral line

Secondary line

**Figure 5.4.** Pressure regulator for the drip irrigation system.

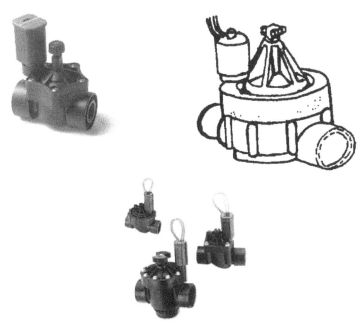

**Figure 5.5.** Solenoid valves for drip irrigation system.

**Figure 5.6.** Air relief valve or vacuum breaker, and pressure release valve.

### The pressure regulator or pressure relief valve

The pressure regulator or pressure relief valve can be fixed or adjustable. When installed in the principal or secondary lines, it can compensate the changes in elevation or frictional losses of lines. The manual valves, the automatic metering valves, the semiautomatic valves, and timers are recommended for the drip irrigation system. The automatic flow valves provide desired amount of irrigation in a specified time. These also reduce pressure variations among the lateral lines in an uneven land. The combinations of pressure regulator and flow control valve are also available.

### Air relief valve or vacuum breaker

This valve removes air from the drip irrigation system. The suction or vacuum is developed when the irrigation system is shut off. This vacuum can obstruct the drippers if the dirty water or dust is suctioned into the system. Vacuum breaker or air relief valve of "one inch for each 25 gpm of flow" is recommended.

### Flushing valve

The flushing valve at the end of each lateral helps in the flushing of the system.

### The Dripper or Emitter

The water emission devices (drippers or emitters or micro sprinklers) are unique in the drip irrigation system. The drippers supply water through small orifices in small amounts near the plant. The pressure through the emitters must be sufficiently greater than the pressure difference due to land topography and friction losses in the system. The orifice must be large enough to avoid serious obstructions. These design considerations have resulted in the manufacture of several types of drippers (Figure 5.7). There are two categories of drippers for installation in the field:

1. In-line emitters/drippers.
2. On-line emitters/drippers.

The in-line emitters are used in the green houses and for row crops such as: vegetables and some fruit orchards. This layout also consists of a series of equally spaced perforations (orifices) throughout the lateral: single or biwall drip laterals. The emitter flow fluctuates from three to four (3–4) liters per minute for 30 m of lateral. The operating pressure of this layout fluctuates from 2 to 30 psi. However, most of these systems operate at a pressure less than 15 psi. The lateral or drip lines can be up to 90 m of length depending on the slope. These laterals are limited by irrigation flow rate and by uneven terrain of the land. The on-line drippers are suitable to leveled lands to maintain acceptable uniformity of emitter flow. Because the operating pressure is low, a moderate variation in elevation causes large variation in the emitter flow. The engineers are the right professionals for design of drip irrigation systems. The flow variations of drippers are due to changes in pressure because of uneven lands. For these situations, compensating pressure drippers provide almost constant flow for wide. The flow variations fluctuate from moderate to high depending on pressure changes because of non uniform orifices in a single chamber and biwall drip lines.

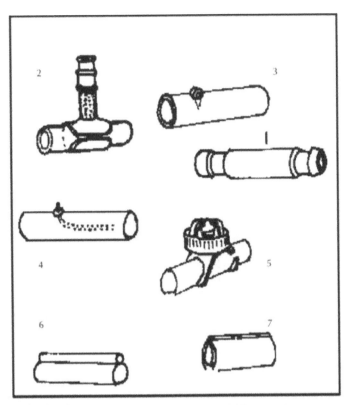

1. In-line emitters.
2. On-line emitters.
3. Micro sprinklers.
4. Micro tube (Spaghetti)

5. Compensating pressure emitter.
6. Low pressure emitter lines.
7. Biwall drip lines.

**Figure 5.7.** Types of emitters.

Micro tubes or spaghetti connected on the lateral line are used for fruit orchards, ornamentals, and flower pots in the nursery (green house). One or numbers of micro tubes irrigate individual plants or pots, and these are generally connected on the plastic or PE hose. In this system, plants are not near the lateral line. The irrigation water enters the dripper at high pressure and leaves the emitter at design pressure. The reduction in pressure is affected by the flow of water through long labyrinths, tortured (zigzag) path, size of orifices, and other design variations by the manufacturer.

Specially, designed emitter can allow free passage of particles of moderate size out of the dripper. These drippers are of self cleaning type at low operating pressures of 5–60 psi with average emitter flow of 2–7.6 liters per hour.

## THE IRRIGATION DURATION

The irrigation duration depends on the following factors:

1. The plant water requirements in gallons or liters/day.
2. Irrigation interval between each application and irrigation frequency.
3. Dripper flow rate and volume of water application.
4. Soil and plant characteristics.

The irrigation duration can be calculated as follows:

$$\text{Hours of irrigation per day} = \frac{[\text{liters of water required by plant per day}]}{[\text{Emitter flow rate in liters per hour}]} \quad \text{......................(1)}$$

## THE CLOGGING AGENTS IN THE IRRIGATION WATER

Depending on the water resource, the quality of water varies considerably depending on physical/chemical and biological composition, water demand, and rainfall. The clogging agents in the water can be of physical, chemical, or biological nature. The physical clogging agents are sand, silt, and clay. The chemical clogging agents include minerals and salts in the water. Many of these clogging agents may stimulate growth of microorganisms. The algae, bacteria, fish, ants, insects, and leaves are biological clogging agents. In general, an adequate filtration system, flushing, chemical treatments, and good management can prevent obstructions of the drippers and the lateral lines.

## THE FLUSHING OF SUB MAINS AND LATERALS

The maintenance and flushing of the main, secondary, and lateral lines are indispensable for good operation of a drip irrigation system. A good filtration system may catch larger size particles. However, smaller size particles may pass through the filter thus arriving at the secondary and lateral lines. These smaller size particles accumulate in the lateral lines and drippers thus causing obstructions. Some particles combined with certain type of slime and bacteria form aggregates that cause emitter clogging. A filter of >100 mesh size is adequate to avoid clogging due to physical agents. The periodic flushing of lateral lines will reduce these obstructions. The main and secondary lines must have a sufficient flow rate to allow flushing. In the lateral lines, the water flow at 30 meters/second is adequate for a good flushing action. The main line is flushed first, then the secondary line, and finally the lateral line. Several lines are opened at the same time when the water pressure is adequate. If the adequate pressure cannot be maintained with all the lines open, then few lines are opened at a time.

The flushing time must be sufficient to allow all sediments out of the lines. A regular program of inspection, maintenance, and flushing helps to reduce clogging of the emitters. The nature of the filtration system, the quality of the water, and the experience of the operator will determine when it is necessary to flush the lines. The maintenance and flushing must be a routine procedure. This practice of maintenance reduces obstructions due to sediments in the lines and drippers. It also prevents formation of slime, bacteria, and algae. The invasion by ants and insects is also reduced.

## THE IRRIGATION DISTRIBUTION SYSTEM

The irrigation distribution includes pump house, main and sub mains, and drip lateral lines. The water from the pump is distributed to the field through a main line. The secondary lines of smaller diameters take the water from the main line to the lateral lines. And the lateral lines supply the water to the drippers, which allow slow application to the plant. The main lines can be of high pressure plastic, galvanized iron, and ployethylene or lay flats. The main lines must be buried at least 0.6 meters (2 feet) to avoid mechanical damage during the field operations. Commonly, PE tubes of 12–16 centimeters (approx. 1/2–2/3 inches) of diameter are used for the lateral lines. These lateral lines provide water to the soil surface by means of the drippers. The PE hoses are available in different sizes, lengths, and strength. Some manufacturers produce weather resistant PE hoses. These are suitable for irrigation application.

### The Uneven Terrain

The unevenness of the agricultural land is an important design criterion. The change in an elevation causes a pressure loss or gain. A change in an elevation of 2.3 feet causes a pressure change of one psi. In a leveled or almost leveled land, the lateral lines with drippers must run throughout crop row. In slopes, drip lines must follow the contour. To install a system in an uneven field, one should consult a specialist. When the lateral lines in uneven field are designed, it is advisable to consider the advantage of the slope. Thus, the energy gain with the decrease in elevation is balanced. To maintain the uniform pressure in slopes of five percent, it is recommended to use shorter length of laterals and pressure compensating emitters; and to install pressure regulators.

### The Fruit Orchards

During dry periods, the fruit trees respond to the drip irrigation with good vegetative and fruit growth. The quantity of fertilizers is reduced to about 30–50% with fertigation. Generally in fruit orchards, drip irrigation system is permanent with the main and secondary lines buried. The lateral lines with drippers can be buried or left on the soil surface. In fruit orchards, the emitters are located within the shaded area of the tree (Figures 5.8–5.11). When a drip irrigation system for fruit trees is designed, more drippers are added as the tree grows. Rule of thumb is to irrigate 60% of the shaded area. When installing the system, it is important to consider the life of the tree (10–20 years). This is the main reason that the sub mains are buried.

## The Vegetable Crops

The drip irrigation system is more beneficial in vegetable crops that are planted in rows. The emitters apply the desired amount of water throughout the row, and the row spaces are left dry. The drip lines are placed on the soil surface along the rows of plants. It can also be located few inches below the soil surface. Also plant row can have one drip line on either side or two drip lines on both sides.

Commonly, PE tubes of 12–16 mm (approx. 1/2–2/3″) of diameter are used for the lateral lines. These lateral lines provide water to the soil surface by means of the drippers. The PE hoses are available in different sizes, lengths, and strength. Some manufacturers produce weather resistant PE hoses. These are suitable for irrigation application. Generally, these lines of distribution are located perpendicular to the direction of the secondary main line.

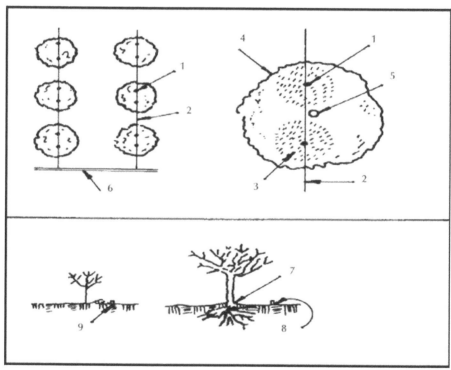

| | |
|---|---|
| 1. Emitter | 7. Do not allow flooding of the base of the trunk |
| 2. Lateral line to avoid diseases | 8. Move the emitter away from the tree trunk |
| 3. Pattern of wetting zone | 9. The emitter is near the trunk for the young tree |
| 4. Leaf coverage depending on the age of a tree | |
| 5. Trunk | |
| 6. Buried main line | |

Figure 5.8. Drip irrigation system in fruit orchards.

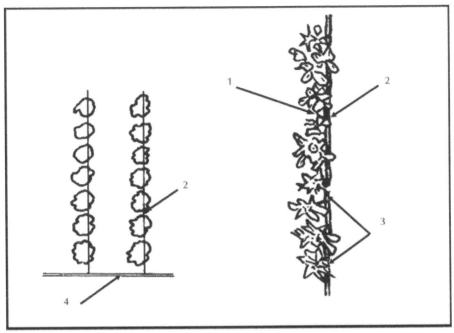

1. Continuous band of wetted zone.
2. Lateral lines with emitters.
3. Emitters.
4. Secondary line or sub mains or lateral lines.

**Figure 5.9.** Drip irrigation system in small fruit orchards (wines).

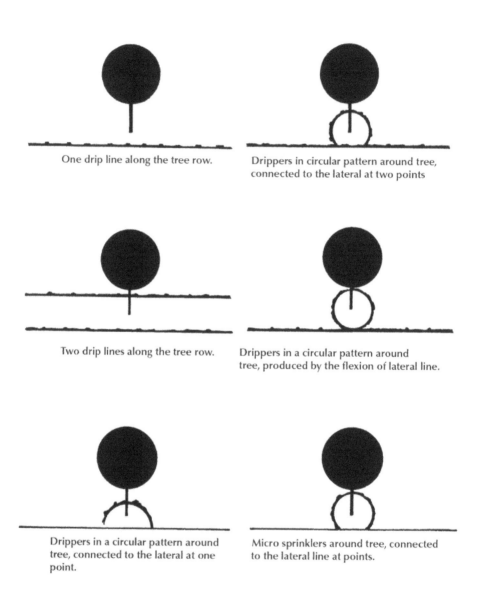

One drip line along the tree row.

Drippers in circular pattern around tree, connected to the lateral at two points

Two drip lines along the tree row.

Drippers in a circular pattern around tree, produced by the flexion of lateral line.

Drippers in a circular pattern around tree, connected to the lateral at one point.

Micro sprinklers around tree, connected to the lateral line at points.

**Figure 5.10.** Different arrangements of drip irrigation laterals for fruit orchards.

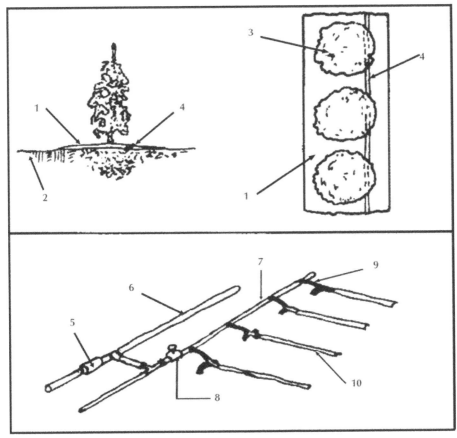

1. Organic matter mulch.
2. Soil surface.
3. Plant.
4. Emitter integrated on the line or in-line drippers.
5. Filter.
6. Principal or main line
7. Secondary line or sub main.
8. Optional: Gate valve.
9. Connecting tube (or spaghetti)
10. Lateral line.

**Figure 5.11.** Drip irrigation system in vegetable crops

## SUBSURFACE DRIP IRRIGATION SYSTEM (SDI)

Subsurface drip irrigation (SDI) is a variation of traditional drip irrigation where the tubing and emitters are buried beneath the soil surface, rather than laid on the ground or suspended from wires. SDI is the placement of permanent drip tape (trickle) below the soil surface, usually at a depth of between 20 and 40 cm. The products being used today in SDI come in three basic configurations: 1) hard hose, 2) drip tape, and 3) porous tubing (see Figure 5.12). Lateral injector to install subsurface laterals in SDI is shown in Figure 5.13. "ASAE S526.1 (1999a): Soil and Water Terminology" defines the SDI as "application of water below the soil surface through emitters, with discharge rates generally in the same range as drip irrigation." Earlier, "subirrigation" and "subsurface irrigation" sometimes referred to both SDI and subirrigation (water table management), and "drip/trickle irrigation" can include either surface or subsurface drip/trickle irrigation, or both. Other definitions of SDI require drip lateral placement below specified depths, such as normal tillage depths or a depth that would ensure use for several years. Generally, SDI has been used to describe drip/trickle application equipment installed below the soil surface only for the past 15–20 years.

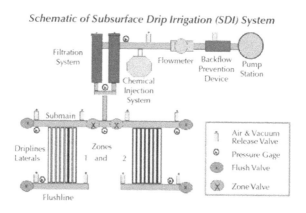

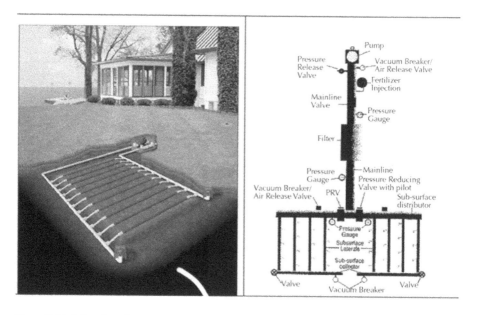

**Figure 5.12.** Subsurface drip irrigation (SDI) system.

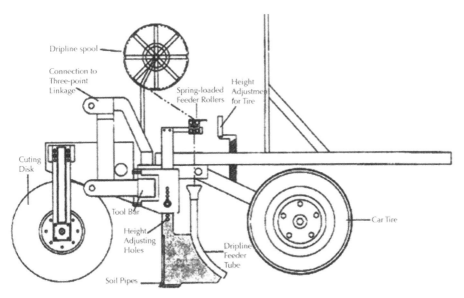

**Figure 5.13.** Subsurface drip irrigation (SDI): Lateral injector.

C. R. Camp, F. R. Lamm, R. G. Evans, and C. J. Phene presented a review paper on "SDI—past, present, and future" at the 4th Decennial National Irrigation Symposium on November 14–16, 2000 in Phoenix AZ. C.R. Camp published a paper on "Subsurface Drip Irrigation: A Review" in Transactions of American Society of Agricultural Engineers [41(5): 1353–1367]. According to them, SDI has been a part of agricultural irrigation in the USA for about 40 years but interest has increased rapidly during the last 20 years. Early drip emitters and tubing were somewhat primitive in comparison to modern materials, which caused major problems, such as emitter plugging and poor distribution uniformity (DU).

As plastic materials, manufacturing processes, and emitter designs improved, SDI became more popular but emitter plugging caused by root intrusion remained a problem. Initially, SDI was used primarily for high-value crops such as fruits, vegetables, nuts, and sugarcane. As system reliability and longevity improved, SDI was used for lower-valued agronomic crops, primarily because the system could be used for multiple years, reducing the annual cost. Design guidelines have also evolved to include unique design elements for SDI, including air entry ports for vacuum relief and flushing manifolds. Specific installation equipment and guidelines have also been developed, resulting in more consistent system installation, improved performance, and longer life. Crop yields with SDI are equal to or better than yields with other irrigation methods, including surface drip systems. Water requirements are equal to or lower than surface drip. Fertilizer requirements are sometimes lower than for other irrigation methods. Interest in the use of wastewater with SDI has increased during the last decade. The future of SDI is very promising, including its use in wastewater systems, and especially in areas where water conservation is important or water quality is poor. SDI is a very precise irrigation method, both in the delivery of water and nutrients to desired locations and the timing and frequency of applications for optimal plant growth.

Hard hose products generally have wall thicknesses of 0.75–1.25 mm (30–50 mils: One mil is a unit of linear measurement equivalent to 0.0254 mm, one thousandth of an inch, often used in measuring the diameter of wires), with nominal inside diameters of 13–16 mm. The emitter is either manufactured as an integral part of the tubing or is inserted later, and is typically placed at a repeated spacing interval of 0.5–1.5 m. The advantages to hard hose products are: strength and resistance to kinking, punctures, and rodent damage. Also, pressure compensating emitters may be incorporated into hard hose products. A disadvantage of hard hose products can be the initial cost.

Drip tapes, have wall thickness of 0.1–0.5 mm (4–20 mils). The thinner materials are more commonly used in single season or throw away applications, typical of strawberry production. The heavier wall materials ranging from 0.35 to 0.50 mm (15–20 mils) are more commonly found in multi-year applications or where the extra strength is required due to stones or other problems identified at the field site. Some tape products have emitter formed by the use of glue or other adhesive material used in the manufacturing process, where the actual water channel is made of the tubing material. Other tape or thin wall products have a pre-manufactured emitter glued or attached to the tape wall during the manufacturing process. Emission outlets are typically spaced along the tubing from 0.2 to 0.6 m, with wider outlet spacing having the larger flow path emitters.

Tape or thin products are initially the most cost attractive of the three basic configurations on a per unit basis. The disadvantage of tape products is that they are the most prone to mechanical or rodent damage due to reduced wall thickness, and thereby reduced structural strength of the tubing wall. However, the use of new generation PE resins has enhanced mechanical properties of tapes. This has lead to new applications and longer field life. Porous pipe products emit water all along the length of the tubing. There are literally thousands of orifices (pores) per meter where water weeps out of the tubing. This design has shown resistance to plugging by roots. However, its flow path is by far the smallest of the three configurations. This increases likelihood of plugging by fine particles. Other design components of SDI such as filtration and accessories are very similar to those found in conventional drip application. Proper filtration protects the emitter from contamination and clogging. Unfortunately, soil particles and other contaminants can be drawn into the emitter from the outside. This generally occurs at system shutdown, when a vacuum is developed in the lines and draws inorganic particles back into the emitter. To keep this from happening, many subsurface irrigation designs incorporate vacuum relief-valves that break the vacuum and keep water and inorganic particles from sucking into the emitter. Water use requirements of subsurface drip may not differ significantly from conventional uses. Evidence suggests that water applied on top of the soil through conventional drip systems has a cooling affect through evaporation on adjacent plant material. The net difference between surface and subsurface drip systems in total requirements is small.

**Advantages of SDI**

Graham A. Harris of Australian Cotton Corporative Research Centre at Toowoomba—indicates following advantages and limitations:

**1. Irrigation:** There is a high degree of water application control with the potential for high uniformity of application. For new systems, DU can be 93% or higher compared with 60–80% for sprinkler and 50–60% for conventional irrigation. The high frequency of irrigation with SDI allows maintenance of optimum soil moisture in the root zone. This is important where salty water is used and with shallow rooted crops, and is an advantage over surface and sprinkler irrigation where the fluctuation in soil water potential is greater, increasing the stress on the crop. Compared with sprinkler irrigation: It is possible to irrigate regardless of wind conditions. Lower pressures are generally needed and lower flow per unit area, requiring smaller diameter mains and laterals. For SDI, land levelling to enable drainage is required. SDI can be installed on a range of paddock sizes and shapes. SDI maintains soil surface structure more effectively than other irrigation types and makes it easier to allow for rainfall events or "catch up," provided there is sufficient system capacity. A well maintained SDI system requires less labour to operate than alternative systems.

**2. Agronomical practices:** The partial soil wetting provided by SDI has several benefits:

- Improved efficiency of nutrient uptake at the fringes of the wetted soil volume.
- Less water lost from soil surface evaporation.
- Less weed germination and growth.
- Unrestricted travel for field operations such as spraying and harvesting.
- Improved access to rainfall infiltration in some row crop situations.
- Maintains dry crop foliage: The benefits include reduced incidence of foliar disease, reduced loss of applied pesticides, reduced evaporation losses direct from the crop canopy, and less leaf burn where saline water is used for irrigation.
- Fertigation can be used with SDI. The possible benefits include savings in labour, more efficient use of nutrients and less risk of nutrient leaching, enhanced timing of nutrient applications to match crop requirements according to development stage and crop condition.

**3. Salinity problems:** Saline irrigation water applied through SDI will have less effect on crops than if applied through surface or sprinkler irrigation. This is due to no foliage absorption of salt. High frequency drip irrigation reduces the effect of increased salt concentration of the soil solution between irrigations compared to the lower-frequency of surface and sprinkler irrigation. There is a leaching of salts from the active root zone to the outer part of the wetted soil volume.

**4. Water use efficiency (WUE):** Water use savings range from 0 to 50% when compared with traditional irrigation systems. In situations where water savings are not made there is often a significant yield increase resulting in improved production per unit of irrigation water—improved WUE. For spray irrigated Lucerne in the Callide Valley of Australia, the long-term WUE is 1.37 t/ML. Trevor and Lyn Stringer have tabulated the WUE for SDI Lucerne grown at Biloela Research Station along with a direct comparison between SDI and spray irrigated Lucerne. They found that the improvement in WUE of SDI compared with spray irrigation was 43% in 1996–97 and 95% in 1997–98. SDI and furrow irrigation comparisons for cotton.

**5. Yield improvement**: Improved yields have been obtained with SDI installations. For Lucerne the yield improvement was 13 and 34% for 1996–97 and 1997–98 respectively. For cotton the yield change ranges from 0 to 21%.

## Limitations of SDI

Graham A. Harris of Australian Cotton Corporative Research Centre at Toowoomba indicates following limitations of SDI:

**1. Emitter clogging**: The high DU inherent in a well-designed SDI system can be readily destroyed through emitter clogging. The orifice diameters of emitters are usually 0.5–1 mm$^2$ and are susceptible to clogging by root penetration, sand, rust, micro organisms, water impurities, and chemical precipitates. The clogging is usually the result of insufficient water filtration, lateral flushing and/or chemical injection. Root penetration has generally only been observed in perennial crops like Lucerne where the soil has dried out around an emitter and the roots penetrate the emitter seeking water.

**2. System shutdown**: Water will flow to the lowest point in the field. If air is not allowed to enter the system by means of an air/vacuum release valve, a vacuum will be created and soil is sucked back into the emitter. On undulating fields, it is possible for this to occur throughout the field resulting in emitter clogging. The soil dries out around these emitters and roots penetrate the emitter seeking out water and further blocking the emitter. This problem can be avoided with an adequate design (including the strategic location of air/vacuum release valves), field preparation (levelling to drain), and high frequency irrigation that produces a permanently saturated soil zone around the emitter.

**3. Chemical treatments:** There are no herbicides registered in Australia to prevent root intrusion. Root hairs that have penetrated an emitter can be burned out with hydrochloric acid. Sand is readily removed from water using centrifugal separators. Suspended organic matter and clay particles can be separated with gravel filters, disk, and screen filters. If the water supply has more than 200 mg/L suspended solids then a settling reservoir is recommended before the water enters the filtration unit—above this level the filter system will be overloaded resulting in excessive back flushing. Bacterial slimes and algae growing on the interior walls of the laterals and emitters can combine with clay particles in the water to block the emitters. Bacterial precipitation of sulphur and iron is a further problem. These must be treated with chlorine. Chemical agents can precipitate to cause emitter clogging and these agents are:

- Iron oxide due to an iron concentration of greater than 0.1 mg/L.
- Iron bacteria further exacerbate the problem.
- Manganese oxide due to a manganese concentration of greater than 0.1 mg/L.
- Iron sulphide due to iron and sulphide concentrations of greater than 0.1 mg/L.
- Carbonates due to bicarbonate levels above 2 meq/L in water with a pH of 7.5 or higher, and Calcium at similar levels or a fertiliser containing calcium.

**4. Salt accumulation:** When saline water is used, salts accumulate at the wetting front. In SDI, this results in an accumulation of salt above and mid-way between the laterals. Where there is insufficient rainfall to move salt below the root zone, this accumulation will affect the growth of the current crop (even a relatively salt tolerant crop like Lucerne) or the establishment and growth of subsequent crops.

**5. Mechanical damage:** SDI laterals must be installed at the required depth below the ground surface along the full length of the field. To achieve this cross-rip, it is ensured that the turn-out area at the end of each lateral run is the same level throughout the field; otherwise the lateral will be installed at different depths at the end of the field.

**6. Mice damage** can be significant on cracking soils used to grow grain crops. Do not let these soils crack open between crops by using some irrigation water to keep the soil about the lateral closed up.

**7. Insect damage** has been a significant problem on some SDI sites. There are a number of possible insects that chew through SDI tape. Damage has resulted from crickets, earwigs, false wireworm larvae, and white-fringed weevil larvae. Care needs to taken in choosing and preparing the field for SDI, in its installation and maintenance in order to minimize the negative effect of insect damage. There are no insecticides registered in Australia for the control of insect pests for SDI systems.

**8. Crop establishment:** Soil type and the depth of placement of the SDI laterals will determine the ability of the system to wet the soil surface to aid the crop establishment. In most situations, crops cannot be established using SDI alone. Where it has been installed on farms with existing sprinkler or furrow irrigation systems, these have been used to establish the crop. In some situations, the SDI system has been used to pre-irrigate the crop and fill the soil profile, with planting following rainfall. Using SDI to fill the profile prior to crop establishment runs the risk of deep drainage as water is being released below the soil surface—this can negate the possible benefits of the system in improving crop WUE.

**9. Soil structural effects:** In certain soils the use of high quality water through SDI has resulted in increasing clay content, exchangeable sodium percentage and calcium—magnesium ratio away from the emitter. The result is a decrease in the lateral spread of water from the emitter during irrigation and a smaller effective root volume.

## SUMMARY

Drip irrigation is an efficient method of water application in agriculture. The drip irrigation system allows a constant application of water by drippers at specific locations on the lateral lines it allows favorable conditions for soil moisture in the root zone and optimal development of plant. Well designed drip irrigation system can increase the crop yield due to: efficient use of water, improved microclimate around the root zone, pest control, and weed control, agronomic and economic benefits. The main components of the drip irrigation system are water source, pump and the energy unit, filtration system, the chemical injection system, the valves and controls, the water distribution system, and the drippers. This complex system requires a more careful high technology management, the high initial investment, and adequate maintenance.

**KEYWORDS**

- Adapter
- Aggregate
- Automation
- Available water
- Cation exchange capacity
- Centrifugal pump
- Chemigation
- Class A pan evaporation
- Clay
- Clogging
- Conductivity
- Conductivity, hydraulic
- Contour
- Coupling
- Crop development stage
- Drainage
- Drip/trickle irrigation
- Dripper or emitter
- Elbow
- Emitter or dripper
- Evaporation
- Evapotranspiration
- Fertigation
- Fertilizer
- Filter
- Filtration system
- Frequency of irrigation
- Infiltration
- Injection system
- Irrigation distribution system
- Irrigation, furrow
- Irrigation, sprinkler
- Irrigation, subsurface
- Irrigation, surface

- Leaching
- Low tension
- Main line
- Nipple or union
- Organic matter
- pH
- Polyethylene (PE)
- Precipitation
- Pump
- Pump house
- Reducer
- Root zone
- Saline water
- Salinity
- Sand
- Slime
- Soil moisture
- Soil nutrient
- Solar Radiation
- Subsurface drip irrigation (SDI)
- Union
- Valve, automatic metering
- Valve, flow control
- Valve, flushing
- Valve, gate
- Valve, one way
- Valve, pressure regulator
- Valve, solenoid
- Valve, volumetric
- Water potential
- Water quality
- Water requirement
- Water source
- Water use efficiency
- Weed

## BIBLIOGRAPHY

Refer to bibliography section 5.1 for the literature cited.

## BASIC ASPECTS IN DRIP IRRIGATION RESEARCH

Basic aspects for consideration in drip irrigation research were compiled by the research team of the Southeastern region of the United States Department of Agriculture [S-143], "Trickle Irrigation in Humid regions." The following aspects should be considered:

### I. Atmospheric Conditions
    A.  Precipitation:
        1.  Total amount.
        2.  Effective amount:
            a.  Measurement of soil moisture.
            b.  Estimation of runoff.
        3.  Frequency.
        4.  Rate.
    B.  Evapotranspiration (ET):
        1.  Pan evaporation:
            a. Class A Pan.
            b. Pan wire mesh.
    2.  Models to estimate ET:
        a. Penman (favorite).
        b. Jensen-Haise.
        c. Blaney-Criddle.
        d. Hargreaves-Samani.
        e. Others.
    C.  Climatic parameters:
        1.  Daily solar radiation.
        2.  Active photosynthetic radiation.
        3.  Maximum and minimum temperatures (daily, monthly, or annual).
        4.  Average speed of the daily wind and direction.
        5.  Dew point temperature (a humidity index).

### II. Soil
    A.  Classification.
    B.  Physical properties:
        1.  Texture.
        2.  Compaction (apparent density / restrictive layer).
        3.  Hydraulic conductivity curve (soil moisture versus soil tension).

4. Infiltration rate.

5. Soil temperature at different depths.

6. Soil depth:

    a. Soil profile.

    b. Water aquifer depth.

    c. Depth of restrictive (hard) layer.

7. Description of the general topography.

C. Chemical properties:

    1. pH.

    2. Electrical conductivity.

    3. Percent of interchangeable sodium.

    4. Fertility:

        a. N, P, K, Ca, Mg.

        b. Micro nutrients: Fe, Zn, Mn, Cu.

        c. Cation exchange capacity.

    5. Applications of fertilizers:

        a. Amount.

        b. Program (frequency).

        c. Method of application.

        d. Source:

            ~ Type of fertilizer.

            ~ Chemical composition.

    6. Organic matter (%).

    7. Toxic ions:

        a. Boron.

        b. Chloride.

        c. Heavy metals.

        d. Other specific ions.

## III. Characteristics of Irrigation System

A. Physical characteristics of the system:

    1. Description of the system:

        a. Type of dripper.

        b. Operating pressure.

        c. Flow rate.

        d. Pattern and number of drippers per plant.

        e. Filtration system.

        f. Method to measure flow.

        g. Description of pumping system.

     h. Water source.
  2. Uniformity of water application and method of evaluation.
  3. Relative volume of the irrigated root zone.
  4. Experimental design:
     a. Treatments.
     b. Replications.
     c. Control.
B. Quality of the water:
  1. Sediments (solid).
  2. Salinity.
  3. Specific ion concentration affecting the soil properties.
     a. Sodium and sodium absorption rate (SAR).
     b. Bicarbonates.
     c. Fe, S.
     d. Ca, Mg.
  4. Toxic ions of the plant:
     a. Boron.
     b. Chloride.
     c. Heavy metals.
  5. pH.
  6. Potential for organic growth:
     a. Iron bacteria.
     b. Sulfur bacteria.
     c. Algae.
  7. Clogging or obstruction problems:

     a. Iron.
     b. Carbonates.
     c. Bicarbonates.

C.  Irrigation programming or scheduling:
1.  Criteria:
   a.  Method/indicator:
      Soil moisture content.
      Drainage capacity of soil.
      Estimated ET.

      Estimated water reserve.
      Simulated model of water reserve.
      Time.
      Time with adjustments for rainfall.
      Plant indicators:
        ~ Leaf temperature.
        ~ Leaf water potential.
        ~ Crop models.
   b.  Location of the measurements:
     i.  Soil measurements in the root zone of the plant.
     ii.  Measurements of leaf water potential.
   c.  Frequency of measurements:
     i.  Continuous measurements.
     ii.  Periodic measurements (specify).
2.  Necessary measurements:
   a.  Amount of water applied by irrigation.
   b.  Irrigation rate.
   c.  Duration of irrigation.
   d.  Frequency of irrigation application.

## IV.  Crop Characteristics
   A.  Type of crop (species, variety).

B.  Factors that affect the growth:
1.  Plant density per hectare.
2.  Plant and row spacing.
3.  Method of sowing or planting.
4.  Method of soil management.
5.  Diseases and location (wet and dry zones).
6.  Insect/diseases and location.
7.  Chemigation:
    a.  Chemical types.
    b.  Method of chemical application.
    c.  Frequency and rate of application.
C.  Crop response:
1.  Growth and development:
    a.  Growth curve.
    b.  Leaf size.
    c.  Plant mass.
    d.  Density.
    e.  Stage of phonological growth.
    f.  Fruit development stage.
    g.  Vegetative development stage.
    h.  Root density and distribution.
2.  Physiological response:
    a.  Leaf water potential.
    b.  Osmotic potential.
    c.  Stomach conductivity.
    d.  Chemical analysis of leaf and fruit (N, P, K, Ca, Mg).
D.  Crop yield:
1.  Amount:
    a.  Total.
    b.  Marketable or commercial (by classification).
2.  Fruit quality:
    a.  Size.
    b.  Heat.
    c.  Soluble solids.
    d.  Firmness or density.
    e.  Other parameters for fruit quality:
    ˜ Classification.
    ˜ Acidity.
    ˜ Brix.
    ˜ pH.

# Chapter 6

## Installation

Michael J. Boswell and The Toro Company

[1]Boswell, Michael J., 1985. Installation Procedures, Chapter XII. In: *Micro-irrigation Design Manual*. Pages 12.1–12.7. USA. James Hardie Irrigation, El Cajon—CA, 92022-22461. USA. Now The Toro Company.

## INTRODUCTION

Once a micro-irrigation system has been designed, it is necessary to translate the design drawings into a functioning irrigation system. Every irrigation system is unique, and therefore no installation guideline can cover all situations. However, experience has shown that there are a great many tasks, and numerous pitfalls, which are common to all systems. The following guidelines are meant to clarify commonly accepted procedures for construction, installation, and testing of micro-irrigation systems. Use of these guidelines will help to synchronize the various steps which are required for a proper and smoother installation, and result in fewer problems.

## GENERAL INSTALLATION OF THE MICRO IRRIGATION

### System Design, Planning, and Procurement of Materials

Accurate topographic maps of the area are obtained. A contour interval of five or ten feet is desirable.

The topographic map should include: water source and other pertinent features of the area, such as an electricity source, utility cables or pipelines, rock piles or rock outcroppings, roads, and so forth.

Irrigation system is designed and drawings are prepared.

Materials take-off list is prepared from the design.

List of materials is adjusted for wastage and breakage.

Sheet for specification of materials is prepared.

List of materials and specifications are distributed to hardware stores to request quotations on price and delivery dates.

Quotations are revised; the final purchase order is prepared.

Materials are ordered from suppliers.

## Requirements for Equipment and Tool

A good stock of shovels, hand saws, hacksaws, files, crowbars, wrenches, screwdrivers, rags, drills, and hoists should be kept in a tool truck.

A pipe trailer, 22 feet long.

A vehicle to carry parts: valves, fittings, glue, and so forth.

A backhoe or trenching machine for digging trenches.

A dozer for backfilling trenches.

A forklift or boom truck for lifting heavy items, including filters, large valves, pumps, tanks, and so forth.

A wire spooler to string control wire or tubing.

Tractors as required by field installation operations.

A small portable sump pump to remove water from flooded areas for making repairs.

## Receiving and Handling of Materials

Prepare the warehouse for materials, including a protected area for drip irrigation lines (reels), glue, and so forth.

Set up receiving procedures for materials. Delivery of materials should be carefully checked against purchase order.

Pipes should be delivered to a specified location near the job site on a planned delivery schedule.

Pipes should not be stacked more than two bundles high in the field unless there is a forklift available.

## Assembly of Fittings and Parts

Designate a location for the assembly of fittings and accessories: gate valves, air and pressure relief valves, risers, and so forth. It is advantageous to do as much of the assembly work as possible in a central, well-equipped facility.

Set up assembly equipments: tables, jigs, tools, bins for ditches (trenches), parts, glue, teflon tape, saws, wrenches, and so forth.

## Surveying, Staking, and Making Ditches (Trenches)

The surveyor stakes out mainline and sub main routes, field boundaries, locations of underground utility cables or pipelines, roads, and so forth.

The surveyor ensures that the pipeline grade enables drainage of the system if this is necessary to avoid damage.

Trenches are dug along the routes where the mains and sub mains will be laid.

## Assembly of Pipelines

Lower the principal and secondary pipe lines carefully to the trenches.

Pipes for mainlines and sub mains are laid out.

Assembly crew team assembles mainlines and sub mains.

Make sure the correct glue is being used with plastic pipe. There are different types of glue available for different sizes of pipe. Make sure primer is used correctly.

Clean the pipes end with a sanitary napkin and acetone.

Put glue at both ends of the pipe.

Join the parts immediately after applying the glue.

Use a rubber hammer to knock and force the entrance of one tube inside the other.

In areas of high humidity, care should be taken to prevent solvent cement from absorbing moisture, as it is hygroscopic and will lose its effectiveness if moisture is absorbed. When in doubt about glue quality, dispose it off rather than risk failures of the glue joint.

In hot weather, pipe exposed to sunlight will expand significantly. At night or after burial, the pipe will then cool and contract. This contraction may cause shifting of the pipe, separation of pipe joints, and various other problems. Smaller sizes of pipe may be "snaked" in the trench. Larger sizes of pipe should be laid in the trench and covered when cool.

In cold weather, additional time should be allowed for glue joints to cure.

Plastic pipe should be cut properly, with a square clean-cut end. Small diameter pipe (4-inch and less) can be quickly and cleanly cut with tubing cutter. Larger pipe should be cut with a power saw. A hacksaw is probably the least desirable tool to use with larger sizes of plastic pipe, since it is difficult to produce a straight cut with a hacksaw. Assembly the elbows, tees, raisers flushing valves, and check valves. The major parts are assembled outside the trenches.

## Installation of Control Wire, Tubing in Trenches, and Hydraulic Tubes for the Automation System

In systems with electronic or hydraulic valve controllers, the wire or hydraulic tubing is laid at the bottom of the trench. Wire should be laid in the trench loosely.

Sufficient slack should be allowed at each sub main riser for making connections to field control valves, and also to allow for contraction. Keep wire or tubing away from sharp edges of pipe line fittings and rocks.

Main and sub main lines are carefully lowered into the trenches. Wire or tubing should be tested before burial.

In hot weather, pipe should be laid into the trenches and covered with the soil in the early morning or late evening. If this is done when the pipe is hot, the buried pipe will tend to contract after buried.

## Installation of Fittings, Valves, Risers

All the remaining pipeline work should be completed at this time. Install all the fittings, valves, and riser assemblies.

Connect the wire or control tubing to the control valves in the field.

Paint the entire exposed plastic pipe and the fittings with a compatible paint, to protect against the sunlight and prevent algae growth.

## Partial Backfilling

Partially backfilling should be done immediately after pipelines are laid in the trench.

Backfill all low spots as soon as possible, especially on rainy days.

When backfilling, sub mains with flexible hose risers, hose kinking may be prevented by first slipping a short length of 1″ plastic pipe over the hose riser. After backfilling, these short pipe lengths may then be removed and used again at the next sub main.

## Installation of Thrust Blocks (construction of thrust walls)

Mainline tees, elbows, reducers, risers, and valves, should be anchored in place with thrust blocks, where significant hydrostatic thrust forces may develop.

Thrust block should be dug out or built after partial backfilling.

Install reinforcement steel and mounting brackets where required.

Pour concrete for thrust blocks. Allow several days for concrete to cure properly.

## Installation of Lateral Lines in Field and Flushing, Pressurizing, and Testing the System

Lateral lines are installed in the field, but not yet connected to sub mains. Care should be taken to ensure that the lateral lines are not contaminated with soil, insects, and so forth. Ends of lateral lines should be kept closed.

Close all sub main control valves.

Fill mainlines with water. Open flushing valves to flush out foreign materials out of mainline.

Once the main lines have been thoroughly flushed, the flush out valves should be closed, and the mainline pressure brought up to test pressure.

Maintain test pressure for 24 hours. If leaks develop in the mainline, immediately shut off the system, repair leaks, flush the main line, and repeat test.

Pressurize sub mains, open the flush out valves to flush out foreign materials out of sub mains.

Ensure that filters are working and that system has been flushed and cleaned before connecting laterals.

## Connecting Lateral lines to Sub mains

With ends of lateral line open, connect lateral lines to sub mains. The connection is more easily made if water is running in the sub mains.

Flush lateral lines thoroughly.

Close ends of lateral line. Bring system pressure up to operating pressure. Adjust all sub main pressures as required. Check for leaks, mark them, shut system off, and fix any leaks in sub main and lateral lines.

## Testing of System Operation and Backfilling Trenches

Check operation of controllers, sub main and control valves, filters and filter controllers, and so forth.

When it has been determined that all underground components, including pipes, fittings, control wires, and tubing, are working properly, backfill all trenches. Care must be taken during backfilling, particularly with large, thin-wall pipe, to prevent collapse or other damage to the pipe.

Once the system is running properly, take pressure and flow readings at strategic points in the system, that is, before and after pumps, filters, main control valves. These readings will be useful in verifying the design, in serving as operating specifications, and in the trouble shooting of the system at some later date.

## INSTALLATION OF BI-WALL® (DOUBLE WALL LATERALS) TUBING

The following recommendations apply to the installation of both Bi-Wall® and Tree-Line® tubing:

### Above Ground Assembly

Install the laterals with the orifices facing upwards. Installation of Bi-Wall® upside down will result in plugging of orifices due to contaminants in the irrigation water.

An air/vacuum relief valve should always be installed at the sub main riser to prevent suction from occurring in the Bi-Wall® tubes when the system is shut down. Suction in buried Bi-Wall® tubing will tend to draw muddy water back into the tubing through the orifices, causing contamination.

Bi-Wall® tubing may be laid on the surface or buried up to 18 inches deep. Burial is preferred where possible, since it protects the tubing from accidental damage, reduces the possibility of clogging due to algae and bacterial slimes and evaporative deposition, and ensures that water is applied at the desired location.

### Below Ground Assembly

Care should be taken during installation to prevent soil, insects, and other contaminants from getting into the tubing. The ends of tubing should be closed off by kinking or knotting until the tubes can be hooked to the system.

Injection equipment used to install Bi-Wall® tubing should be free of sharp edges, burrs, and areas where the tubing could be damaged. Bends, rollers, and other points of contact with the tubing should be kept to a minimum to reduce both the possibilities for damage and the tension on the tubing as it is injected. A simple Bi-Wall® injection tool is shown in Figure 6.1.

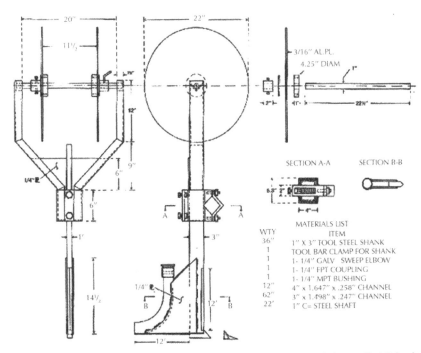

**Figure 6.1.** Tool for irrigation of the Bi-Wall tubing below the ground. [Boswell, Michael Micro Irrigation Design Manual. Hardie Irrigation. Page 12.7].

It is strongly recommended that the Bi-Wall® tubing be monitored as it is buried in the soil. Someone should be watching to ensure that the tubing maintains its orifices inward, orientation, to assist in case the tubing becomes tangled in the injector, and to signal the tractor driver when the Bi-Wall® real is empty and must be replaced.

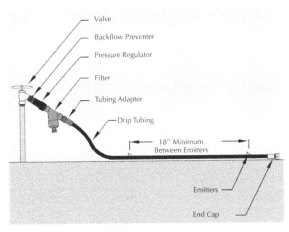

Field manifold

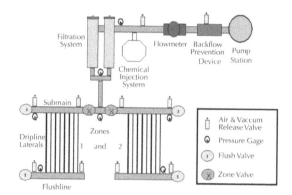

Components of a typical injection system for a drip irrigation system.

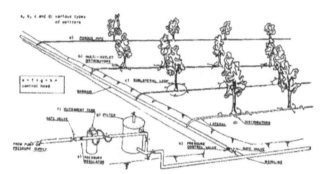

Installation of a drip irrigation system

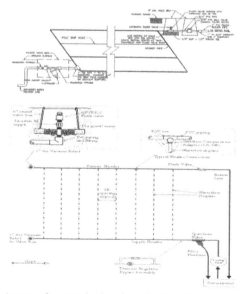

Layout of a typical subsurface drip irrigation system

a. Success of drip irrigation depends on the support from specialist for installation and maintenance (left top).
b. Flow meter to measure the volume of water applied (right top).
c. Water source, pump and check valve assembly (right bottom).

## SUMMARY

After micro-irrigation system has been designed, it is necessary to translate the design drawings into a functioning irrigation system. The procedures for construction, installation, and evaluation of micro-irrigation systems consist of planning and procurement of materials, requirements for equipment and tools, receiving and handling of materials, assembly of fittings and parts, surveying, staking, and making ditches (trenches), assembly of pipelines, installation of control wires or tubing in trenches and hydraulic tubes for the automation system, installation of fittings, valves, risers, partial backfilling, installation of thrust blocks, installation of lateral lines in field and flushing, pressurizing and testing the system, connecting lateral lines to sub mains, testing and operation, and backfilling the trenches. The lateral lines tubings may be laid on the surface or buried up to 45 cm depth. For a below ground assembly, care should be taken during installation to prevent soil, insects, and other contaminants from getting into the tubing.

**KEYWORDS**

- Automation
- Bi-Wall® tubing
- Drainage
- Elbow
- Filter
- Flow meter
- Micro-irrigation system
- Polyvinylchloride (PVC) pipe
- Pump
- Reducer
- Slime
- Trenches
- Valve, check
- Valve, flushing
- Valve, gate
- Valve, hydraulic

# Chapter 7

## Automation

[1]This chapter is modified and translated from: "Goyal, Megh R, E.A. Gonzalez, G. Fornaris y J.V. Pagan. 1990. Automatización. Capitulo VIII en: *Manejo de Riego por Goteo,* editado por Megh R. Goyal, paginas 241–280." Rio Piedras, PR: Servicio de Extensión Agrícola, UPRM. For more details one may contact Dr. Megh R. Goyal by E-mail: goyalmegh@gmail.com

## INTRODUCTION

Drip irrigation is an artificial method to apply the essential water for the plant growth that the nature has failed to provide [1]. Typically the irrigation water is applied to supply moisture to root zone when most of the "water available" to the plant has been used. There are several methods of pressure irrigation, such as: Sprinkler irrigation, center pivot and LEPA; micro jets, drip/micro or trickle irrigation, surface or subsurface irrigation. These help to maintain the soil moisture that is adequate for the plant growth. Among these systems, drip irrigation is the most efficient in terms of water use efficiency. Drip irrigation system is used extensively in humid, arid and semi-arid regions of the world. Any interruption or disturbance in an irrigation scheduling will cause a water stress to the crop. Therefore, the scheduling of drip (high frequency) irrigation should be automated so that it is able to respond to slower and faster changes in the soil moisture, the plant water or evapotranspiration. Automation of drip irrigation system has several advantages: Economy, saving of manual labor, increase in crop yield, conservation of energy and effective control of irrigation. This chapter presents basic concepts for automation of drip irrigation system, different methods of automation and irrigation programming [11].

## PRINCIPLE OF AUTOMATION

Current technologies of irrigation programming consider several factors such as [2, 3]: Duration and stage of crop growth, allowable plant water stress, soil aeration, soil water potential, soil salinity, soil moisture available to the plant, class A pan evaporation and evapotranspiration. In most cases, programming of drip irrigation has been limited to a control system that uses duration or depth of irrigation. The irrigation controller is programmed to operate solenoid valves in sequence and to verify operating pressure and flow rates, wind, temperature and other indirect variables. To obtain the minimum cost-benefit and high efficiency of water use, it is necessary to achieve high crop yield. The water loss due to several processes (control of salinity, requirement of infiltration, evaporation, irrigation losses and runoff), must be reduced to a minimum so that the accurate application of the irrigation is limited only to the crop requirements. Four methods for automation of irrigation systems are based on: (1) Soil moisture, (2) Plant water stress, (3) Estimation of evapotranspiration and (4) Combination of one or more of these methods.

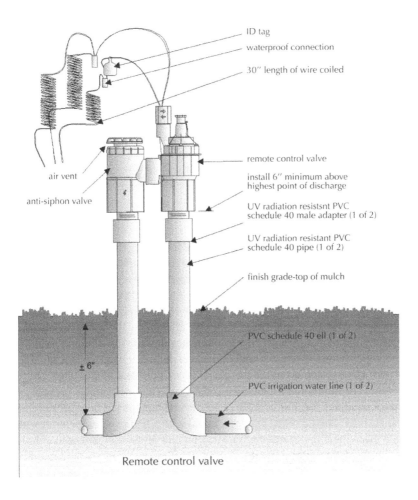

ID tag

waterproof connection

30″ length of wire coiled

remote control valve

air vent

install 6″ minimum above
highest point of discharge

anti-siphon valve

UV radiation resistsnt PVC
schedule 40 male adapter (1 of 2)

UV radiation resistant PVC
schedule 40 pipe (1 of 2)

finish grade-top of mulch

PVC schedule 40 ell (1 of 2)

± 6″

PVC irrigation water line (1 of 2)

Remote control valve

## Soil Moisture Method

Irrigation based on soil water potential is perhaps the oldest method to program irrigation. Microprocessors along with sensors, tensiometers, heat transfer psychrometric
methods, gypsum blocks and thermocouples have been used successfully for irrigation scheduling. The sensors can provide quick information to make decisions for
application of irrigation depth. The microprocessor circuits combined with a computer
programming can help to estimate the irrigation duration on the basis of field data,
matrix potential of soil; and to calculate the number of days between two successive
irrigation events.

A thermal method measures the matrix potential of soil, independent of soil texture, temperature or salinity. It is based on frequent measurements of ability of a porous ceramic sensor to dissipate a small amount of heat. With a good calibration, the
sensor can be used in any soil to automatically watch the matrix potential of the soil
and for irrigation scheduling. For closed circuit automated irrigation, the soil sensor

is placed in the root zone. For an automatic control of an irrigation system based on matrix potential of a soil, we need equipments for the:

1. Automatic sampling from several sensors in sequence,
2. Comparison of the reading of each sensor at which the irrigation begins at a predetermined matrix potential of the soil, and
3. The operation of irrigation controller to control the irrigation depth. Desktop computers in combination with microprocessors have been successfully used. There is also a commercial equipment to measure the matrix potential of the soil and for an automatic control of a drip irrigation system.

### Water Content in the Plant
The water is frequently one of the limiting factors in agriculture. Transpiration loss occurs from the plant surface due to an evaporative demand of the atmosphere. Less than one percent of the absorbed water is retained by the plant. This small fraction of water is often used to replace the deficit between water use and transpiration. Thus any water deficiency can cause a plant water stress. The total water potential (the sum of turgor, matrix and osmotic potential) is used to indicate the condition of the plant water. The plant development and growth (cellular enlargement and photosynthesis), pollination, fruit formation, crop yield and fruits quality are affected by the water deficit. Probably, the cellular growth is most sensitive to the water deficit. There are several methods to estimate the condition of plant water. These include determination of relative water content, diffusive conductivity of the plant, water potential of the plant and surface temperature. The indirect or direct measurement of water potential is probably a good indicator of the plant water stress. There are several methods to measure the plant water stress such as: The total leaf water potential with a leaf psychrometer; temperature of leaf surface with an infrared thermometer, and the leaf water potential indirectly on the basis of the diameter of the stem.

### Leaf Water Potential
The leaf water potential can be measured by psychrometer or by adhering thermocouples to the leaves. Although the psychrometric measurements are taken routinely for research purpose, yet the instruments are expensive and not feasible for commercial purposes.

### Temperature of the Leaf
Measurements of leaf temperature can indirectly indicate status of a water stress [10]. Plant water stress index can be used to automate the irrigation system, and to indicate when to irrigate. The operating system can be easily automated to take the data, to calculate the index of plant water stress, to make comparisons with pre-determined values of irrigation depth and to make decisions for irrigation scheduling. Leaf temperature is measured with a non-contact infrared thermometer. The accuracy of temperature of the surface of leaf depends on the precision of calibration. The measurements are sensitive to changes in the ambient temperature, interactions with surrounding surfaces (such as soil), and leaf area index. Measurements of leaf area index of a crop vary from plant to plant. There is no standard value.

## Stem Diameter

The diameter of stem and the leaf water potential are closely related to one another. The measurements of stem diameter can be used for continuous recording of the stem growth and the condition of plant water. The periodic calibration of the changes of diameter of stem versus leaf water potential can be conducted for each phenological stage of a plant. This technique can be used for the purpose of automation.

## Evapotranspiration Estimations

To program the irrigation, the evapotranspiration models have been successfully used throughout the world. The following information is needed for the evapotranspiration estimations and the criteria to decide when to irrigate.

1.  Evapotranspiration of a reference crop, potential ET, etc.
2.  Crop growth curve, crop coefficient and consumptive use of a crop.
3.  Index to estimate the additional evaporation from the soil surface when the soil is wet or dry.
4.  Index to estimate the effect of soil water loss in relation to ET.
5.  Estimation of available soil moisture used by a crop: Consumptive water use.
6.  Relation between expected crop yield and crop water use.

To estimate the ET, many of the variables are not well defined and must be estimated. Although the ET models can be useful to accurately estimate the irrigation needs, yet these are not viable for irrigation scheduling as available weather data are limited for a particular location.

## Direct Measurement of Essential Evapotranspiration

The weighing lysimeter in a given crop can serve as a guide to provide an adequate irrigation depth for the crop need. A water tank is connected to a lysimeter so that the weight of the irrigation depth is included in the daily weight of lysimeter. Whenever one millimeter of ETc is registered, lysimeter is automatically watered by drip irrigation system to maintain the soil water potential. The tank is automatically filled daily to a constant depth. Therefore, the daily changes in the weight of lysimeter represent the crop growth. The water potential of the soil is almost maintained constant by the drip irrigation system (see Figure 7.1) [8, 11].

## INSTRUMENTATION AND EQUIPMENTS

The automation of a drip irrigation system at an operating pressure can potentially provide an optimum crop yield and optimum water use. A system of controls in an automated irrigation system must use sensors to measure variables, such as: Depth and frequency of irrigation, flow rate, operating pressure; and environmental conditions such as wind speed, ambient temperature, solar radiation, rain fall, soil moisture, leaf temperature, leaf area index, etc. Maximum irrigation efficiency is possible with the continuous monitoring and control of the operation of the system with measurements of flow (solenoid valves) and operating pressure (pressure regulators) at strategically important locations in the field. The data or control of functions can be transmitted by electrical cables, laser or hydraulic lines, rays, radio frequency signals, remote control or by satellites. A

wide variety of instrumentation and equipments with characteristics are available commercially. These can be subdivided in six categories: (1) Controls, (2) Valves, (3) Flow
meters, (4) Filter, (5) Chemical injectors, and (6) Environmental Sensors.

### Controls

The controls receive feedback about the volume of water for the field, pressure in the
line, flow rates, climatic data, soil water, plant water stress and from the field sensors.
This information is then compared with the predetermined values and the irrigation
is reprogrammed to adjust for the new values, if necessary. The controls, volumetric
valves, hydraulic valves, fertilizer or chemical injectors, flushing of filters, etc., can be
operated automatically or manually.

### Valves

Automatic valves (Figures 7.2 to 7.9) can be activated electrically, hydraulically or
pneumatically and these are used to release or to stop the water in the lines; to flush the
mains and laterals; to continue the water from one field to another field and to regulate
flow or pressure in main, sub-main or lateral lines. The type of valve will depend on
the desired purpose. Valves receive feedback to verify the precision of operation.

### Automatic Volumetric Valve: Flow meters

The flow metering valve (Figure 7.10) allow programming the predetermined values.
Usually these meters are calibrated to measure applied volume of water or to measure
the flow rate.

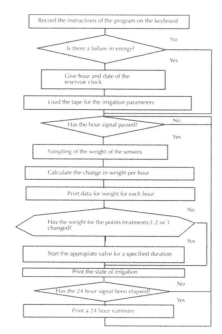

**Figure 7.1.** Logic diagram to measure weight of lysimeter sensors and to control the irrigation
sequence with three depths of irrigation.

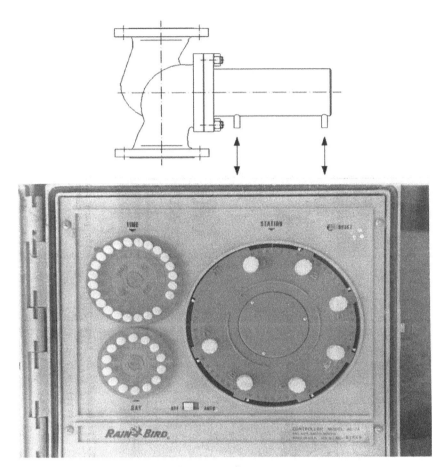

**Figure 7.2.** Automatic irrigation controller (Rain Bird).

**Figure 7.3.** Logic hydraulic valve.

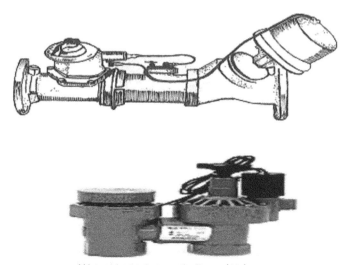

Water Master Automatic Control Valve

**Figure 7.4.** Automatic metering valve along with a hydraulic valve.

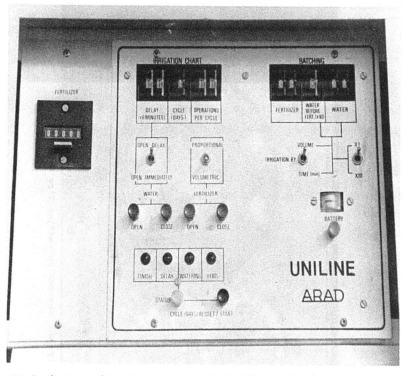

**Figure 7.5.** Fertilization and irrigation programmer for six different valves (for green house or field).

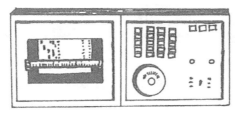

**Figure 7.6.** Automatic controller (Nirim electronics), using a programmer with a perforated tape or card.

**Figure 7.7.** Fertigation and chemigation equipments.

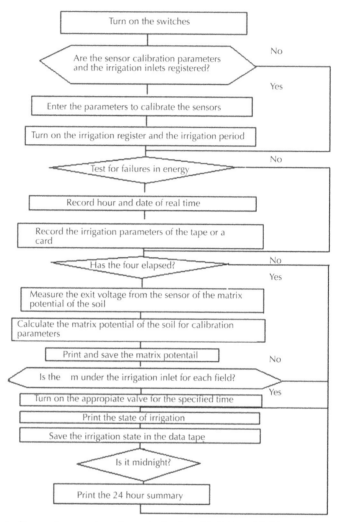

Turn on the switches

Are the sensor calibration parameters and the irrigation inlets registered?    No / Yes

Enter the parameters to calibrate the sensors

Turn on the irrigation register and the irrigation period

Test for failures in energy    No

Record hour and date of real time

Record the irrigation parameters of the tape or a card

Has the four elapsed?    No / Yes

Measure the exit voltage from the sensor of the matrix potential of the soil

Calculate the matrix potential of the soil for calibration parameters

Print and save the matrix potentail    No

Is the    m under the irrigation inlet for each field?    Yes

Turn on the appropiate valve for the specified time

Print the state of irrigation

Save the irrigation state in the data tape

Is it midnight?

Print the 24 hour summary

**Figure 7.8.** Logic diagram for an automatic controller in a drip irrigated field.

http://www.bermad.com/

**Figures 7.9 and 7.10.** Bermad automatic volumetric valve.

### Ambient Sensors

Various types of instruments are available to determine the soil moisture (ceramic densitometry, ceramic cup, heat dissipater sensor, soil psychrometer); to measure climatic parameters (weather station, automated evaporation tank, etc.), plant water stress or leaf temperature of the crop (leaf psychrometer, porometer for stomate diffusion, infrared and sensorial thermometer to measure stem diameter). These can be used as feedback for the management of irrigation. If the soil at a particular field station is wet, the sensor opens the circuit of the hydraulic or solenoid valve and this station is bypassed. If the soil at this field station is dry, the closer the circuit and the field at this station are irrigated for a specified duration.

### Filters

The obstruction in the drippers caused by clogging agents (physical, chemical or biological) is a common problem and is considered a serious problem in the maintenance of the drip irrigation systems. The suspended solids may finally clog or reduce the filtration efficiency. The automatic flushing valve is available for different types of filters. The flushing is done by means of back flow of water [4, 8, 11], thus allowing the water to move through the filter in an opposite direction (Figure 7.11).

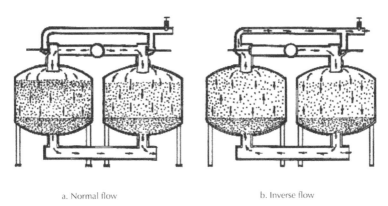

a. Normal flow                              b. Inverse flow

**Figure 7.11.** Automatic flushing of filters by inverse or back flow.

### Chemical Injectors

The chemigation methods to inject the fertilizers, pesticides and other inorganic compounds are: (1) Pressure differential, (2) Venturi meters and (3) Injection Pumps. In all these cases, digital flow meters can be used for the chemigation by allowing a known amount of chemicals in a known amount of water to maintain a constant concentration of chemicals-in-the-irrigation-water [8, 11].

### AUTOMATIC SYSTEMS

With the exception of a volumetric metering valve that operates according to the time or the discharge rate, the automatic irrigation systems can be divided in three groups on the basis of operation: (1) Sequential hydraulically operated system, (2) Sequential electrically or hydraulically-electrically operated systems and (3) Non-sequential electrically operated system with or without programming: With the possibility of using information of the field (feedback) by remote control.

### Sequential Hydraulically Operated System

This system controls the valves in sequence (Figure 7.12). The valves open and close based on the water pressure in the line. The pressure arrives at the valve by means of a flexible hydraulic tube (micro tube: polyethylene tube of small diameter) to provide a required pressure. The diameter of the micro tube is generally between 6 and 12 mm and is connected to the hydraulic valve at one end and the other end is connected to the automatic control or the line of water.

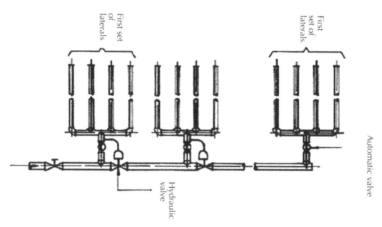

**Figure 7.12.** Sequential hydraulically operated system for green houses, gardens, nurseries, and fruit orchards.

Some hydraulic systems can be connected to the main valve of the line or to the system that replaces the water. In this case, the main valve is connected automatically to open when the system in series is in operation and to close at the end of the irrigation cycle. Electrically operated automatic system activates the pump and deactivates the pump, when the irrigation cycle is over.

Sequential hydraulically operated system is controlled by a predetermined amount of water. The amount of water can be different for each valve and can be adjusted by a regulator mounted in same valve. The hydraulically sequential system can be used to water fruit orchards, gardens, green houses and nurseries, establishing low flow rates through tubes of small diameter and for flow rates in any diameter of tube. The system includes automatic metering valve, hydraulic valve and hydraulic tube.

### Sequential System: Operated Electrically or Operated Hydraulically-Electrically

These systems supply an electrical current through cables for the remote control of the valves (Figure 7.12). The current from the "control panel" to the valves, usually passes through a step down transformer to supply a voltage of 24 V. For safety reasons, a current of 220 V should not be used when the subsurface cables extend to the field valves. The regular solenoid valves are mainly used for low flow rates. For pipes of larger diameters, the solenoid valves are used only as controls to activate the hydraulic valves and all the automation process is hydraulically-electrical. The control of the second valve is always hydraulic. In the hydraulic sequential system, the opening is controlled electrically by a timer mounted next to the main valve. In such cases, the current source is direct and not alternating.

### Programming irrigation with solenoid valves

The solenoid valves can be used to program the irrigation (Figure 7.13). In order to calculate the crop water use, a computer program can be used with the information such as: The soil moisture, evapotranspiration, the date of the next irrigation and the amount of water to be applied. The irrigation programs are based on evapotranspiration estimations, complex water budgets in several dimensions or crop growth models. The ET models use crop and climatic data such as crop coefficient, root zone depth, allowable depletion, drainage rates, air temperature, sun radiation, precipitation and constants in the evapotranspiration equations, and so forth.

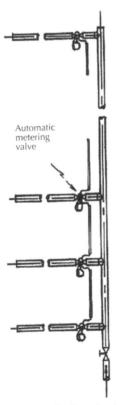

Automatic
metering
valve

**Figure 7.13.** Electrically operated tensiometer and solenoid valve.

Then the model incorporates the climatic information to calculate the evapotrans-piration rates and to adjust the water balance in the soil as the water is being used. The evapotranspiration model requires an irrigation criteria based on the allowable depletion or the irrigation interval. Actual field data after the irrigation can be helpful to compute the infiltration and immediate drainage for correction of estimated soil moisture. The rate of computed ET can be used to indicate the required amount of ir-rigation or to specify the time for irrigation interval. This method is more practical for drip irrigation than for other irrigation methods. The records of field data are kept in the office files for the irrigation programming, so that the data can be used to-update-the-inputs-in-the-program.

**Automatic valves**

The automatic valves are commonly used for the pump house and filters; for regu-lating the pressure in the main line; to control the flushing cycles in the filters, or to control the volume of water through the secondary or lateral lines. The solenoid valves can be used in the secondary or lateral lines to control the volume of water to the individual blocks. The primary function of a solenoid valve is to switch on or switch off the system. However these valves can be equipped with pressure regula-tors and check valves. The solenoid valves are operated electrically from the "Cen-tral Control Panel." Automatic control valves can also be equipped with manual valves for better efficiency. Automatic valves require periodic maintenance to assure a satisfactory operation. The maintenance program depends on the use of the valve and the flushing operations.

At least, it is recommended that all the diaphragm valves are disarmed and cleaned at least once a year. It is important to clean the deposits on the stem of the valve. Almost all the manufacturers provide a service or fast replacement of most of the components. This can usually be done without removing the valve from the irrigation line. A number of auxiliary controls can be adapted to the diaphragm valves to provide flexibility and convenience.

*Pressure reducing valve*

This valve responds to changes in the pressure at the exit of the main valve and adjusts to the pressure in the cap or valve cover to compensate for any change. A trouble in the operation can be caused by contamination, obstructions, incorrect assembly, and damages or worn out parts.

*Pressure-regulator-valve*

This valve is used to separate the system from the pressure in the main line. It must be open during the normal operation. Whenever the pressure exceeds a preset value, the valve releases the excess pressure.

*Controls-to-adjust-the-velocity-of-the-main-valve*

These are small adjustable controls in the pilot control system. These regulate the speed of opening and closing of the main valve by blocking or strangling the flow that enters or leaves the casing. These can be subjected to obstructions by fine sediments if tightly fit.

## Controls

Several electromechanical and electronic controls in the drip irrigation system are automated. The controls with mechanical time clocks open and close only a single valve at one time. These are programmed based on series of climatic and soil sensors: to decide when to begin and to end the irrigation cycle; start and to put off pump; to open and close the valves to supply an irrigation depth and to remember how much water and fertilizer was applied to each block within the field. The controls are also available to diagnose operation and identify the troubles and to take remedial steps. Others put off the system during rainfall and restart the system when necessary. A timer uses a clock to program the beginning and sequence of irrigation. The control is a source of electric or hydraulic signal to activate by remotely located valves to allow or to stop the flow.

The communication between the irrigation controller and the valves is by means of electrical wires, hydraulic lines or radio signals. The microprocessors and microcomputers also can be programmed using data of tensiometers, pan evaporation, thermocouples, soil moisture tension gages, anemometer, flow meter, pressure transducer, etc. These controls are based on the climatic and soil sensors or according to the program specified by the irrigator. Using these data, the controller uses a program to compute irrigation requirements for each crop and block within a field.

The data from the flow meters and pressure gages is used to determine the flushing time and to detect any troubles in the system. In most of the cases, the controller has a calendar programmer, so that the cycle of irrigation begins automatically on a particular day of the week and at a particular time of the day. Most of the controllers can be programmed for fourteen days, while others are only limited to seven days. Practically all automatic controllers have a station selector on the outer surface of the panel (Figures 7.4–7.7). This station selector shows a green light to show the station in operation. In addition, it can also be set manually so that the irrigation operator can start and put if off whenever desired.

## Sequential System: Electrically Operated

In these systems, the amount of water distributed to the different blocks is determined by a flow meter. A timer determines the duration of operation: 14 days and 24 hours per day. Sensors based on tensiometers or pan evaporation can activate these. Although this type of system was developed mainly to water green houses, yet it can be used for the drip irrigation system.

## Non-sequential System

These systems are completely automatic and are controlled electrically. These non-sequential systems are controlled by hydraulic or electrical valves that can operate the valve in the desired block at random, and can supply known amount of water for a known duration to a desired block. Each unit can supply a known flow at different hours during the day, in response to soil moisture status in each block. The "Control Panel" consists of electrical circuits that operate the pump, main valve, adds fertilizer according to a pre-established schedule and measures the soil moisture to estimate the crop irrigation requirements. This system usually operates by a remote control system and is designed to provide feedback of field data, so that the automatic adjustment can be made and adjustments for changes in pressure and flow rates can be made to the discharge flow in the distribution lines.

### Central panel

The central panel controls all the operations of the field, sending instructions to the valves and receiving continuous data on the operation of the irrigation system. It consists of a programmed unit of irrigation, a unit for transmission of information, a unit for the control of flow in the laterals and a unit for warning signals.

### Field panel

The field panel is placed centrally in the field and operated by remote control unit. The signals of the main panel are sent by an individual communication channel and these are transmitted to individual field panel. The field panel can collect the data on water meters, operating pressures and warning signals. Then the data can be transmitted to the main panel (control panel).

## Use of Sensors to Program Irrigation

In addition to the above mentioned instruments, sensors are available to determine the soil moisture tension or the soil moisture. Tensiometers and gypsum blocks are simple and economical to use. Another method is a neutron scattering method, but it is quite expensive and is used for research purpose only.

## Use of Gypsum Blocks and Tensiometers

The gypsum blocks can measure the soil moisture tension in the range of 1–15 atmospheres. There are two electrodes inserted in each block and the changes in the soil moisture are calibrated with variations in the resistance. The precision of this method is based on the temperature, salt concentration in the soil solution, physical characteristics of the gypsum block and the electrical resistivity of the soil. For tensions of 80 cbars, a tensiometer is recommended instead of a gypsum block. Tensiometer (Figure 7.14) measures the tension and the reading is given in cbars. The main disadvantage of a tensiometer is a relatively low critical tension of 85 cbars after which the air enters the plastic stem of a tensiometer. The soil moisture by any method will show variations in the soil moisture within the same field. A sample of the soil in a given location represents only the soil condition of that location. Therefore several observations of soil moisture at various locations in the field are desirable.

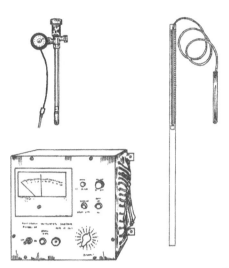

**Figure 7.14.** Automatic unit for control of irrigation based on the gypsum blocks (sensors).

## Neutron Scattering Method

The neutron scattering method consists of a neutron radiation source of high energy and a neutron detector. Neutrons travel through the soil medium, loose energy, and the speed is reduced when these hit the elements that are present in the soil. The hydrogen, a component of the water, is dominant in the reduction of the speed of fast neutrons. Due to other factors that can affect the reading, the calibration of this method is done in a location where the equipment will be installed and used. The use of the neutron scattering method requires the installation of access tubes at the beginning of planting and removal of these tubes after the last harvest.

It is recommended to install one sensor at each 30 cm depth. Periodically the operator will obtain the readings of the tube at the desired depth. A minimum of three readings are taken: at shallow root depth, at middle depth, and at a deeper depth. The water content of these readings is added and the water content at field capacity is deducted from the sum. The difference between these two estimates will be the amount of water that should be applied. The readings can be recorded automatically and are stored in the memory of the neutron scattering equipment. Then these can be downloaded on the computer of the Control Panel. With this information, the computer will give the necessary commands to the drip irrigation system so that the crop water requirements are met in the desired block.

## Class A Pan Evaporation to Automate the System

The relationship between pan evaporation and the water loss have been well established. Both are exposed to similar climatic conditions in the same field. This correlation can be used to schedule the irrigation. If electrodes in the tank can be installed at a depth (based on previous experience), the irrigation can be controlled automatically. The irrigation will begin when the surface of the water in the class A pan lowers to a predetermined level and will stop when the level raises to certain level in the tank [7].

## PREVENTIVE MAINTENANCE

### Preparation after the Last Harvest

1. Clean the controllers, valves and sensors.
2. Examine the condition of the control panel and store it well.
3. Remove and store batteries.
4. Flush and drain the hydraulic tubes.
5. Disconnect the electrical wires in the field.
6. Examine for possible breakage and defects in electrical conductors.

### Preparation for the Start of a Crop Season

1. Be sure that all the electrical connections are cleaned and adjusted well.
2. Make sure that the electrical contacts are free of corrosion and dirt.
3. Inspect all the hydraulic lines and pneumatic lines for leakage or breakage.
4. Verify that the equipments, accessories and sensors operate properly.

### During the Crop Season

1. Visually examine all external components weekly.
2. Disconnect the electrical wires in the field during electric storms.
3. Disconnect the batteries when the control is out of service for one week or more than one week.

## TROUBLE SHOOTING

| Trouble | Cause | Remedy |
|---|---|---|
| **Controls** | | |
| 1. The cycle of irrigation does not work at the pre-established time. | The clock of the control panel is outside pre-established calibration for the schedule of the cycle. | Calibrate clock at the pre-established time. |
| 2. Some stations do not operate. | Control station for time is off. Cables of the valves are not connected properly. Hydraulic tube is broken or missing. | Place ignition control in "on" position. Check connections between valves. Replace the hydraulic tube. |
| 3. Danger signal is in "on-position" | Battery is dead. Program-of-emergency-is-in operation due to bad operation of the system. | Recharge battery. To locate the source of the problem in the system and to correct it. |
| **Filters** | | |
| 4. Poor filtration | High difference in pressure is due to obstruction of filters by the clogging agents. | Flushing of filters by inverse (back) flow. |
| 5. Pressure difference at the entrance and exit of a filter -exceeds the recommended -values. | Depth of filter media is not adequate. Valves are obstructed. | Add more media until it is at recommended level. Verify valves for obstruction. |

## SUMMARY

Principle of Automation includes factors such as: duration and stage of crop growth, allowable plant water stress, soil aeration, soil water potential, soil salinity and evapotranspiration. Leaf water potential can be measured by a psychrometer or by adhering thermocouples to the leaves. Leaf temperature is measured with a non-contact infrared thermometer. The accuracy of temperature of the surface of leaf depends on the precision of calibration. Measurements of leaf area index of a crop vary from plant to plant. The diameter of the stem can be used for continuous recording of the stem growth and the condition of plant water stress for each phenological stage of a plant. This technique can be used for the purpose of automation. The automation of a drip irrigation system provides an optimum crop yield and optimum water use. The system uses sensors to measure depth and frequency of irrigation, flow rate, operating pressure, wind speed, ambient temperature, solar radiation, rain fall, soil moisture, leaf temperature, leaf area index, etc. The instrumentation and equipments for automation can be subdivided in six categories: (1) Controls, (2) Valves, (3) Flow meters, (4) Filter, (5) Chemical injectors, and (6) Environmental.

There are three types of automatic irrigation systems. In sequential hydraulically operated system, the valves open and close in response to the application or elimination of water pressure. In sequential electrically or operated hydraulically-electrically, the system supplies an electrical current for remote control of the valve. The automatic valves are commonly used for the pump house and filters; for regulating the pressure in the main line; to control the flushing cycles in the filters, or to control the volume of water through the secondary or lateral lines. Solenoid valves are used in the secondary or lateral lines to control the volume of pressure-regulator-valves are used to separate the system from the pressure in the main line. Whenever the pressure exceeds a preset value, the valve releases the excess pressure. The controls with mechanical time clocks open and close only a single valve at one time.

The communication between the irrigation controller and the valves is by means of wires, hydraulic lines or radio signals. In Electrically Operated Sequential System the amount of water distributed to the different blocks is determined by a flow meter. The non-sequential systems are controlled by hydraulic or electrical valves that can operate the valve in the desired block at random, and can supply known amount of water for a known duration to a desired block. The central panel allows control all the operations of the field. The field panel is operated by a remote control unit. The signals of the main panel are sent by an individual communication channel and are transmitted to individual field panel. Sensors are available to determine the soil moisture tension or the soil moisture. Tensiometer and gypsum blocks are simple and economical to use. Another method is a neutron scattering method but it is quite expensive and is for research purpose only. Preventive maintenance and trouble shooting of the system are also presented.

**KEYWORDS**

- Atmosphere
- Automatic metering valve
- Automatic system
- Automation
- Back flow
- Bar
- Cap or plug
- Check valve
- Chemigation
- Class A pan
- Clogging
- Consumptive water use
- Control system
- Crop coefficient
- Crop water requirement
- Diffusion
- Drainage
- Dripper or emitter
- Evaporation
- Evaporation tank
- Evapotranspiration
- Fertigation
- Fertilizer
- Field capacity
- Filter
- Flow meter
- Flushing valve
- Gypsum blocks
- Hydraulic system
- Hydraulic valve
- Infiltration
- Irrigarion, drip/trickle
- Irrigation frequency
- Irrigation requirement

- Irrigation, depth
- Irrigation, duration
- Irrigation, sprinkler
- Irrigation, subsurface
- Irrigation, surface
- Lysimeter
- Main line
- Main valve
- Microtube
- Neutron scattering
- Pan evaporation
- Photosynthesis
- Polyethylene (PE)
- Precipitation
- Psychrometer
- Pump
- Pump house
- Root zone
- Sequential electrically operated system
- Soil moisture
- Soil texture
- Solar radiation
- Solenoid valve
- Station selector
- Tensiometer
- Transpiration
- Volumetric valve
- Water content
- Water potential

## BIBLIOGRAPHY

Refer to bibliography Section 7.1 for the literature cited.

# Chapter 8

## Chemigation

## INTRODUCTION

Several decades ago, it began the idea of applying fertilizers through irrigation system. During the recent years, the application of chemicals through drip irrigation was

---

[1]This chapter is modified and translated from "Goyal, Megh R. y Luís E. Rivera Martínez, 1990. Quimigación. Capítulo X en: *Manejo de Riego Por Goteo* editado por Megh R. Goyal, páginas 289–310. Río Piedras, PR: Servicio de Extension Agrícola, UPRM." For more details one may contact by E-mail: goyalmegh@gmail.com.

adopted. Thus, new technical terms were defined such as: Chemigation, chloration, pestigation, insectigation, fungigation, nemagation, and herbigation [2, 3, 5]. The term chemigation was adopted to include the application of chemicals through the irrigation systems. The chemigation offers following economic advantages compared to other conventional methods [5]:

1. Provides uniformity in the application of chemicals allowing the distribution of these in small quantities during the growing season when and where these are needed.
2. Reduces soil compaction and the chemical damage to the crop.
3. Reduces the quantity of chemicals used in the crop and the health hazards/risks during the application.
4. Reduces the pollution of the environment.
5. Reduces the costs of manual labor, equipment, and energy.

To obtain a better efficiency and to diminish the clogging problems in the lateral lines, filters, drippers or any other part of the system, it is recommended to conduct a chemical and mechanical analysis of the water source [1]. If the chemical analysis shows a high concentration of salts, it can cause clogging problems. Therefore, it is recommended to avoid the use of chemicals that can cause precipitates. For a better efficiency in application, the chemicals should be distributed uniformly around the plants. The uniformity of chemical distribution depends on:

1. The efficiency of the mixture.
2. The uniformity of the water application.
3. The characteristics of the flow.
4. The elements or chemical compounds that are present in the soil.

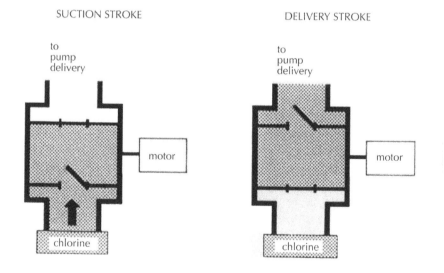

## CHEMICAL INJECTION METHODS

### Selection of a Pump for the Injection of a Chemical

While selecting an injection pump, one must consider the parts that will be in direct contact with the chemical substances. These parts must be of stainless steel or of material that is corrosion resistant. The injection method consists of basic components such as: Pump, pressure regulator, the gate valve, the pressure gage, the connecting tubes, check valve, and the chemigation tank. The injection pump should be precise, easy to adjust for different degrees of injection, corrosion resistant, durable, rechargeable, with availability of spare parts. The recommended materials are: Stainless steel, resistant plastic, rubber, and aluminum. Bronze, iron, and copper are non acceptable materials. The injection pump must provide a pressure on the discharge line greater than the irrigation pump. Therefore, the operating pressure of the irrigation system exerts a minimum effect in the endurance of the injection pump.

### Injection Methods

The efficiency of chemigation depends on the capacity of the injection tank, solubility of a chemical in water, dilution ratio, precision of dilution, the potability, the costs and the capacity of the unit, the method of operation, the experience of the operator, and the needs of the operator. The chemical compounds for the chemigation process must be liquid emulsions or soluble powder. The injector must be appropriate to introduce the substances in the system. In addition, it must be of an adequate size to supply necessary amount of chemical at a desired flow rate. Normally, the injection is carried out with an auxiliary electric pump or by interconnecting the injection pump to the irrigation pump. The most common methods for chemigation are described in Figures 8.1 and 8.2.

Fertilizer Injector.using Venturi system.

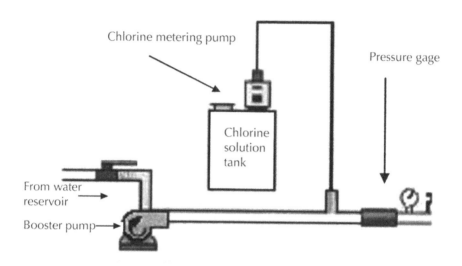

Chlorine metering pump

Pressure gage

Chlorine
solution
tank

From water
reservoir

Booster pump

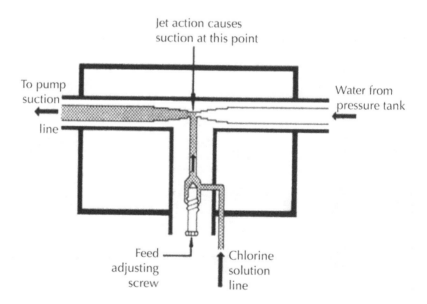

Jet action causes
suction at this point

To pump
suction

line

Water from
pressure tank

Feed
adjusting
screw

Chlorine
solution
line

Chemigation by Venturi injection

a. Chemigation
   pumping method

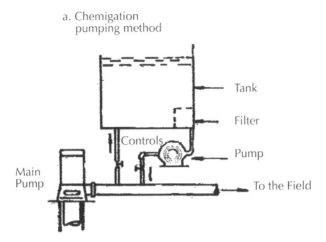

b. Venturi Type Injector

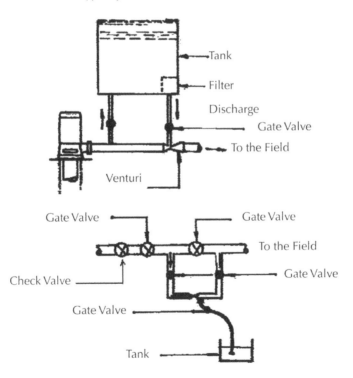

**Figure 8.1a.** Chemical injection methods: (a) Pressure pump and (b) Venturi type injector.

c. Pressure difference method

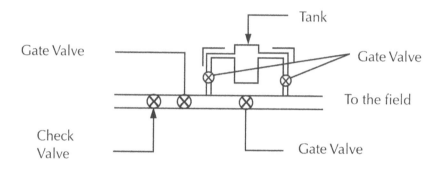

d. Injection method on the
   suction line of a pump

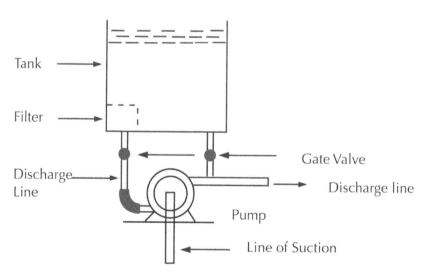

**Figure 8.1b.** Chemical injection methods: (c) Pressure difference method, and (d) Using the suction line of an irrigation pump.

**Figure 8.2.** Chemigation process. Top: The installation of a check valve (the nitrogen and pesticide tanks are behind the check valve). Bottom: Bypass Venturi system.

## *Injection by a pressure pump*

A rotary, diaphragm or piston type pump can be used to inject the chemicals to flow from the chemical tank towards the irrigation line. The chemigation pump must develop a pressure greater than the operating pressure in the irrigation line. The internal parts of the pump must be corrosion resistant. This method is very precise and reliable for injecting the chemicals in the drip irrigation system.

## Injection by pressure difference

This is one of the easiest methods to operate. In this method a low-pressure tank is used. This tank is connected with the discharge line at two points: one which serves as a water entrance to the tank and the other is an exit of the mixture of chemicals. A pressure difference is created with a gate valve in the main line. The pressure difference is enough to cause a flow of water through the tank. The chemical mixture flows into the irrigation line. The concentration of chemical in the water is difficult to calculate and control. Therefore, it is recommended to install an accurate metering valve to maintain a precalibrated injection rate.

## Injection by Venturi principle

A Venturi system can be used to inject chemicals into the irrigation line. There is a decrease in the pressure accompanied by an increase in the liquid velocity through a Venturi. The pressure difference is created across the Venturi and it is sufficient to cause a flow by suction of the chemical solutions from a tank.

## Injection in the suction line of the irrigation pump

A hose or tube can be connected to the suction pipe of the irrigation pump to inject the chemicals. A second hose or tube is connected to the discharge line of a pump to supply water to the tank. This method should not be used with toxic chemical compounds because of possible contamination of the water source. A foot valve or safety valve at the end of a suction line can avoid the contamination.

## FERTIGATION

All fertilizers for the chemigation purpose must be soluble (see Table 8.1). The partially soluble chemical compounds can cause clogging and thus can create operational problems [2, 4].

**Table 8.1.** Solubility of commercial fertilizers.

| Fertilizer | Solubility (grams/liter) |
|---|---|
| Ammonia | 97 |
| Ammonium Nitrate | 1185 |
| Ammonium Sulfate | 700 |
| Calcium Nitrate | 2670 |
| Calcium Sulfate | Insoluble |
| Di-Ammonium Phosphate | 413 |
| Di-Calcium Phosphate | Insoluble |
| Magnesium Sulfate | 700 |
| Manganese Sulfate | 517 |
| Mono-Ammonium Phosphate | 225 |
| Mono-Calcium Phosphate | Insoluble |
| Potassium Chloride | 277 |
| Potassium Nitrate | 135 |
| Potassium Sulfate | 67 |
| Urea | 1190 |

## Nitrogen

Nitrogen is an element that is most frequently applied in the drip irrigation system. The principal sources of nitrogen for chemigation are: anhydrous ammonia, liquid ammonia, ammonium sulfate, urea, ammonium nitrate, and calcium nitrate. The anhydrous ammonia or the liquid ammonia can increase the pH of the irrigation water, thus a possible precipitation of calcium and magnesium salts. If the irrigation water has high concentration of calcium and magnesium bicarbonates, then these can result in precipitation of chemical compounds. This enhances the clogging problems in the drippers, filters, and laterals. The ammonium salts are very soluble in water and cause less problems of clogging, with the exception of ammonium phosphate. The phosphate salts tend to precipitate in the form of calcium and magnesium phosphates, if there is an abundance of Ca and Mg in the irrigation water. Ammonium sulfate causes little obstruction problems or changes in pH water. The urea is very soluble and it does not react with the irrigation water to form ions, unless the water contains the enzyme urease. This enzyme can be present if the water has large amounts of algae or other biological agents.

The filtration system does not remove urease. This can cause hydrolysis of the urea. Because the concentrations of the enzyme are generally low compared to those in the soil, the urea will not hydrolyze to a significant degree in the irrigation water. The nitrate salts (e.g., calcium nitrate) are relatively soluble in water and do not cause large changes in the pH of the irrigation water. The nitrogen fertigation is more effective than the conventional methods of application, especially in sandy soils. In addition, the nitrogen fertigation is more efficient than the conventional methods in fine textured soils.

## Phosphorus

The phosphorus can be applied in the irrigation system, as an organic phosphate compound and glycerophosphates. The organic phosphates (orthophosphates) and urea phosphate are relatively soluble in water and can easily move in the soil. The organic phosphates do not precipitate, and the hydrolysis of an organic phosphate requires large lapse time.

The glycerophosphates react with calcium to form compounds of moderate solubility. The application of phosphorus can enhance obstructions in the drip irrigation system. When phosphoric fertilizers are applied in the irrigation water with high concentrations of Ca and Mg, then the insoluble phosphate compounds are formed that can obstruct the drippers and lateral lines. The phosphorus moves slowly in the soil and the root zone. In addition, the moist soil particles absorb phosphorus to form insoluble compounds. It is not recommended to fertigate phosphorus fertilizers during the growth period of a crop. Instead it should be applied before seeding, during seeding, and during fruit formation. The plant uses phosphorus early in its growth. In the drip irrigated crops, the fertigation of phosphorus can be combined with traditional methods of application. The drip irrigation system is efficient in the application of soluble phosphorus compounds, because the water is applied in the root zone; that facilitates the availability of phosphorus.

## Potassium

The potassium can be fertigated in the form of potassium sulfate, potassium chloride, and potassium nitrate. Generally, potassium salts have good solubility in water and cause little problems of precipitation.

## Micro Nutrients

The micro nutrients are supplied in the form of chelates. Thus, its solubility in the water is increased and these do not cause any problems of obstruction and precipitation. If the micro nutrients are not applied as recommended, then iron, zinc, copper, and magnesium can react with the salts in the soil causing precipitation. This enhances the clogging of the emitters. The chelates should be dissolved before fertigation. The chemigation of micro nutrients benefits the plant to accomplish a good development and growth. In addition, the operational cost is lower compared to the foliage application.

## PESTIGATION

Although, sufficient information is available on the application of pesticides through the drip irrigation system in different regions of the world, yet the data does not necessarily adapt to the weather and soil conditions of all regions of the world.

## Insectigation

When dispersed emulsions or formulations are used in water, these comprise of the liquid phase and therefore can be distributed uniformly. In order to control foliage insects, the insecticides applied through the drip irrigation system must be systematic and of high solubility.

## Fungigation

The fungi dragged by the wind are difficult to control. These fungi produce numerous spores that often are located on the leaves which are not easy to control. By means of injection of soluble and systematic fungicides through the drip irrigation, the effectiveness of certain fungicides can be increased.

## Nemagation

The application of fumigants and nematicides through the drip irrigation system is convenient and safe. Using this method, the soil can be fumigated in an efficient way to control nematodes and other detrimental organisms. The success of the operation will depend on many conditions such as: Temperature, soil moisture, soil aeration, content of organic matter, and uniformity of irrigation.

## Herbigation

In places where the rainfall is limited, the application of herbicides through drip irrigation serves to activate the applied herbicides. This action eliminates the need for mechanical incorporation and reduces the cost of weeding operation. As the drip irrigation system directly takes the water with herbicides to the place where the weeds are to be controlled, a uniform distribution is obtained. There is a significant reduction of losses of non-available chemicals as a result of the inactivation by rubbish or organic matter. Thus, the irrigation efficiency can be maximized and the efficiency of the herbicide application can be increased.

## CHLORATION OR CHLORINATION

The chlorine is the cheapest and effective treatment for the control of bacteria, algae, and the slime in the irrigation water. The chlorine can be introduced at low concentrations (1 ppm) at necessary intervals, or at high concentrations (10–20 ppm) for few minutes. The chlorine can be injected in form of chlorine powder (solid) and chlorine gas. The gas treatment is expensive and dangerous for the operator. The calcium hypochlorite can also be used, but the calcium tends to precipitate. The chlorine also acts as a biocide to iron and sulfur bacteria.

## INSTALLATION, OPERATION, AND MAINTENANCE

### Installation

To be effective, all types of chemigation equipments must be suitable, reliable, and precise. All the electrical devices must resist risks due to bad weather. In addition, all the valves, accessories, fittings, and pipes must resist the operating pressure. Some chemicals can cause corrosion problems. The chemical compounds and the concentrations of chemicals must be compatible with the injection system. The materials of the system must be corrosion resistant. These should be washed with clean water after each fumigation process.

### Operation

The chemigation procedure must follow a pre-established order. Irrigation system is operated until the soil saturates to a field capacity. Then the chemicals are injected. Once the chemigation has finished, water is allowed to flow free in the system for a sufficient time to remove all the sediments (salts) of chemicals from the laterals and drippers.

### Maintenance

The maintenance is a routine procedure. One should inspect all the components of the injection system after each application. It is recommended to replace the defective components before these will stop working altogether. In order to clean the injection system, it is convenient to have the water accessible near the system. After each application, it is recommended to clean and wash exterior of all the parts with water and detergent, and to rinse with clean water. The chemigation system can be cleaned in two ways:

1. By pressurized air.
2. By using acids or other chemical agents.

The pressurized air is used to clean the laterals of accumulation of the organic matter. Also, the lines can be cleaned with a commercial grade hydrochloric acid, phosphoric acid, or sulphuric acid at a concentration of 33–38%. When the acid is used, it is convenient to use protective clothing to avoid risks and accidents due to burns. Before using acid, it is recommended to allow the water flow through the system for 15 minutes. It is safe to fill the tank to 2/3 parts of its capacity and make sure that all components are in good condition. Now add the acid to the tank. The system will operate at a pressure of 0.8–1.0 atmospheres to apply acid. The acid treatment will avoid precipitation of salts and the formation of slime in the system. When the treatment has been completed, allow the water to flow to remove the residues of acid. Other practices for a good operation are:

1. To lubricate the movable screws and parts, after using the system.
2. To lubricate the movable screws and parts if the fertigation system was inactive during a prolonged period. It is convenient to activate the system and to make sure that all the parts are in good condition.

Rule of thumb is to chemigate during the middle of the irrigation cycle.

## Calibration

The calibration consists of adjustment of the injection equipment to supply a desired amount of chemical. The adjustment is necessary to make sure that the recommended dosage is applied. An excess of chemical is very dangerous and hazardous, whereas a small amount will not give effective results. The use of excess fertilizers is not economical. The amounts less than recommended dosages cause reduction in the crop yield. The precise calibration helps us to obtain accurate, reliable, and desirable results. A simple calibration consists of the collection of a sample of a chemical solution that is being injected to cover the desired area during the irrigation cycle. During the chemigation process, the rate of injection of a chemical compound can be calculated with the following equation:

$$g = (f \times A)/(c \times t_2 \times t_r) \tag{1}$$

where: g = Rate of injection (liters/hour)
F = Quantity of chemical compound (Kg/hectare)
A = Irrigation area (hectares)
C = Concentration of the chemical in the solution (Kg/liter)
$t_2$ = Chemigation time (hours)
$t_r$ = Irrigation duration (hours)

## SAFETY CONSIDERATIONS

The safety during the chemigation process is of paramount importance. It is recommended to use specialized equipment to protect the water source, the operators of the system and to avoid risks, health hazards, and accidents. It is necessary to install a safety valve (check valve or one way valve) in the main line between the irrigation pump and the injection point. A manual gate valve is not enough to avoid contamination of the water source. The safety valve allows the water flow in a forward direction. If it is properly installed, it avoids back flow of chemicals towards the water source.

It is safer to install an air-relief valve (vacuum breaker) between the irrigation pump and safety valve.

The air-relief valve allows escape of air from the system. It will avoid suction of chemical solution towards the water source. The pump to inject the chemicals and the irrigation pump can be interconnected. The pumps are interconnected in such a way that if one is shut off; the other is shut off automatically. This is convenient when the two pumps are electrical. A safety valve must be installed in the line of chemical injection. This arrangement avoids back flow towards the tank. This back flow can cause dilution of chemical, causing spills and breakage of the system.

The spills of pesticides are extremely dangerous because these can contaminate the water source and can cause health hazards. It is recommended to locate the chemical tank away from the water source. In the case of deep well, the pesticides can wash through the soil and contaminate the well. In addition, the operator and the environment are exposed to the danger of the contamination. The safety valve generally has spring and requires pressure so that the water will flow through those. This valve allows the flow when only an adequate pressure exists in the injection pump. When the injection pump is not in operation, there is no escape of liquid due to small static pressure in the tank. A gate valve at the downstream of a chemigation tank will help to avoid flow of irrigation water towards the tank when the chemigation is not in progress. This valve can be a manual gate valve, ball valve, or solenoid metering valve. The valve should be installed close to the tank. It must be open only during the chemigation. Also it must be corrosion resistant.

The automatic solenoid valve is interconnected electrically to the injection pump. This interconnection allows automatic closing of the valve in the supply line of chemical. Thus, it avoids flow of water in both directions when the chemigation pump is not in operation. The top of a chemical tank should be provided with wide openings that will allow easy filling and cleaning. In addition, the tank must be provided with a sieve or a filter. The tank must be corrosion resistant such as: Stainless steel or reinforced plastic with fiber glass. The flow rate from of the tank should correspond to the capacity of the pump. The tank should be equipped with an indicator to register the level of the liquid. The centrifugal pump provides high volume at low pressure. The pumps of piston and diaphragm provide volumes between moderate and high flows at high pressure. The pumps of roller and gear type provide a moderate volume at low pressure. If a pump is allowed to operate dry, then it can be damaged. It is recommended to follow the instruction manual of the manufacturer for a long life of a pump. Maintain all protectors in place. The injection pump of low volume can inject the concentrated formulation of pesticide. In that way, the problem of constantly mixing the solution in the tank is avoided. In addition, the calibration becomes easier.

One must select the hoses and synthetic or plastic tubes that can resist the operating pressure, climatic conditions, and the solvents in some chemical compounds. Do not allow the bending or kinking of hoses and tubes with another object. Wash the exterior and interior of hoses frequently so that these can last longer. These must be cleaned, washed, and stored well when are not being used. If it is possible, avoid exposure to sun. Due to climatic changes the hoses or tubes show deteriorations on the outer surface. The areas where chemigation is in progress, should display a sign that chemical compounds are being applied through irrigation system. The operator of the chemigation equipment must take all precautions: To use protective clothes, boots, protective goggles, and gloves. The waiting period to enter the field is necessary to avoid health hazards and risks. The precautions are necessary so that persons or animals do not enter the treated area during the application of pesticides and toxic substances.

## TROUBLE SHOOTING

| Cause | Remedy |
|---|---|
| **Uniformity of application is not adequate** | |
| 1. Drippers are clogged with precipitates or clay particles. | Replace drippers. Inject HCL according to the instructions. Use dispersing agents such as: Na and Al. |
| 2. Drippers are clogged with microorganisms. | Use biocides, algaecides, and bactericides. |
| 3. Lines are clogged. | Flush the lines by opening the ends. |
| 4. Filters are clogged. | Clean the filters. Open the corresponding gate valves. |
| **Chemical tank is overflowing** | |
| 5. Gate valve between two injection points is closed. Gate valve of injection line is closed. | Open the valves. |
| 6. Filters are totally obstructed. | Flush filters. |
| **Signs of chlorosis** | |
| 7. Adequate dosage of fertilizers is not used. | Use recommended dosages. |
| 8. Lack of uniformity of application. | Improve the uniformity. Flush and change drippers if necessary. |
| 9. Formulations of nitrogen that precipitate. | Use acid to clean the lines. Do chemical analysis to check the calibration. |
| **Components are leaking** | |
| 10. Corrosion of the components. | Use anti-corrosive components. |
| 11. Purple color in young leaves: Lost of phosphorus due to precipitation in line. | Use formulations of phosphorus that do not precipitate. Apply the phosphorus in bands not through the irrigation system. |

## SUMMARY

The chemigation is an application of chemicals through the irrigation system. The chemigation provides uniformity in the application of chemicals allowing the distribution of these chemicals in small quantities during the growing season when and where these are needed; reduces soil compaction and the chemical damage to the crop;

reduces the quantity of chemicals used in the crop and the hazards/risks during the application; reduces the pollution of the environment; and reduces the costs of manual labor, equipment, and energy. The chemical injection unit consists of chemigation pump, pressure regulator, gate valve, pressure gage, connecting tubes, check valve, and chemigation tank. All fertilizers for the chemigation must be soluble. The chlorine can be introduced at low concentrations at necessary intervals, or at high concentrations for few minutes. The chlorine can be injected in form of chlorine powder and chlorine gas. All chemigation equipments must be corrosion resistant, reliable, and precise. The chemigation procedure must follow a pre-established order. Irrigation system is operated until the soil saturates to a field capacity. Then the chemicals are injected. Once the chemigation has finished, water is allowed to flow free in the system for a sufficient time to remove all the sediments from the laterals and drippers. Rule of thumb is to chemigate during the middle of the irrigation cycle. Maintenance is a routine procedure. One should inspect all the components of the injection system after each application. It is recommended to replace the defective components. The calibration consists of adjustment of the injection equipment to supply a desired amount of chemical. It is recommended to use specialized equipment to protect the water source, the operators of the system and to avoid risks, health hazards, and accidents.

## KEYWORDS

- Acid treatment
- Air relief valve
- Atmosphere
- Automatic solenoid valve
- Back flow
- Centrifugal pump
- Check valve
- Chemigation
- Chlorination or chloration
- Chlorosis
- Clay
- Clogging
- Dripper or emitter
- Fertigation
- Fertilizer
- Field capacity
- Filter
- Filtration system

- **Fungigation**
- **Gate valve**
- **Glycerophosphates**
- **Herbigation**
- **Injection system**
- **Insectigation**
- **Manual gate valve**
- **Nemagation**
- **Nitrogen**
- **Nitrogen fertigation**
- **Organic matter**
- **Phosphorus**
- **Potassium**
- **Precipitation**
- **Pump**
- **Root zone**
- **Sand**
- **Slime**
- **Soil moisture**
- **Solenoid valve**
- **Solubility**
- **Urea**
- **Venturi system**
- **Weed**

**BIBLIOGRAPHY**

Refer to bibliography section 8.1 for the literature cited.

# Chapter 9

## Chloration

### INTRODUCTION

Drip irrigation is an efficient method of irrigation application and is extensively used in arid and semi-arid regions of the world. The system consists of applying water slowly and frequently into the soil with emitters or drippers distributed throughout the laterals (drip lines). The obstruction of the filters, the distribution lines, or the emitters is a main problem associated with the operation and management of a drip irrigation

---

[1]This chapter is modified and translated from "Goyal, Megh R., L.E. Rivera y Antonio Poventud, 1990. Cloración. Capítulo XI en: *Manejo de Riego por Goteo*, editado por Megh R. Goyal, páginas 311-330. Rio Piedras, Puerto Rico: Servicio de Extensión Agrícola, UPRM." For more details one may contact by E-mail: goyalmegh@gmail.com

system [2, 9, 11]. These obstructions are caused by physical agents (solid particles in suspension), chemicals agents (precipitation of insoluble compounds), or biological agents (macro and microorganisms). The preventive maintenance is the best solution to reduce or to eliminate the obstructions in the emitters or components of the system [1]. The chlorination is an addition of chlorine to water [11]. The chlorine, when it dissolves, acts as an oxidation agent and attacks the microorganisms, such as the algae, fungi, and bacteria. This procedure has been used for many decades to purify the drinking water [3]. Chloration is an injection of chlorine compounds through the irrigation system. The chloration solves effectively and economically the problem of obstruction of the emitters or drippers due to biological agents. Caution: Do not use any chemical agent through the drip system without consulting a specialist [9, 10, 11, 12].

## QUALITY OF WATER

### Water Source

It is necessary to conduct the physical and chemical analysis of the water before designing a drip irrigation system and choosing an appropriate filtration system [12]. For the chemical analysis, it is important to take a representative sample of the water. If the water source is subsurface (e.g., deep well), the sample must be taken an hour and a half after the pump begins to work. When the water source is from a lake, river, pool, or open channel, the samples must be taken at the surface, at the center, and at the bottom of a water source.

It is important to analyze the sample for suspended solids, dissolved solids and the acidity (pH), macro organisms, and microorganisms [5]. The acidity of the water must be known, since it is a factor that affects the "chemigation directly" and therefore the chloration. For example, the chloration for the control of bacteria is ineffective for a pH > 7.5. Therefore, it is necessary to add acid to lower the pH of irrigation water and to optimize the biocide action of the chlorine compound. If a chemical analysis of the water is known, we can predict the obstruction problems and take suitable measures. In addition, a program of adequate service and maintenance can be developed. The physical, chemical, and biological agents are classified in Table 9.1. The factors are classified in order of the risk: from low to severe. When the water contains amount of solids, salts, and bacteria within the acceptable limits, then the risk of clogging is reduced.

Also the particles of organic matter can combine with bacteria and produce a type of obstruction that cannot be controlled with filtration system. The fine particles of organic matter are deposited within the emitters and are cemented with bacteria such as: Pseudomonas and Enterobacter. This combined mass causes clogging of the emitters. This problem can be controlled with super chlorination at the rate of 1000 ppm (mg/L). However, the chloration at these high rates can cause toxicity of a crop. The obstruction caused by the biological agents constitutes a serious problem in the drip irrigation system that contains organic sediments with iron or hydrogen sulfide. Generally, the obstruction is not a serious problem if the water does not have organic carbon, which is a power source for the bacteria [Promotes the bacterial growth]. There are several organisms that increase the probability of the obstructions when there are ions of iron ($Fe^{++}$) or sulfur ($S^{--}$).

**Table 9.1.** Water quality: Criteria that indicates risk of obstruction of the emitters.

| Type of Problem | Risk of Obstruction | | |
| --- | --- | --- | --- |
| | Low | Moderate | Severe |
| **Biological Agents** | | | |
| Bacteria population** | <10,000 | 10,000–50,000 | >50,000 |
| **Physical Agents** | | | |
| Suspended solids* | <50 | 50–100 | >100 |
| **Chemical Agents** | | | |
| Acidity (pH) | <7.0 | 7.0–8.0 | >8.0 |
| Dissolved solids* | <0.2 | 0.2–1.5 | >2000 |
| Iron* | <0.1 | 0.1–1.5 | >1.5 |
| Hydrogen sulfide | <500 | 500–2000 | >2.0 |
| Manganese* | <0.2 | 0.2–2.0 | >1.5 |

**Maximum concentration of the representative sample of water. Given in ppm (mg/L).
*Maximum number of bacteria per milliliter. Obtained from field samples and laboratory analysis.

Algae in surface water can add carbon to the system. The slime can grow on the inner surface of the pipes. The combination of fertilizer and the heating of the polyethylene pipes (black) due to sunlight can promote the formation and development of these microorganisms. Many of the water sources contain carbonates and bicarbonates that serve like an inorganic power source to promote slime growth; also autotrophic bacteria (that synthesize their own food) are developed. The algae and the fungi are developed in the surface waters. Besides obstructing the emitters, the filamentous algae form a gelatinous substance in the pipes and emitters, which serve as a base for the development of slime. Another type of obstruction can also happen when the filamentous bacteria precipitate the iron into the insoluble iron compounds ($Fe^{+++}$).

## Growth of Slime in the Drip Irrigation System
The bacteria can grow within the system in absence of light and produce a mass of the slime or cause the precipitation of the iron or sulfur dissolved in water. The slime can act like an adhesive substance that agglutinates fine clay particles sufficiently large enough to cause clogging [2, 10, 11].

## Growth of Algae in the Water Source or in the Irrigation System
One of the most frequent problems is the growth of algae and other aquatic plants in the surface water that can be used for drip irrigation. The algae grow well in the surface water. The problem becomes serious if the water source contains nitrogen, phosphorus, or both. In many cases, the algae can cause obstructions in the filtration system. When the screen filters are used, the algae can be entangled in the sieves (screen) of the filter. In high concentration, these aquatic microorganisms can create problems in the sand filters. This requires a frequent flushing and cleaning of the filters.

**Types of Algae**
The main groups of organisms in surface water are classified like protists, plants, and animals. The protists include bacteria, fungi, protozoa, and algae. Algae are unicellular organisms or multicellular autotrophic and photosynthetic, and require organic compounds to reproduce. The major nutriments are carbon dioxide, nitrogen, and phosphorus. The minor elements like iron, copper, and manganese are also important for the development of these organisms.

It is important to note that algae problems usually occur due to ponds being neglected. Often, people think you can simply "dig a hole" and then let the pond take care of itself. Unfortunately, this is not the case. Healthy ponds require proper aeration, bacterial treatments, and adequate pond weed management. The use of copper sulfate has NOT been recommended for algae control as research and field usages have shown a high potential for detrimental environmental effects. In certain waters, copper sulfate is quite toxic to fish and other organisms. Overuse of this product is common due to its short-term effectiveness. This can result in copper build-up in the sediments leading to a sterile bottom, killing important beneficial bacteria, microorganisms, snails, and other beneficial "creatures." Many large pond or lake owners are concerned about "toxic algae." Death and sickness to pets, livestock, wildlife, and even man have been attributed to the presence of certain algae, mostly blue–green forming species, in water supplies. Lethal substances produced by these algae are retained within the cells and released after death or are secreted from living cells. Many unattended farm ponds and other waters contain some of these toxic forms, posing a threat to human health and the environment. Medical case histories, biologist reports, and laboratory tests show some of the possible effects of toxic algae. A list complied by the U.S. Department of the Interior Federal Water Pollution Control Administration summarizes medical case histories of algal poisonings for a 120 year period. Exposure to and ingestion of algae caused a variety of "discomforts" including: skin rashes, headaches, nausea, vomiting, diarrhea, fever, muscular pains, and eye, nose, and throat irritation. California State Water Resources Control Board states in the *Water Quality Criteria Handbook* (Second Edition): "There have been reports of rapid deaths of a great variety of animals after drinking water containing high concentrations of blue–green algae such as Microcystis, Aphanizomenon, Nostoc rivulare, Nodularia, Gleotrichia, Gomphosphaeria, and Anabaena. Fatal poisonings have occurred among cattle, pigs, sheep, dogs, horses, turkeys, ducks, geese, and chickens. It is believed that such algae may be toxic to all warm-blooded animals." To find out how one can avoid algae problems in large bodies of water and create a healthier pond, please read an article from: http://www.pondsolutions.com/blue-green-algae.htm

Algae are primitive plants closely related to fungi (Figure 9.1). These exhibit no true leaves, stems, or root systems and reproduce by means of spores, cell division, or fragmentation. These "live" from excess nutrients in the water and sunlight for growth. Over 17,400 species of algae have been identified and thousands more probably exist. The simplest algae are single cells (e.g., the diatoms); the more complex forms consist of many cells grouped in a spherical colony (e.g., *Volvox*), in a ribbon like filament (e.g., *Spirogyra*), or in a branching thallus form (e.g., *Fucus*). The cells of the colonies are generally similar, but some are differentiated for reproduction and for other functions. Kelps, the largest algae, may attain a length of more than 200 ft (61 m). *Euglena* and similar genera are free-swimming one-celled forms that contain chlorophyll but that are also able, under certain conditions, to ingest food in an animal like manner. The green algae include most of the freshwater forms. The pond scum, a green slime found in stagnant water, is a green alga, as is the green film found on the bark of trees. The more complex brown algae and red algae are chiefly saltwater forms; the green color of the chlorophyll is masked by the presence of other pigments. Blue–green algae have been grouped with other prokaryotes in the kingdom *Monera* and renamed *cyanobacteria*. There are four classes of algae for the irrigation system:

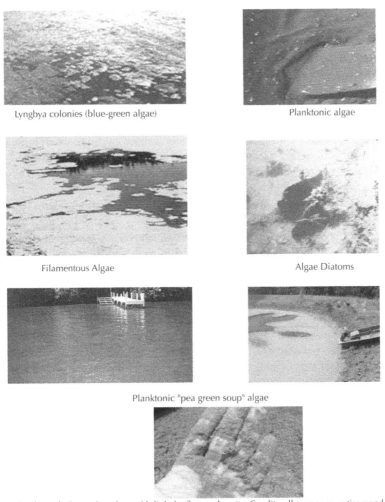

Lyngbya colonies (blue-green algae)

Planktonic algae

Filamentous Algae

Algae Diatoms

Planktonic "pea green soup" algae

Duckweed. Super tiny plant with little leaflets and roots. Can literally cover an entire pond. Commonly transferred from pond to pond by waterfowl

**Figure 9.1.** Types of algae.

### Green algae [Chlorophyta, klōrof'*utu*]

These are commonly known as photosynthetic organisms: Phylum (division) of the kingdom *Protista*. The organisms are largely aquatic or marine. The various species can be unicellular, multicellular, coenocytic (having more than one nucleus in a cell), or colonial. Those that are motile have two apical or subapical flagella. A few types are terrestrial, occurring on moist soil, on the trunks of trees, on moist rocks, and even in snow banks. Cells of the Chlorophyta contain organelles called chloroplasts in which photosynthesis occurs; the photosynthetic pigments chlorophyll *a* and chlorophyll *b*, and various carotenoids, are the same as those found in plants and are found in similar proportions. Chlorophytes store their food in the form of starch in plastids and, in many, the cell walls consist of cellulose. Unlike in plants, there is no differentiation into specialized tissues among members of the division, even though the body, or thallus, may consist of several different kinds of cells. According to "*The Columbia Electronic Encyclopedia,* 6th edition 2006, Columbia University Press." there are four evolutionary lineages of green algae.

1. **Euglenophyta** (yOO"gl*u*nof'*utu*), small phylum (division) of the kingdom *Protista*, consisting of mostly unicellular aquatic algae. Most live in freshwater; many have flagella and are motile. The outer part of the cell consists of a firm but flexible layer called a pellicle, or periplast, which cannot properly be considered a cell wall. Some euglenoids contain chloroplasts with the photosynthetic pigments chlorophyll *a* and *b*, as in the phylum Chlorophyta; others are heterotrophic and can ingest or absorb their food. Food is stored as a polysaccharide, paramylon. Reproduction occurs by longitudinal cell division. The most characteristic genus is *Euglena,* common in ponds and pools, especially when the water has been polluted by runoff from fields or lawns on which fertilizers have been used. There are approximately 1000 species of euglenoids.

2. **Dinoflagellata** (dī'nōflăj"*u*lät'*u*, –lā'tu), phylum (division) of unicellular, mostly marine algae, called dinoflagellates. In some classification systems, this division is called Pyrrhophyta. There are approximately 2000 species of dinoflagellates. Most have two flagella that lie perpendicular to one another and cause them to spin as they move through the water. Most have walls, or thecae, that are rigid and armorlike and sometimes take on fantastic shapes. The plates that make up these walls are actually located inside the plasma membrane rather than outside, as cell walls are. Some species are heterotrophic, but many are photosynthetic organisms containing chlorophyll *a* and chlorophyll *c*. The green of these chlorophylls may be masked by various other pigments. Still other species are symbionts, living inside such organisms as jellyfish and corals. Food reserves are largely starch. Reproduction for most dinoflagellates is asexual, through simple division of cells following mitosis. They are unusual in that in each cell, the chromosomes remain compact between divisions, instead of stretching out into slender threads, as in most other organisms. The chromosomes are constricted at regular intervals and do not have centromeres, or fiber-attachment centers. There is no spindle, yet the very numerous chromosomes are divided equally at the time of mitosis.

3. **Chrysophyta** (kr*u*sof'*utu*), phylum (division) of unicellular marine or freshwater organisms of the kingdom Protista consisting of the diatoms (class Bacillariophyceae), the golden, or golden–brown, algae (class Chrysophyceae), and the yellow–green algae (class Xanthophyceae). In many chrysophytes the cell walls are composed of cellulose with large quantities of silica. Some have one or two flagella, which can be similar or dissimilar. A few species are ameboid forms with no cell walls. The food storage products of chrysophytes are oils or the polysaccharide laminarin. Formerly classified as plants, the chrysophytes contain the photosynthetic pigments chlorophyll *a* and *c*; all but the yellow–green algae also contain the carotenoid pigment fucoxanthin. Under some circumstances, diatoms will reproduce sexually, but the usual form of reproduction is cell division. The diatoms and golden–brown algae are of great importance as components of the plankton and nanoplankton that form the foundation of the marine food chain.

4. **Phaeophyta** (fēof'*utu*), phylum (division) of the kingdom Protista consisting of those organisms commonly called brown algae. Many of the world's familiar seaweeds are members of Phaeophyta. There are approximately 1500 species. Like the chrysophytes, brown algae derive their color from the presence, in the cell chloroplasts, of several brownish carotenoid pigments, including fucoxanthin, in addition to the photosynthetic pigments chlorophyll *a* and *c*. With only a few exceptions, brown algae are marine, growing in the colder oceans of the world, many in the tidal zone, where they are subjected to great stress from wave action; others grow in deep water. Among the brown algae are the largest of all algae, the giant kelps, which may reach a length of over 100 ft (30 m). *Fucus* (rockweed), *Sargassum* (gulfweed), and the simple filamentous *Ectocarpus* are other examples of brown algae. The cell wall of the brown algae consists of a cellulose differing chemically from that of plants. The outside is covered with a series of gelatinous pectic compounds, generically called algin; this substance, for which the large brown algae, or kelps, of the Pacific coast are harvested commercially, is used industrially as a stabilizer in emulsions and for other purposes. The normal food reserve of the brown algal cell is a soluble polysaccharide called laminarin; mannitol and oil also occur as storage products. The body, or thallus, of the larger brown algae may contain tissues differentiated for different functions, with stemlike, rootlike, and leaflike organs, the most complex structures of all algae.

## Motile

These form colonies when mature. It has flashy green color and it is unicellular and flagellated. The flagella are those that make motile in the water. Filamentous algae, or commonly referred to as "pond scum" or "pond moss" forms greenish mats upon the surface of water. This alga usually begins its growth along the edges or bottom of the pond and "mushrooms" to the surface. Individual filaments are a series of cells joined end to end which give the thread-like appearance. They also form fur-like growths on bottom logs, rocks, and even on the backs of turtles. Some forms of filamentous algae are commonly referred to as "frog spittle" or "water net."

### Yellow–green algae [Synonyms for *yellow–green algae*]

The diatoms are most important in this group. It is a group of common unicellular and colonial algae of the phylum Chrysophyta, having mostly yellow and green pigments, occurring in soil and on moist rocks and vegetation and also as a slime or scum on ponds and stagnant waters. These are found in fresh water and salt water.

### Blue–green algae

Popular name for those microorganisms that are now more properly called cyanobacteria [sī"*u*nōbăktir'ē*u*, sī-ăn"ō–] or photosynthetic bacteria that contain chlorophyll. Cyanobacteria are familiar to many as a component of pond scum. Despite their name, different species can be red, brown, or yellow; blooms (dense masses on the surface of a body of water) of a red species are said to have given the Red Sea its name. These are unicellular organisms with flagella. These can form big masses in the surface water. In addition, these can use nitrogen from the atmosphere. Nitrogen-fixing cyanobacteria need only nitrogen and carbon dioxide to live.

### PRINCIPLE OF CHLORATION

The principle of chloration for treating the water by drip irrigation is similar to the one that is used to purify the water for drinking purpose. Table 9.2 includes basic reactions of chlorine and its salts. When the chlorine in gaseous state ($Cl_2$) dissolves in water, the chlorine molecule is combined with water in a reaction called hydrolysis. The hydrolysis produces hypochloric acid (HOCL; Reaction (1)). Following this reaction, the hypochloric acid enters an ionization reaction as shown in Reaction (2). The hypochloric acid (HOCl) and the hypochlorite (OCL) are known as free available compounds and are responsible for controlling the microorganisms in the water. The equilibrium of these depends on the temperature and pH of the irrigation water. When the water is acidic (low pH) the equilibrium moves to the left, resulting in an increase of HOCl. When the water is alkaline (high pH), the chlorine increases in the form of OCl–. The efficiency of HOCl is 40–80 times greater than OCl–. Therefore, the efficiency of the chloration depends greatly on the acidity (pH) of the water source. Reaction (1) produces hydrogen ions (H+) that can increase the acidity. The basicity depends on the amount of added chlorine and the buffer capacity of the water. The sodium hypochlorite [NaOCl] and the calcium hypochlorite [$Ca(OCl)_2$] hydrolyze and produce OH– ions that tend to lower the acidity of the water (Reactions (3) and (4)). If the pH is extremely low, the gaseous chlorine ($Cl_2$) predominates and can be dangerous. Therefore, it is recommended to store the sources OCL compounds separate from solids. Also the available free chlorine reacts with oxidizing compounds (like iron, manganese, and hydrogen sulfide) and produces insoluble compounds, which must be removed from the system to avoid clogging.

**Table 9.2.** Basic forms of chlorine reactions and its salts.

| Reactions | Reaction Number |
|---|---|
| $Cl_2 + H_2O = H^+ + Cl^- + Col$ | (1) |
| $HOCl = H^+ + OCl^-$ | (2) |
| $NaOCl + H_2O = Na^+ + OH^- + Col$ | (3) |
| $Ca(OCl)_2 + 2H_2O = Ca^{2+} + 2OH^- + 2HOCl$ | (4) |
| $HOCl + NH_3 = NH_2Cl + H_2O$ | (5) |
| $HOCl + NH_2Cl = NHCl_2 + H_2O$ | (6) |
| $HOCl + NHCl_2 = NCl_3 + H_2O$ | (7) |
| $HOCl + 2Fe^{2+} + H^+ = 2Fe^{3+} + Cl^- + H_2O$ Ferrous to ferric | (8) |
| $Cl_2 + 2Fe(HCO_3)_2 + Ca(HCO_3)_2 = 2Fe(OH)_3$ (insoluble) $+ CaCl_2 + 6CO_2$ | (9) |
| $HOCl + H_2S = S^-$(insoluble) $+ H_2O + H^+ + Cl^-$ | (10) |
| $Cl_2 + H_2S = S^-$(insoluble) $+ 2H^+ + 2Cl^-$ | (11) |

The chlorine has two important chemical properties: At a low concentration (1–5 mg/L), it acts as a bactericidal. At a high concentration (100–1,000 mg/L), it acts as an oxidizing agent which can disintegrate particles of organic matter. It is necessary to watch, because the chlorine at these high levels can affect the growth of some plants.

## SOURCES OF COMMERCIAL CHLORINE

The most common chorine sources [7] used in a drip irrigation system are sodium hypochlorite, calcium hypochlorite, and gaseous chlorine.

### Sodium Hypochlorite, NaOCl

Sodium hypochlorite is liquid and is commonly used as whitener for clothes. It can be easily decomposed at high concentrations, in the presence of light and heat. It must be stored at room temperature in packages resistant to corrosion. This compound is easy to handle. The amounts can be measured precisely and causes few problems.

### Calcium Hypochlorite, Ca (OCl)$_2$

Calcium hypochlorite is available commercially as dust, granulated, or in pellets. It is well soluble in water and is quite stable under appropriate storage conditions. It must be stored at room temperature in a dry place and in packages resistant to corrosion. When this compound is mixed in a concentrated solution, it forms a suspension that contains calcium oxalate, calcium carbonate, and calcium hydroxide. These compounds can obstruct the drip irrigation system.

### Gaseous Chlorine, Cl$_2$ gas

It is available in liquid form at high pressure in cylinders from 45 kg to 1,000 kg. The $Cl_2$ is very poisonous and corrosive. It must be stored in a well ventilated place. Table 9.3 shows equivalent amounts of chlorine for different commercial sources and the required amount to treat 1,233 m³ (1 acre-foot) of water to obtain one ppm of chlorine.

The NaOCl is safer than $Cl_2$ and avoids calcium precipitates in the emitters, which can happen when using Ca $(OCl)_2$. It is more economical to use than the $Cl_2$ in large systems. In small systems, it is appropriate to use sodium or calcium hypochlorite. The use of $Cl_2$ is preferred in situations where the addition of sodium and calcium can be detrimental to the crop. It is necessary to observe that $Cl_2$ is dangerous under certain conditions. Thus, the instructions on the label must be followed. It is recommended to install a security valve (one way or check valve) in the tank that is used for injecting the chlorine.

**Table 9.3.** Equivalent amounts of the commercial sources of chlorine and required amounts to treat one acre-foot of water to obtain one ppm of chlorine.

| Commercial Source of Chlorine | Equivalent Amount to Obtain 454 g (1 lb.) of Chlorine | Required Amount to Treat One Acre-foot (1233 m³) of Water and Obtain one ppm of Chlorine |
|---|---|---|
| Gaseous chlorine ($Cl_2$) | 454 g (1.0 lb) | 1226 g (2.7 lb.) |
| **Calcium Hypochlorite, Ca(OCl)₂** | | |
| 65–70% of available chlorine | 681 g (1.5 lb.) | 1816 g (4.0 lb.) |
| **Sodium Hypochlorite, NaOCl** | | |
| 15% of available chlorine | 2.54 liters (0.67 gallons) | 6.81 liters (1.8 gallons) |
| 10% of available chlorine | 3.78 liters (1.0 gallons) | 10.22 liters (2.7 gallons) |
| 0.5% of available chlorine | 7.57 liters (2.0 gallons) | 20.44 liters (3.4 gallons) |

## CHLORATION METHOD

The chloration in a drip irrigation system can be continuous or in intervals, depending on the desired results. Application at intervals is appropriate, when the objective is to control the growth of microorganisms in lateral lines, emitters, or in other parts of the system. The continuous treatment is used when we want to precipitate the iron dissolved in the water, to control algae in the system, or where it is not reliable to use the treatment at intervals.

### General Recommendations

1. Inject chlorine before the filters. This controls the growth of algae or bacteria in the filters that otherwise would reduce the filtration efficiency. This also allows the filtration of any precipitate caused by the injection of chlorine.
2. Calculate the amount of chlorine to inject. It is necessary to know the volume of water to be treated, the active ingredient of the chemical compound to be used and the desired concentration in the treated water.
3. The chlorine should be injected when the system is in operation.
4. One should take samples from the water at the nearest drippers and most distant drippers to determine the chlorine level at these points. Allow sufficient time so that the lines are filled with the chlorine solution.
5. Adjust the injection ratio. Repeat steps 4 and 5 until the desired concentration in the system is obtained.

## Recommended Chlorine Concentrations

1. Continuous treatment (in order to prevent the growth of algae or bacteria): Apply from 1 to 2 mg/L continuously through the system.
2. Treatment at intervals (in order to eliminate the algae or bacteria): Apply from 10 to 20 mg/L for 60 minutes. The frequency of the treatment depends on the concentration of these microorganisms in the water source.
3. Super chloration (in order to dissolve the organic matter and in many cases the calcium precipitated in the drippers): Inject chlorine at a concentration from 500 to 100 mg/L, depending on the case. After this, close the system and leave it for 24 hours, to clean all the secondary and lateral lines. It helps to clean the obstructions in the secondary and lateral lines. We have to be careful while applying these amounts, since these chlorine levels can be toxic to certain crops. Table 9.4 shows typical dosages of chlorine.

**Table 9.4.** Typical dosages of chlorine.

| Problem due to | Dosage |
|---|---|
| Algae | 1 to 2 ppm continuous, or 10–20 ppm for 30–60 min. |
| Ferro bacteria | 1 + ppm: Varies with the amount of bacteria |
| Slime | 0.5 ppm |
| Precipitation of iron | $0.64 \times$ [content of $Fe^{++}$] |
| Precipitation of Manganese | $1.3 \times$ [content of manganese] |
| Hydrogen sulfide | 3.6 a $8.4 \times$ [the content of $H_2S$] |

## Chlorine Requirements

Chlorine requirements must be known before the chloration. The $Cl_2$ or NaOCl is a biocide that must be applied in the amounts and at recommended concentrations. The excess of chlorine in the irrigation water can cause damage to the young plants or young trees. On the other hand, the low levels do not solve the problems associated with the growth of microorganisms in the irrigation water. The chlorine is a very active and toxic agent at high concentrations; therefore it must be handled carefully. When it is injected in the irrigation lines, some chlorine reacts with inorganic compounds and organic substances of the water or it adheres to them. In most wells and water sources, from 65 to 81% of the chlorine is lost by this type of reaction. The chlorine (like hypochlorous acid) that adheres to the organic matter or that reacts with other compounds does not destroy microorganisms. For this reason, it does not have value as a biocide agent. The free chlorine (the excess of hypochlorous acid) is the agent that inhibits the growth of bacteria, algae, and other microorganisms in the water. Therefore, it is indispensable to establish the chlorine requirements before the chloration. In this way, we can maintain the desired concentrations of available chlorine.

In order to inhibit the growth of microorganisms, a minimum contact time of 30 minutes is required (45 minutes of injection). It also requires a minimum concentration of 0.5–1.0 mg/L of available chlorine measured at the end of the drip line and 2.0–3.0 mg/L of available chlorine at the injection point. The following equations are used to calculate the gallons per hour (gph) of NaOCl that must be injected to obtain the desired concentration of chlorine per minute (gpm):

1. Formula for gpm of 10% NaOCl: (1)
= [0.0006 × (gpm desirable chlorine) × (Discharge of the pump, gpm)]

2. Formula for gph of 5.25% NaOCl: (2)
= [0.000114 × (ppm desirable chlorine) × (Discharge of the pump, gpm)]

3. Formula for pounds by hectare of $Cl_2$ (gas): (3)
= [0.000998 × (ppm desirable chlorine) × (Discharge of the pump, gpm)]

4. Gallons of liquid chlorine per hour: (4)
= [0.06 × ppm of desirable chlorine × discharge of the pump in gpm]/
[Percentage of chlorine in the material]

5. Dry chlorine in pounds per hour: (5)
= [0.05 × ppm × gpm]/[percentage of chlorine in the material]

6. Dry chlorine in pounds per 1000 gallons of water: (6)
= [0.83 × ppm]/[percentage of chlorine in the material]

7. For chlorine gas: (7)
= Take percentage of chlorine = 100; and calculate as dry chlorine.

## Methods to Measure Chlorine with the D.P.D. Method [N,N, dietil P-femilenediamina]

It is essential to measure the chlorine when using liquid chlorine as bactericidal and algaecide in irrigation systems of low volume [5, 6]. Most of the methods of measuring chlorine that are used in the swimming pools are not adequate for irrigation systems. This is because many of these equipments measure only total chlorine, but not the residual free chlorine. Equipment "D.P.D." of good quality can measure total chlorine and the free available chlorine. The test equipment D.P.D. is very simple. The directions and procedures come with the equipment. The equipment is used to measure each type of chlorine. When applying these compounds, the water becomes pink in the presence of the chlorine. The more intense is the color, higher is the chlorine concentration. In order to know the chlorine concentration, the color of the water is compared with that of a calibrated chromatic chart. One must remember that the free chlorine is the one that determines the biocide action. If there is not sufficient free chlorine available, the bacteria continue growing even though chlorine has been injected into the system. In other words, if the amount of total chlorine is not sufficient to maintain chlorine free in solution, the treatment gets is of no value. The test equipment D.P.D. can be purchased from the sellers of irrigation equipments or from chemical agents who are specialized in water treatment.

## EXAMPLES FOR CALCULATING THE AMOUNT OF CHLORINE

### Example # 1

A farmer wishes to use a cloth whitener (NaOCl—5% active chlorine or available chlorine) to reach a concentration of one ppm of chlorine at the injection point. The flow rate for the system is 100 gpm. In what ratio the chlorine must be injected?

$$IR = [Q \times C \times M]/S \tag{8}$$
$$= (100 \times 1 \times 0.006)/5$$
$$= 0.21 \text{ gph}$$

where: IR = Rate of chlorine injection (gallons/hour).

Q = Flow rate of the system (gallons/minute).

C = Desired concentration of chlorine (ppm).

S = Percent of active ingredient (%).

M = 0.006 for the liquid material (NaOCl), or 0.05 for the solid material $Ca (OCl)_2$

## Example # 2

A farmer wants to inject $Cl_2$ through the drip irrigation system at a concentration of 10 ppm. What will be the rate of injection of the chlorine gas? The flow rate of the system is 1,500 gpm.

$$IR = Q \times C \times 0.012 \tag{9}$$

where: IR = Rate of injection of chlorine (pound/day).

Q = Flow Rate of the system (gallons/minute).

C = Desired chlorine concentration (ppm).

IR = $1500 \times 10 \times 0.012$

= 180 pounds per day

## SUMMARY

Chloration is a process of injection of chlorine compounds through an irrigation system to prevent obstruction. The chloration solves effectively and economically the problem of obstruction of the emitters or drippers due to biological agents. The chloration can be continuous or in intervals, depending on the desired results. The most common chlorine sources are sodium and calcium hypochlorite and chlorine gas. The NaOCl is safer than the $Cl_2$. The chapter discusses the quality of water, principle of chloration, commercial sources of chlorine, methods of chloration, and examples to calculate the injection rates.

## KEYWORDS

- **Atmosphere (atmyósfera)**
- **Biological agents**
- **Chemigation (quimigación)**
- **Chloration (cloración)**
- **Chlorophytes**
- **Chloroplast (cloroplastos)**

- Clay (arcilla)
- Clogging (obstrucción)
- Cyanobacteria
- Diatoms
- Dissolve (disolver)
- Distribution lines (línea de distribución)
- Drip irrigation (riego por goteo)
- Dripper (gotero)
- Emitters (surtidores o goteros)
- Fertilizer (fertilizante, abono)
- Filter (filtro)
- Filter, sand (filtro de arena)
- Filtration system (sistema de filtración)
- Line, distribution (línea de distribución)
- Maintenance (mantenimiento)
- Motile
- Nitrogen (nitrógeno)
- Nutrients (nutrientes)
- Organic matter (materia orgánica)
- pH
- Photosynthesis (fotosíntesis)
- Polyethylene (polietileno)
- Precipitation (precipitación)
- Pump (bomba)
- Root system (sistema radical)
- Sand (arena)
- Slime (limo)
- Solid particles in suspension (partículas sólidas en suspensión)
- Term (plazo)
- Water quality (calidad del agua)
- Water source (fuente de agua)
- Weed (yerbajo)

## BIBLIOGRAPHY

Refer to bibliography Section 9.1 for the literature cited.

# Chapter 10

## Filtration Systems

---

[1]This chapter is modified and translated from "Rivera, Luis E. y Megh R. Goyal, 1990. Sistemas de Filtración. Capítulo VII en: *Manejo de Riego Por Goteo* editado por Megh R. Goyal, páginas 207–240. Río Piedras, PR: Servicio de Extensión Agrícola, UPRM." For more details one may contact by email: goyalmegh@gmail.com

## INTRODUCTION

Drip irrigation is a novel technique which has extended in different regions of the world. This system provides many benefits compared to other irrigation systems. Also, the drip system involves application of new technology for a guaranteed operation. The success of drip irrigation depends on the filtration of water. The purpose of any filtration system is to reduce any clogging by removing any suspended solids or particles from the irrigation water. Clean water free of contaminants is of vital importance for a better operation of the drip irrigation. An adequate filtration system is an indispensable requirement to eliminate free suspended solids. Partial or total obstructions by different agents may render the drip irrigation system out of order. Thus clogging or obstruction causes economic loss to the farmer. Therefore, the selection of an adequate filtration system is one of the considerations for designing and installing a drip irrigation system [3, 5, 7].

## CLOGGING AGENTS

The presence or development of particles causes the obstruction problem, thus reducing the emitter flow. The problem is progressive, if no corrective measures are taken at a proper time. Once the emitter flow rate decreases, the obstruction is accelerated leading to a complete clogging. The obstruction may occur in any place in the system but predominantly in the emitters/drippers. The problem can be avoided by using clean water and avoid the injection of agents that may precipitate. The clogging agents are grouped in three categories: Physical, chemical or biological. In some cases, the obstruction may be due to two or more of these factors at the same time.

### Physical Agents

Different types of soil particles are present in the irrigation water. These particles may consist of soil material of different sizes (sand, silt and clay), sedimentation that builds up in the reservoirs, and small pieces of tubing from pipe lines, etc. These particles can be large enough to pass through the filters or irrigation lines.

Clay particles in combination with salts can accumulate on the internal walls of drippers or the filters, and thus can reduce performance of drippers or the filters. Silt and clay with other chemicals may form aggregates to cause clogging of drippers. During the planning stage of a filtration system, it is important to test the water source for chemical and physical analysis.

### Chemical Agents

Irrigation water contains many water soluble salts that can precipitate on the surface of the drippers as the water evaporates between the irrigation intervals. If the salts cannot be dissolved, then mineral deposits can be formed to obstruct the drippers. A high concentration of calcium, magnesium and bicarbonate ions in the irrigation water promotes deposits of calcium and magnesium carbonates. High concentrations of calcium and sulfate ions may cause formation of calcium sulfate on the surface of the dripper. Generally the subsurface water contains iron and manganese salts. On contact with atmospheric air, the insoluble oxides of iron and manganese precipitate. These may cause obstructions of the drippers. Water rich in sulfite ions may produce insoluble sulfur compounds. Besides naturally occurring inorganic compounds, the precipitates can also be formed due to injection of liquid fertilizers or other chemical substances in the irrigation water. Soluble fertilizers may be used with success. There are simple qualitative laboratory tests to identify most of the insoluble compounds. Testing of irrigation water is essential to know the probability of clogging by chemical deposits. It is recommended to identify the insoluble compounds before chemigation; and to use a good filtration system.

## Biological Agents

The irrigation system can also be obstructed by macro organisms and microorganisms. The irrigation water that comes from the surfaces open to atmospheric air may contain great amounts of organic material. Organic matter content in water may consist of partially decomposed organic matter (mostly of vegetative origin) and microorganisms (algae, bacteria and protozoa). The environmental conditions may favor development and rapid formation of different types of microorganisms. Some species like Sphaertilius and other microbiological cells multiply in darkness and in the presence of iron and sulfur in water. The oxides of iron, manganese and sulfur can be formed in the presence of microorganisms and these oxides may increase the obstruction. Crustaceans, fishes and other microorganisms may cause obstruction problems in the filtration system and irrigation lines. Spiders, ants and other insects may obstruct the drippers. A visual inspection normally helps to identify these organisms. To identify appropriately bacteria, algae and protozoa, a microscopic examination is needed. It is not necessary to obtain an exact identification of the microorganisms, but it is important to identify the presence of non organic compounds in the irrigation water.

## PREVENTION OF OBSTRUCTION OR CLOGGING

The need to prevent obstruction by means of an adequate filtration unit should not be underestimated. This importance can be illustrated with the following example: If water is pumped from a pond to a drip irrigation system, this water might contain organic matter, weeds and suspended particles. In the absence of a filtration system, the drippers will be clogged quickly. If the problem is not corrected, the complete system may become useless. Under any circumstances, untreated water should not be used in drip irrigation. The water may look clean. It must be filtered with an adequate filtration system. Depending on the type of impurities and suspended solids, one may use screen filters, disk or ring filters, sand filters and hydrocyclone filters. Other treatments include chloration; acidification and use of compressed air.

## TYPES OF FILTERS

### Gravity Filters

Ponds, lakes, open ditches, irrigation channels and water reservoirs are good candidates for the gravity filter. Gravity filters can separate part of the suspended solids from the water. The method is not very reliable if the water has high concentration of microorganisms.

### Pressure Filtration System

#### *Screen filters*

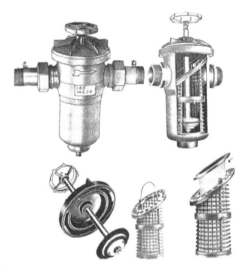

**Figure 10.1.** Screen filter commonly used in a drip irrigation system.

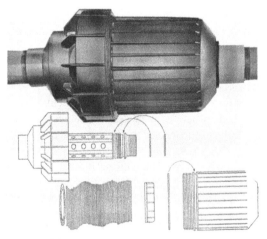

**Figure 10.2.** The components of disk (ring) filter commonly used in a drip irrigation system.

**Figure 10.3.** Disk filter: Flushing operation.

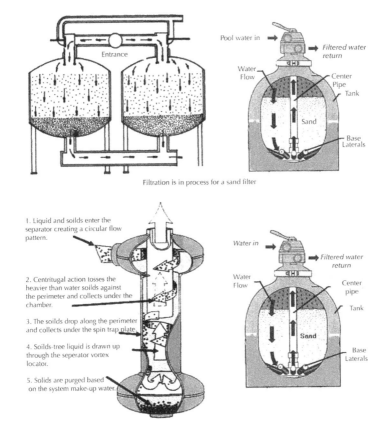

**Figure 10.4.** Cut section of a typical sand filter.

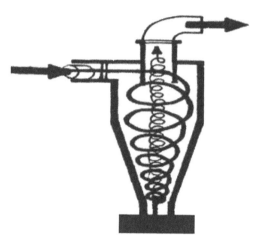

**Figure 10.5.** Hydrocyclone filter.

The screen filter is the simplest of all filtering methods. The principal component is a sieve made of metal, plastic or synthetic fabric enclosed in a special casing. The sieves are classified by number of squares per square inch, with a standard mesh for each size of sieve. The mesh size of a screen increases as the squares become smaller. For a mesh size of 200, the standard caliber is 75 micrometers (0.0029 inches). Most manufacturers recommend sieves of 75–150 micrometers to avoid clogging in the drippers. However, filters with a mesh size of 30 (600 micrometers) can be used, if water quality is good.

A double screen filter is shown in Figure 10.6. It consists of two cylindrical screens [3, 4, 5]. The external screen is made of a sieve of 80 mesh size to separate the suspended particles that are larger than $2 \times 10^5$ nanometers. The internal screen is made of a sieve of 120 mesh size to separate the particles up to $1.25 \times 10^5$ nanometers. Particles smaller than $1.25 \times 10^5$ nanometers are not considered harmful to cause clogging of drippers. The double screen filter can be easily dismantled, cleaned and the sieves can be replaced quickly. These filters are available with sieves of different mesh sizes depending on the quality of water. A simple screen filter may consist of one cylindrical sieve. In each type of a screen filter, the sieve separates solid particles. The presence of biological agents in the irrigation system may cause obstructions in the sieve and reduce the filtration capacity. The screen filter can be used in series with other types of filtration devices to make sure that the water is free from sediments and is suitable for the drip irrigation system.

### Disk or ring filter

The type of filter is similar to a screen filter. It consists of a number of disks encapsulated inside a special cover. The disks have small pathways at the surface by which water is filtered as it passes through these pathways (Figures 10.7 and 10.8). It is important that the disks are tightly fitted together for an efficient filtration and to avoid obstruction.

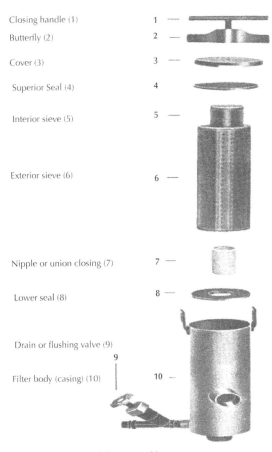

Closing handle (1)

Butterfly (2)

Cover (3)

Superior Seal (4)

Interior sieve (5)

Exterior sieve (6)

Nipple or union closing (7)

Lower seal (8)

Drain or flushing valve (9)

Filter body (casing) (10)

**Figure 10.6.** Components of the double screen filter.

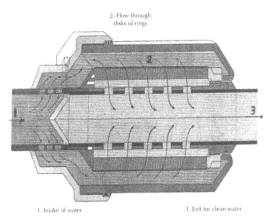

2. Flow through disks of rings

1. Intake of water          3. Exit for clean water

**Figure 10.7.** The direction of flow of water through a disk filter: 1. Intake of water; 2. Flow through disks or rings; 3. Clean water at the outlet (Exit).

**Figure 10.8.** Flushing method for the disk filter.

### Sand filter

In the past, sand filters have been used for domestic and industrial purposes. Before the introduction of the trickle irrigation technique, the use of these filters was limited because of high cost. Sand filter (Figure 10.4) consists of a fine gravel and sand of selected size. Sand filters are not easily obstructed by algae and can separate high quantities of suspended solids. Sand filters may separate particles from 25 to 100 micrometers. In general, the flow rate through these filters cannot exceed 14 lps (20 gpm) per surface area of filtration. The depth of filtration medium (sand) in the tank should be greater than 500 mm (18 inches). Secondary filters (screen or disk) should follow the sand filters. A flushing valve helps in the cleaning of the sand filters.

### Centrifugal filter

The centrifugal filter (hydrocyclone or a sand separator) separates the solids from the water. It is a simple device to separate particles of $2 \times 10^3$ to $2 \times 10^5$ nanometers from water, at a flow rate greater than 600 lph. Basically, the centrifugal filters can separate solid particles that have a specific gravity greater than the water. These are primary filters that are effective to separate particles before the water enters the drip irrigation system. These can also be installed at the intake of the pump (as pre-filter) to minimize the wear of the pump. Screen filter (secondary filter) must follow the hydrocyclone filter to avoid clogging. Centrifugal filters are available in different sizes for different discharge rates. These filters are inefficient to separate most of the organic matter and substances.

## PRINCIPLE OF OPERATION OF FILTERS

### Gravity Filter

The storage of water in a tank for a specified period promotes suspended solids (like gravel, sand, silt and clay) to settle down at the bottom of the tank. Therefore, the solids with a specific gravity greater that water are separated.

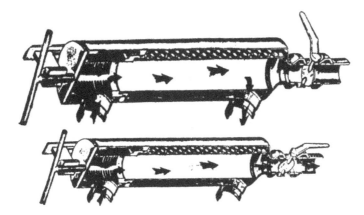

**Figure 10.9.** Screen filter: Draining operation (up) and the filtration operation (down).

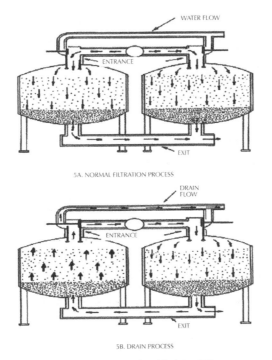

**Figure 10.10.** Sand Filter: Filtration (5A) and principle of flushing (5B).

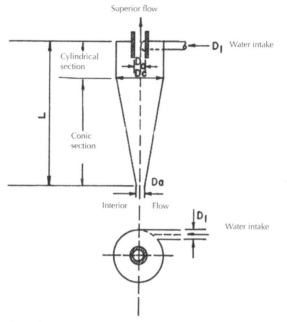

**Figure 10.11.** Hydrocyclone (centrifugal) filter showing water intake, superior and inferior flow.

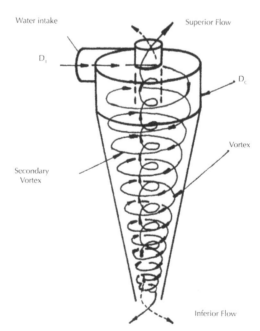

**Figure 10.12.** Vortex formation with superior and inferior flow in a centrifugal filter and quality of the suspended solids to be removed; the chance of chemical o biological obstructions; and quality of water after filtration.

## Pressure Filters
### Screen and disk filter
The principle of operation for both types of filters is similar. The water enters a closed chamber and is forced to pass through sieves or pressed disks (or rings) as shown in Figure 10.7. In this way, the sediments are trapped in the filtering unit and the clean water is forced out of the filters.

### Sand filters
Sand filter is a most efficient method to separate organic matter in suspension and organic solids. Contaminated water (without filtration) enters the filter through superior part of the filter and flows (under pressure) through the sand where the suspended particles are trapped. The filtered water flows to the main line.

### Hydrocyclone filter
The unfiltered water enters tangentially to the cylindrical section. It creates an angular momentum to cause a rotational movement of the water in the cylindrical section (Dc). A conical form of the filter allows the separation of solid particles from the water. This movement of water is called a vortex.

The solid particles, heavier than water, are thrown at the periphery of a circular path. Due to gravity, the particles go down to a flushing tank that is located at the lowest part of a cylinder. Later the sediments are removed from the flushing tank by opening a gate valve, when it is necessary. Part of the water movement is separated from the principal vortex to form a secondary vortex. The secondary vortex produces an upward movement in the middle and the clean water flows out through the superior exit (Di).

## SELECTION OF FILTER
An appropriate filtration system is selected to provide the required filtration with minimum cost and at maximum efficiency. The selection of a filter is based on the following factors:

1. Calculate the size of the irrigation system (the flow rate, operating pressure and volume of water). The capacity of the filter should exceed the total demand of irrigation.
2. Determine the physical, chemical and biological quality of the irrigation water; the size
3. Ask the following questions:
   a. How complex is the filtration system? What are the limitations for flushing?
   b. Is the manual labor available to perform service and maintenance?
   c. Is the location of the filter a problem? Where is the location of the water source?
   d. Is the filtration system flexible? What are possibilities for future modifications?

The filters are available in different sizes depending on the flow rate (Table 10.1). The selection of an adequate filtration system is an important consideration in the design of drip irrigation system.

**Table 10.1.** Guide for the selection of a filter.

| Flow Rate | Concentrations of solids | | **Recommendations |
|---|---|---|---|
| | *Organic | *Inorganic | |
| Less than 11.4 m³/hr. | L | L | A |
| (50  gpm) | L | M | A |
| | L | H | A |
| | M | L | A |
| | M | M | B+A |
| | M | H | B+A |
| | H | L | B+A |
| | H | M | B+A |
| | H | H | B+A |
| 11.4 to 45.6 m³/hr | L | L | A |
| (50–200 gpm) | L | M | A |
| | L | H | A |
| | M | L | A+B |
| | M | M | A+B o A+C |
| | M | H | A+B o A+B+C |
| | H | L | A+C |
| | H | M | A+C o A+B+C |
| | H | H | A+B+C |
| >45.4 m³/hr | L | L | A |
| (200 gpm) | L | M | A+C |
| | L | H | A+C |
| | M | L | A+C |
| | M | M | A+B+C |
| | M | H | A+B+C |
| | H | L | A+C |
| | H | M | A+C+C |
| | H | H | A+B+C |

¹The purpose of this guide is to select a correct size of the filter. The specific and individual requirements of the design must be evaluated.
*Key for concentration of solids: L = Less than 5 ppm; M = 5–50 ppm; H = Greater than 50 ppm
**Key for recommendations: A = Disk or screen filter; B = Hydrocyclone filter; C = Sand filter

## SERVICE AND MAINTENANCE OF THE FILTERS

The filtration system is a pulmonary system of a drip irrigation system. Therefore, we must make sure that the filters are flushed and these perform efficiently. It is necessary to install pressure indicators (gages) at the entrance and at the exit of a filter. When the filters are clean, the gages should indicate the same pressure.

The pressure difference across the filters will increase due to obstruction problems. When this pressure difference is from 10 to 15 psi (depending on the type of filter), then the operator must clean or flush the filters according to the specifications of a manufacturer.

### Flushing of Filters

Filters should be flushed before each irrigation, or when indicated by the pressure gages. In case of water with large quantities of sediments or contaminants, the filters should be cleaned or flushed frequently. Entrance of dust into the system should be avoided.

### *Disk filter*

The flushing or the cleaning is achieved by moving the disk (rings) in the direction of the flow. These are then washed with pressurized water (Figures 10.3, 10.7, and 10.8). After the flushing operation the disks are placed and pressed together. The filter casing is then adjusted and closed tightly. If the disks are not tight enough, the filtration efficiency decreases and the problems of obstruction may occur.

### *Screen filter*

The screen filter is commonly equipped with a cleaning brush or a valve in the inferior part of the filter. If the pressure difference between the gages at the intake and the exit is greater than the allowable difference in pressure, then the flushing is required. Open the drain or flush valve so that the pressurized water escapes through this valve (Figure 10.9). Move the cleaning brush up and down, giving it a torsional movement (if it is provided with the unit). This will help to eliminate the sediments through the flushing valve. If the gages still show a need for flushing, shut off the irrigation system and dismantle the filtering assembly. Check if the screen is not obstructed with sediments. Wash the screen and internal parts with pressurized water to make sure that it is in good condition. Broken and defective parts should be repaired or replaced. Assemble the filter to make sure all the parts are installed properly.

### *Centrifugal filter*

The sediments accumulated in the tank can be removed in two ways:

1. When the system is not in operation, open flushing valve and let the sediments escape out.
2. When the filter is in operation, open the flushing valve and let the sediments escape out.

## Sand filter

Sand filters are flushed by a back flow. For this type of flushing, the two adjacent filters must be interconnected (Figure 10.10). The flow is inverted to one of the filters by means of a three way valve (Figures 10.13), while maintaining the second filter in operation under normal conditions. This causes turbulence in the filter, the particles and matter trapped in the sand are escaped out through a flushing pipe (Figure 10.10). The water flow for flushing is supplied through a superior part of filter and taken out through a separate flushing line. By using two filters in series, the irrigation operation is not interrupted during the flushing process. This way, two or more filters in series are flushed successively.

13a. Manual valve:

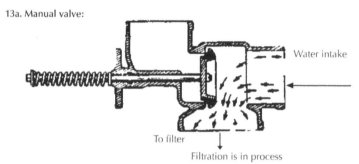

Water intake

To filter

Filtration is in process

13b. An automatic three-way valve:

Flushing flow exit

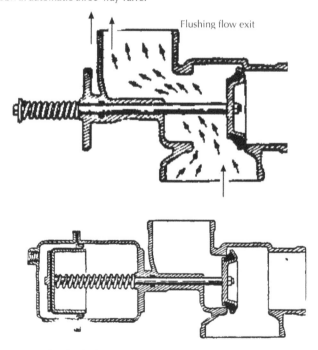

**Figure 10.13.** Three way valve for flushing of sand filters.

An excessive back flow will expand the sand to a point that the sand itself is expelled out back from the tank during the inverse flow. An insufficient flow will not expand the sand enough to flush the trapped contaminants. To obtain the maximum flushing efficiency, adjust the back flow.

## PREVENTION OF CLOGGING

To avoid obstructions in the drip irrigation system [1, 2, 3, 4, 5, 6], one must use an efficient and adequate filtration system. If this is not taken seriously, then it will cost more at later stage and it will be necessary to do modifications in the system. Simple treatments are not always successful to avoid obstructions, because it may be due to a specific condition. Therefore, a particular situation needs a specified solution.

### Physical Agents

Besides an adequate filtration system, lines and drippers should be cleaned regularly to prevent obstructions by physical agents. For flushing, open the ends of the principal (main) line and allow expulsion of the accumulated sediments. Then repeat the same process with the secondary (sub main) and lateral (drip) lines. Under extreme conditions of obstructions in the lines and drippers, one can correct these by a pressurized air. Before this process is initiated, start the system and let it run for approximately 15 minutes. Once the system is charged with (or full of) water, apply pressurized air (approx. 7 bars). The compressed air will force out accumulated material from the lines or drippers.

### Chemical Agents

Most of the causes of obstruction by chemical agents may be resolved by chloration or treatment with acid. By this process, chemical deposits caused by water and fertilizers are dissolved. In severe cases, the drippers are immersed in a diluted acid solution (approximately one percent). In some cases, these are flushed individually. For less severe cases, it is adequate to inject the acid to lower the pH of water. The process should be repeated till a normal flow rate is obtained through the dripper. The quantity of acid required to lower the pH is determined by testing a small volume of water.

Acids are extremely corrosive, so extreme caution should be taken. Commonly used acids are commercial grade sulfuric, hydrochloric and nitric (muriatic) acids. "Do not use or apply any chemical agent through the drip system" without consulting a specialist. The alternative between one method and another will depend on the availability and the cost. The acids may corrode the metal accessories, tubing and casings. Surfaces in contact with acids should be made of stainless steel and plastic. All these parts should be well washed after being in contact with the acid.

### *Quantity of acid required to lower the pH of irrigation water*

The quantity of acid required to lower the pH of irrigation water to a given value can be calculated using equation (1) for gallons applied per hour of water flow and equation (2) for gallons per each 100 gallons of water. In both equations, the acidity factor and the normality of the acid should be known, for example: Concentrated sulfuric acid ($H_2SO_4$) = 36N; Concentrated hydrochloric acid (HCl) = 12N. Acidity factor is a milliequivalents of acid per liter of water to lower the pH to a desired level. It is determined by titration of water sample with acid, in a laboratory.

Gallons of acid per hour = [0.06 × acidity factor × gpm]/[Acid normality]   (1)

Gallons of acid per 100 gallons of water = [Acidity factor]/[Acid normality]   (2)

## Biological Agents

When the silt or algae obstruct the emitters, the common treatment is an injection of a biocide (chlorine) followed by complete flushing of the lines to remove any organic matter. The common biocides are chlorine gas ($Cl_2$) and solutions of hypochlorite ($HOCl^-$). An efficient algae and bacterial control in a drip irrigation system may be maintained by administration of 10–20 ppm during the last 20 minutes of the irrigation cycle. If the irrigation water has a concentration of one ppm of chlorine, then the obstruction due to biological agents is under control.

## SUMMARY

We must select an adequate filtration system to provide water free of clogging agents. This chapter discusses clogging agents, prevention of clogging, types of filters, and principle of operation of filters, selection of filters, service and maintenance of filters, trouble shooting and procedure to solve problems of clogging.

## TROUBLE SHOOTING

| Causes | Remedies |
|---|---|
| **Poor filtration** | |
| A.   **Screen or disk filter:** | |
| 1. Obstruction caused by one of the following: | |
| a. Sand or silt. | a. Use a centrifugal filter. |
| b. Algae or organic matter. | b. Use a sand filter or a biocide. |
| c. Salt or chemical compounds. | c. Wash with acids. |
| d. Use of an inadequate filtration method. | d. Use an adequate filtration system. |
| B.   **Sand filter:** | |
| 2. Excess flow through media causes formation of a cone. | 2. Reduce flow capacity or add more filters. |
| 3. Air inside the tank causes disturbances in the sand and formation of a cone. | 3. Install manual or automatic air relief valves. |
| 4. Use of an incorrect type of media (sand) | 4. Use recommended media. |
| 5. A high pressure difference caused by the sediments. | 5. Flush frequently. |
| C.   **Hydrocyclone filter:** | |
| 6. Organic matter in water | 6. Use a sand filter |

| Pressure variation is constant | |
|---|---|
| A. | **Screen or disk filter:** | |
| | 7. Filter media is obstructed. | 7. Flush or use an adequate media. |
| B. | **Sand filter:** | |
| | 8. Low level of sand in filter causes an inadequate flush flow. | 8. Add sand to an adequate level. |
| | 9. Blocked filter: there is not enough water through the filter for flushing. | 9. Remove cover and scrape the superior crust of sand. Place cover and wash the tanks alternatively at short intervals till water is clean. |
| | 10. Insufficient flushing flow. | 10. Readjust back flow valve. |
| Dripping of a drain valve | |
| A. | **Sand filter:** | |
| | 11. Valve is obstructed. | 11. Remove the obstructions. |
| | 12. Valve is defective. | 12. Replace the valve. |
| Air shocks | |
| | 13. Air in tank. | 13. Install manual or automatic devices to remove air. |
| | 14. Flushing line causes suction. | 14. Add an air relief valve in flushing line. |
| | 15. Inappropriate flushing | 15. Increase flow or duration of flushing. |
| | 16. Low level of sand in filter. | 16. Add more sand to a recommended depth |
| | 17. Water is too dirty. | 17. More filters are required. |

## KEYWORDS

- **Acidity factor**
- **Back flow**
- **Bar**
- **Chemigation**
- **Chlorination or chloration**
- **Clay**
- **Clogging**
- **Cylindrical section (Dc)**
- **Drip irrigation**
- **Dripper**
- **Emitter**
- **Emitter flow rate**
- **Fertilizer**

- **Fertilizer, liquid**
- **Filter**
- **Filter, double screen**
- **Filter, hydrocyclone**
- **Filter, mesh**
- **Filter, sand**
- **Filter, screen**
- **Filtration system**
- **Flushing valve**
- **Gate valve**
- **Main line**
- **Nipple or union**
- **Organic matter**
- **pH**
- **Pump**
- **Soluble salts**
- **Water quality**
- **Water, source**
- **Weed**

## BIBLIOGRAPHY

Refer to bibliography Section 10.1 for the literature cited.

# Chapter 11

## Service and Maintenance

Success of drip irrigation depends on the support from specialist for installation and maintenance

---

[1]This chapter is modified and translated from, "Goyal, Megh R. y Luis E. Rivera-Martínez, 1990. Servicio y Mantenimiento. Capítulo XII en: *Manejo de Riego Por Goteo*. Editado por Megh R. Goyal. Páginas 331–338. Río Piedras- PR: Servicio de Extensión Agrícola, UPRM." For more details one may contact by E-mail: goyalmegh@gmail.com

Flow meter to measure the volume of water applied.

Water source, pump, and check valve assembly.

## INTRODUCTION

The orifices in the drip lines or the emitters emit water to the soil. The emitters allow only the discharge of few liters or gallons per hour. Because the emitters have small orifices, these can be easily obstructed with clogging agents (physical, chemical, and biological). The obstruction can reduce degree of emission, the uniformity of water distribution, and therefore, this can reduce plant growth. Once the system has been obstructed, it becomes more difficult to restore the normal water flow. Therefore, we must prevent the obstructions in the filters, laterals, and emitters. The clogging can be prevented with a good maintenance and periodic service of the system. To operate and to maintain a drip irrigation system in a good working condition, the following considerations are important for an adequate operation [1, 2, 3, 4]:

1. Pay strict attention to the filtration and flushing operation.
2. Maintain an adequate operating pressure in the main, sub main, and lateral lines.
3. Flushing and periodic inspection of the drip irrigation system.

## MAINTENANCE OF FILTERS AND FLUSHING OPERATION

For effective filtration efficiency, we must make sure that the system is maintained in good condition and it is not obstructed by the clogging agents. For this purpose, pressure gages are installed at the entrance and the exit of a filter. The pressure difference between these two gages should vary from 2 to 5 psi when the filter is clean and the mesh is free from obstructions. The filtration system should be cleaned and flushed, when the pressure difference is from 10 to 15 psi. The filters must be flushed before each irrigation operation. If the water contains high percentage of suspended solids, then the filters should be flushed more frequently. Entrance of dust and foreign material should be avoided, when the filters are open. Filters may not be able to remove the clay particles and algae.

## FLUSHING METHOD

The frequency of flushing depends on the water quality. For flushing of irrigation lines, the following procedure can be adopted:

1. Open the ends of the distribution and lateral lines. Allow the flow of water through the lines until all the sediments are thrown out of the lines.
2. Close the ends of the distribution lines. Begin to close the lines one after another, from one block to second, and so on. There must be a sufficient pressure to flush out all the sediments.

### Cleaning with Pressurized Air

The clogging can be caused due to presence of organic matter in water. It may be necessary to use pressurized air to clean the drippers. Before beginning this process, the water is passed through the lines for a period of 15 minutes. When adequate operating pressure has been established, then the air at 7 bars of pressure is allowed through the system. The compressed air will clean the lines, laterals, and drippers from the accumulated organic matter.

### Cleaning with Acids and Chlorine

The clogging may also be caused due to precipitation of salts. The cleaning with acids will help to dissolve the chemical deposits. This process is not effective to remove the organic matter. Sodium hypochlorite (at the rate of one ppm) can be injected on the suction side of a pump for 45–90 minutes before shutting off the pump. The best time of injection is after flushing the sand filters, because the chlorine prevents the growth of bacteria in the sand. The surface water containing iron can be treated with chlorine or commercial bleaching agent for 45 minute for lowering the pH to <6.5. At pH > 6.5, certain reactions in combination with the precipitates of iron may gradually obstruct the irrigation lines.

One may use commercial grade phosphoric acid or hydrochloric acid. Before using the acid, the water is allowed to pass through the system at a pressure greater than the operating pressure. Fill the fertilizer tank up to two third (2/3) parts of the capacity of a tank. Add the acid at the rate of one liter per cubic meter per hour of flow rate. Inject the diluted acid into the system, as one will inject the fertilizer, in a normal process.

**Remember**: When using the chemigation tank, first pour the water and then add the acid.

### METHODS TO REPAIR TUBES OR DRIP LINES

1. The orifices of Bi-wall tubing may be obstructed due to salts, and so forth. A polyethylene tubing of small diameter is used as a bypass method to repair these drip lines.
2. If the line is broken or there is an excessive escape of water, the pipe, or the tube is cut down and is connected with a union or a coupling.
3. If the main line is made of flexible nylon flat and is leaking, then use a small piece of plastic pipe of same diameter to insert into the flexible nylon tubing. The both ends are sealed with the use of pipe clamps.

### SERVICE BEFORE THE SOWING SEASON

1. Clean and flush all the distribution system and the drip lines, with water.
2. Wash with water and clean the pump house system. Lubricate all valves and accessories.
3. Turn on the pump and activate the system. Check the pipes and drip lines for leakage. Repair if necessary.
4. If the system has been used previously, then cleaning and flushing should be carried out for a longer period of time. It is particularly important in sandy soils, as the sand can penetrate into the pipe during the removal of lines.

### SERVICE AT THE END OF CROP SEASON

At the end of a crop season, following steps should be taken:
1. Flush the pipes. Clean the filters and other components of the system.
2. Lubricate all the gate valves and accessories.

3. If the pipes are permanently installed in the field and cannot be removed at the end of a crop season, keep these free of soil and weeds that can grow nearby.

4. If the system can be moved from one place to another (according to the season), the following procedure is adequate:

   a. Flush and clean the system.

   b. Remove the drip lines and collect these carefully.

   c. It is best to leave the main lines in place. If it is not possible or if there is a need for transfer to an area, then these should be rolled. Close both ends and store in the shaded area.

   d. It is advisable to label the hoses with tags. Distance between orifices and frequency of use should be indicated on the tag.

## TROUBLE SHOOTING

| Causes | Remedies |
|---|---|
| **Pressure difference > Recommended value** | |
| 1. Filters are obstructed. | Flush the filters. |
| 2. Lines are broken. | Repair or replace lines. |
| 3. Pump is defective. | Repair or replace the pump. |
| 4. Gate valve is blocked. | Fix or replace the gate valve. |
| 5. Pressure regulator is defective. | Remove and replace the regulator. |
| **Laterals (or drip lines) and drippers are clogged** | |
| 6. Sand is being accumulated in the drippers and lines. | Open ends of laterals and leave open for more than two minutes so that water at pressure passes through. |
| 7. Formation of algae and bacteria. | Wash with chlorine. Paint the PVC pipes or install the lines below soil surface. |
| 8. Sediments are being accumulated. | Wash with acid. |
| 9. Precipitation of chemical compounds due to chemigation. | Wash with acid and conduct the chloration process. |
| 10. Obstruction due to nest of insects. | Wash with insecticide. |
| **Pressure is increased** | |
| 11. Orifices in the drip lines or drippers are clogged. | Flush the drip lines or laterals. |

## SUMMARY

The orifices in the drip lines or the emitters emit water to the soil. The emitters allow only the discharge of few liters or gallons per hour. The emitters have small orifices and these can be easily obstructed. For a trouble free operation, one should follow these considerations: Pay strict attention to filtration and flushing operation. Maintain an adequate operating pressure in the main, sub main and lateral lines. Flushing and periodic inspection of the drip irrigation system is a must.

For effective filtration efficiency, we must maintain the system in good condition and it is not obstructed by the clogging agents. For this, pressure gages are installed at the entrance and the exit of a filter. The frequency of flushing depends on the water quality. Some recommendations for an adequate maintenance are cleaning with pressurized air, acids, and chlorine. This chapter includes methods to repair tubes or drip lines. Also there is a procedure for the service before the sowing season and the service at the end of the crop season.

## KEYWORDS

- Bar
- Check valve
- Chemigation
- Clay
- Clogging
- Dripper
- Emitter
- Fertilizer
- Fertilizer tank
- Filter
- Filter, sand
- Filtration system
- Flow metering valve
- Gate valve
- Leaching
- Line, distribution
- Line, main
- Maintenance
- Orifices
- Poly Vinyl Chloride (PVC) pipe
- Polyethylene (PE)
- Precipitation
- Pressure regulator valve
- Pump
- Pump house
- Sodium hypochlorite
- Union
- Water quality
- Weed

## BIBLIOGRAPHY

Refer to bibliography Section 11.1 for the literature cited.

# Chapter 12

## Design of Trickle Irrigation Systems

I. Pai Wu and Harris M. Gitlin

[1]This chapter is modified and compiled from, "Wu, I.P. and H.M. Gitlin. Drip Irrigation Systems Design. Bulletin No. 144 and 156 of the Cooperative Extension Service of the University of Hawaii."

## INTRODUCTION

A drip irrigation system consists of main lines, secondary lines, and lateral lines [1]. There are other important components such as filters, regulators of pressure, indicators of pressure, valves, injectors of fertilizer, and so forth. The lateral lines can be polyethylene tubes combined with drippers, or simply of low pressure plastic pipe with orifices. These are designed to distribute water to the field with an acceptable degree of uniformity [8, 9, 10]. The secondary line acts as a control system, which can adjust the pressure of water to supply the quantity of flow required in each lateral line. Also it is used to control the time of irrigation in individual field. The main line serves as a system of transportation to supply the total quantity of water required in the irrigation system. This chapter is a combined version of *"Drip irrigation systems design"* bulletins # 144 and 156 by the Coopertative Agricultural Extension Service at the University of Hawaii [11].

## BASIC CONCEPTS OF HYDRAULICS

Plastic pipes of different sizes can be considered like smooth conduits. The Blasius formula can be used to determine the turbulent flow in a smooth conduit [5, 6]. The empirical equation of Williams and Hazen [3] for C = 150 is given in Table 12.1. The equation one is used to determine the loss or fall in energy for the section of main line. In equation (1) and in Table 12.1:

$\Delta H$ = The energy loss by friction in feet or meters.
$\Delta L$ = Length of the pipe section in feet or meters.
$Q$ = Total discharge in the pipe in gallons per minute (gpm) or liters per second (lps).
$D$ = Internal diameter of the pipe in inches or centimeters.

**Table 12.1.** Equations for basic concepts of the hydraulics.

| | | |
|---|---|---|
| $\Delta H = 9.76 \times 10^{-4} \times (Q^{1.852}/D^{4.871}) \times \Delta L$ | F.P.S. | (1a) |
| $\Delta H = 15.27 \times (Q^{1.852}/D^{4.871}) \times \Delta L$ | M.K.S. | (1b) |
| $\Delta H = 3.42 \times 10^{-4} \times (Q^{1.852}/D^{4.871}) \times L$ | F.P.S. | (2a) |
| $\Delta H = 5.35 \times (Q^{1.852}/D^{4.871}) \times L$ | M.K.S. | (2b) |
| $R_i = 1-[1-i]^{2.852}$ | — | (3) |
| $(dh/dl) = -S_f \pm S_o$ | — | (4) |
| $q = C \times (\sqrt{h})$ o $q = C \times (h)^{0.5}$ | — | (5) |

| | | |
|---|---|---|
| $q_{var} = 1-[1-h_{var}]^{0.5}$ | — | (6) |
| $q_{var} = [q_{max} -q_{min}]/q_{max}$ | — | (7) |
| $h_{var} = [h_{max} -h_{min}]/h_{max}$ | — | (8) |

The condition of the flow in the secondary or lateral line is constant and varies especially with the flows of the lateral line. Since the discharge in the line decreases with the length, the loss will be less than is given by equation (1). The loss of energy due to friction for the lateral or secondary lines is shown in equation (2). In equation (2) and in Table 12.1:

$\Delta H$ = Total energy loss by friction at end of the lateral line (or secondary) in feet or meters.

Q = Total discharge at the entrance of the lateral line (or secondary) in gpm or lps.

D = Internal diameter of the lateral line (or secondary) in inches or centimeters.

L = Total length of the lateral line (or secondary) in feet or meters.

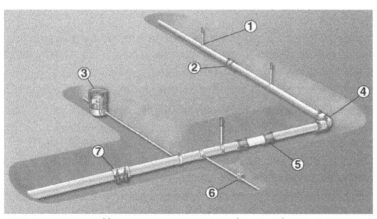

Components a trickle irrigation system: 1. Main line, 6. Submain, 3. Pump.

When the discharge in the lateral line decreases with the length, the gradient line of energy is an exponential curve [2, 5] instead of a straight line. The form of the gradient line can be expressed as the gradient line of energy without dimensions as shown in equation (3). In equation (3) and in Table 12.1:

$R_i = \Delta H_i /\Delta H$, and it is known as a energy drop ratio.

$\Delta H$ = Total loss of energy determined by equation (2).

$\Delta H_i$ = Total loss of energy expressed in feet or meters, for length ratio of i, (i = 1/L).

L = Total length in feet or meters.

L = Given length, measured since the final section of the line in feet or meters.
i = 1/L.

The equation (3) can be used to determine the pattern of energy loss along a lateral line (or secondary), when the total loss in energy is known. The variation in pressure (the change in pressure along the length) can be determined like a lineal combination of slope of the energy and the slope of the line, assuming that the change of velocity head is small or insignificant. This is shown in equation (4). This relation can be used for non uniform and uniform slopes [6]. In equation (4) and in Table 12.1:

$S_f$ = Energy slope or the energy gradient line.
$\pm S_0$ = Slope of the line, with positive sign when the line is down the slope, and with negative sign when the line is above the slope.

Hydraulically, the variation in pressure along a lateral line will cause a variation in the flow of the dripper along the lateral line. A variation in pressure along the secondary line would cause a variation in the flow of the dripper along the lateral line (toward each lateral line) and along the secondary line. For most common dripper types, and assuming a turbulent flow in the lateral line, the discharge of the dripper (or the lateral lines flow for the secondary one) and of the pressure head can be expressed as a simple function by equation (5). In Table 12.1 and in equation (5):

q = Flow of dripper (or flow toward the lateral line).

h = Pressure head.

The equations (6) and (7) describe the relation between the variation in pressure and the variation in the dripper flow. The variation in pressure is given by the equation (8). In equations (6), (7), and (8) and Table 12.1:

$q_{var}$ = Dripper flow variation for lateral line (or secondary line).

$q_{max}$ = Maximum dripper flow for lateral line (or secondary line).

$q_{min}$ = Minimum dripper flow for the lateral line (or secondary line).

$h_{max}$ = Maximum pressure head along the lateral line (or secondary line).

$h_{min}$ = Minimum pressure head along the lateral line (or secondary line).

$h_{var}$ = Variation in pressure head along the lateral line (or secondary line).

The design criteria [6] is a variation in the dripper flow of less than the 20% (approximately 40% for variation in pressure) for the design of lateral line and a variation in the flow of the lateral line less than 5% (approximately 10% for variation in pressure) for the design of the secondary line.

## CHARTS AND DESIGN PROCEDURES

The design charts were based on the hydraulics of the drip lines and were developed with a computer simulation [4, 6, 7] for the design of lateral line, secondary line, and main line. The design charts and its procedures are described below:

### Design Charts for the Lateral Line on Uniform Slopes

The design charts for lateral line of 0.5 inch were developed as shown in Figures 12.1, 12.2, 12.3, and 12.4. The Figures 12.5 and 12.6 show design charts for lateral line of 12 mm and 16 mm, respectively. The procedure is:

**Step 1:** Establish along one of the lateral lines: the length of lateral line (L), the operational pressure head (H), the ratio L/H, and the total discharge Q in gpm or lps.

**Step 2:** Move vertically from L/H in the third quadrant to the total given discharge (gpm or lps) in the second quadrant; then establish a horizontal line toward the first quadrant.

**Step 3:** Move horizontally from L/H (third quadrant) to the percent slope in the fourth quadrant; then establish a vertical line toward the first quadrant.

**Step 4:** The point of intersection of these two lines in the first quadrant determines the acceptability of the design.

**Desirable:** A coefficient of uniformity (CU) over 98%; equivalent to a variation in the dripper flow less than 10% or a variation in pressure less than 20%.

**Acceptable:** CU between 95 and 98%; equivalent to a variation in the dripper flow around 10–20% or variation in pressure of 20–40%.

**Not Acceptable:** CU less than 95%; equivalent to a variation in the dripper flow over 20% or a variation in pressure greater than 40%.

### General Design Charts for the Lateral and Secondary Lines on Uniform Slopes

The design charts were developed for the lateral lines (Figures 12.1–12.4 for the English system, and Figures 12.5 and 12.6 for metric system). The Figures 12.7 and 12.8 are dimensionless charts for all sizes of the line; and these can be used for the secondary and lateral lines. The Figure 12.9 is for the English system and the Figure 12.10 is for the metric system. These figures are used to determine H/L from the total discharge and the size of the line.

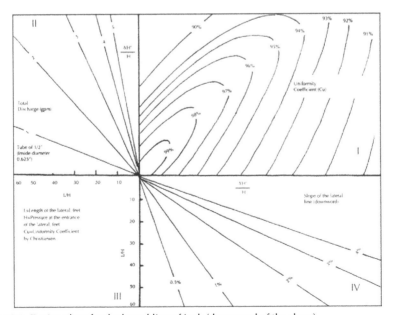

**Figure 12.1.** Design chart for the lateral line of inch (downward of the slope).

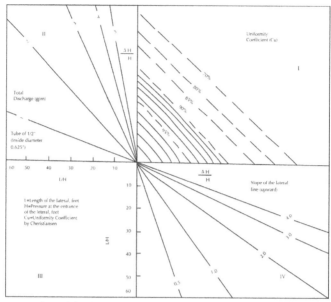

**Figure 12.2.** Design chart of the lateral line of inch (upward of the slope).

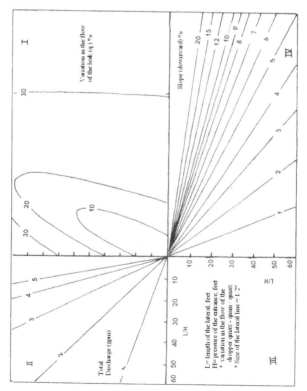

**Figure 12.3.** Design chart of the lateral line of  inch (downward of the slope).

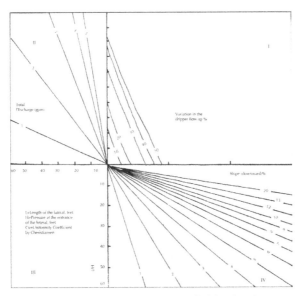

**Figure 12.4.** Design charts of the lateral line of inch (upward of the slope).

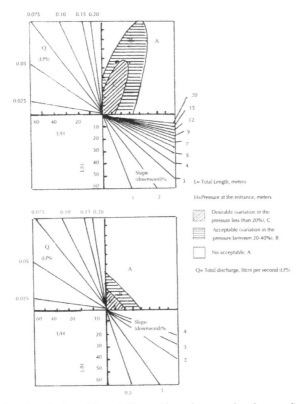

**Figure 12.5.** Design chart for lateral line of 12 mm (slope downward and upward).

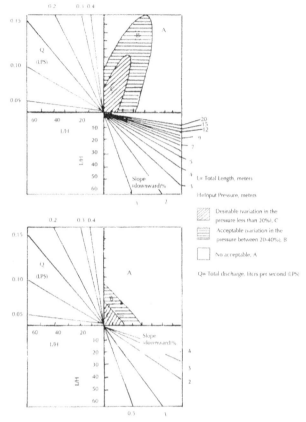

**Figure 12.6.** Design chart for lateral line of 16 mm (slope downward and upward).

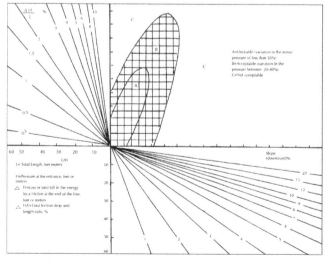

**Figure 12.7.** Dimensionless design chart for lateral and secondary lines (downward of the slope).

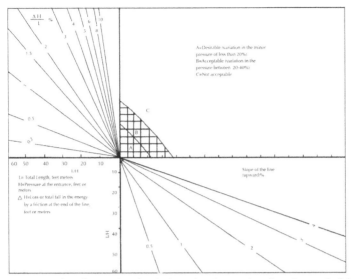

**Figure 12.8.** Dimensionless design chart for the lateral and secondary lines (upward of the slope).

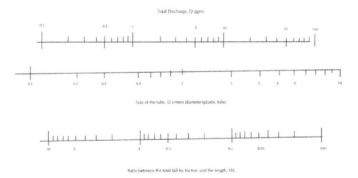

**Figure 12.9.** Nomograph for the design of lateral and secondary lines in F.P.S. units.

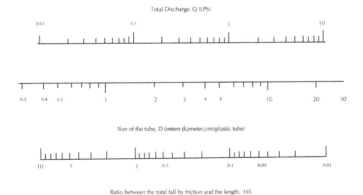

**Figure 12.10.** Nomograph for the design of lateral and secondary lines in metric units.

These design charts can be used to revise the acceptability of the design, if the size of the lateral lines is given, or for selecting an appropriate size of lateral lines to comply with the design criteria. The design procedure is:

**A. To revise the acceptability of the design (if we know the size of the lateral or secondary lines)**

**Step 1:** Establish a trial L/H and the total discharge.

**Step 2:** One should use the total discharge and the size of the lateral line to determine $\Delta H/L$, in Figure 12.9 or 12.10.

**Step 3:** Move vertically from L/H (third quadrant) toward the specific value of $\Delta H/L$ in the second quadrant (Figures 12.7 and 12.8); then establish a horizontal line toward the first quadrant.

**Step 4:** Move horizontally from L/H toward the slope of the line in the fourth quadrant; then establish a vertical line toward the first quadrant.

**Step 5:** The point of intersection of these two lines in the first quadrant will determine the acceptability of the design:

   **Zone A:** Desirable, variation in the dripper flow of less than 10%.

   **Zone B:** Acceptable, variation in the dripper flow from 10–20%.

   **Zone C:** Not acceptable, variation in the dripper flow of greater than 20%.

**B. To select the appropriate size of the lateral or secondary line.**

**Step 1:** Establish a trial value of L/H and the total discharge.

**Step 2:** Move horizontally from L/H toward the slope of the line (up or downward) in the fourth quadrant. From this point, establish a vertical line toward the first quadrant.

**Step 3:** Establish a point along this line in the first quadrant in the upper margin of the acceptable region depending on the design criteria. From this point, establish a horizontal line toward the second quadrant.

**Step 4:** Establish a vertical line toward the second quadrant for the value L/H in such a way to intercept the point in the horizontal line indicated in the step 3.

**Step 5:** Determine the value of $\Delta H/L$ in the second quadrant for this point.

**Step 6:** From Figures 12.9 and 12.10, the total discharge is calculated. The values of $\Delta H/L$ and the minimum size of the lateral lines are established in agreement with the design criteria.

### Design of Lateral Line on Non Uniform Slopes

A simple design chart for lateral line on non uniform slopes [9] was developed as shown in the Figure 12.11. The chart is dimensionless, so that it can be used for English and metric units. The design procedure is:

**Step 1:** Divide the profile of the non uniform slope in several sections, so that each section can be considered as a uniform slope. Determine the slope of each section; calculate the gain (or the loss) of energy in each section due to its slope, and find the total energy gain by the slopes of any section along the line ($\Delta H'_i$).

**Step 2:** Plot the pattern of the non uniform slope, using dimensionless design chart in Figure 12.11: $1/L$ vs. $\Delta H'_i/L$ in the first quadrant.

**Step 3:** Determine the total energy loss by friction ($\Delta H$), using equation (2). Calculate the ratio of total loss in energy ($\Delta H$) and the operational pressure head, $\Delta H/H$. One may also use Figure 12.9 or 12.10.

**Step 4:** Determine the ratio of the length of the lateral lines (L) to the operational pressure head (H): L/H.

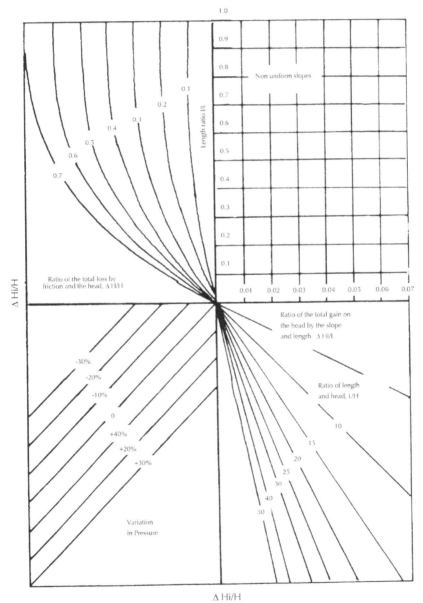

**Figure 12.11.** Dimensionless design chart for lateral line for non uniform slopes.

**Step 5:** Select any point on the profile of the non uniform slope in the first quadrant: Usually a point between two slopes or the middle of the section.

**Step 6:** From this point, draw a vertical line downward of the specific value of L/H in the fourth quadrant, and establish a horizontal line toward the third quadrant.

**Step 7:** Also from the point as indicated in step 5, draw a horizontal line at specific ΔH/H in the second quadrant, then establish a line toward the third quadrant.

**Step 8:** Locate the point of intersection of these two lines in the third quadrant. It will give the variation in the pressure.

**Step 9:** Repeat the same procedure for various other points in the profile of the non uniform slope (Step 1). Revise the variations in pressure for these points of the operational pressure head.

## Design of Trickle Irrigation System for Different Sizes of Lateral and Secondary Lines [10]

The most of the lateral and secondary lines are designed for a pipe of selected size. The energy gradient line for a lateral line of a simple size is an exponential curve, which is used as a basis to design lateral or secondary line in drip irrigation for uniform slopes and non uniform slopes. However, the length of one of the lateral and secondary lines can be relatively large under specific conditions and for non uniform slopes. The design of the lateral and secondary lines is a series of pipes of different sizes.

If the lateral or secondary line of individual size is designed so that the total energy loss by friction ($\Delta$ H) is balanced by the energy gain (H) at the end of the line, then the maximum variation in pressure head will be (0.36 $\Delta H_i$) or (0.36 S $\circ$ L), where:

$S_o$ = Slope of the lateral or secondary lines.

L = Length of the line.

This is caused by the shape of curve of the energy gradient line. The maximum variation in pressure will occur near the middle of the individual section of the line. When a series of different sizes in the design of lateral and secondary lines can be used, the maximum variation in pressure head can be reduced. If one can use three, four or more different sizes, the variation in pressure head along the line will be less, as shown in Figure 12.12. The slope of the line of each section can be used as the energy to design the size of the line. This procedure can be used for uniform and non uniform slopes. If a secondary or lateral line can be divided into different sections and different sizes for each section can be designed, then it has been shown that the energy gradient line of each section can be approximated as a straight line. It has also been shown that the discharge of each section can be used to calculate the total energy loss by friction, which is a basis for the design of size of tubing. With these variables, the engineer can design sections of lateral lines using Figure 12.13. This procedure is valid for uniform and non uniform slopes.

## Simplified Design Chart for Main and Secondary Lines

Design charts similar to Figures 12.1 and 12.2 for English system and Figures 12.5 and 12.6 for metric system can be developed for the design of secondary lines. The design chart for non uniform slopes can be utilized for the design of secondary lines. However, for shorter length of the secondary line, one can assume a uniform slope. Two design charts for secondary lines are developed: One for lines with slopes equal or greater than 0.5%, and another for secondary lines with slopes less than 0.5%. The design charts are given in Figures 12.14 and 12.15 for English units and 12.16 and 12.17 for metric units. Generally, the length of the secondary line is short, from 66–200 feet (= 20–60 meters). If the operational pressure head is from 20 to 30 feet (= 6–9 meters), the two simplified design charts will give a variation in flow of less than 5% for lateral line. The design procedure is:

**Step 1:** Determine the total discharge Q, for the secondary line.

**Step 2:** Determine the ratio of the length of the line to the pressure head: L/H.

**Step 3:** Determine the slope of the secondary line. If the slope is equal or greater than 0.5%, use Figure 12.14 or 12.16 to design the size of the secondary line.

**Step 4:** If the slope is less than 0.5%, use Figure 12.15 or 12.17 to determine the size of the secondary line.

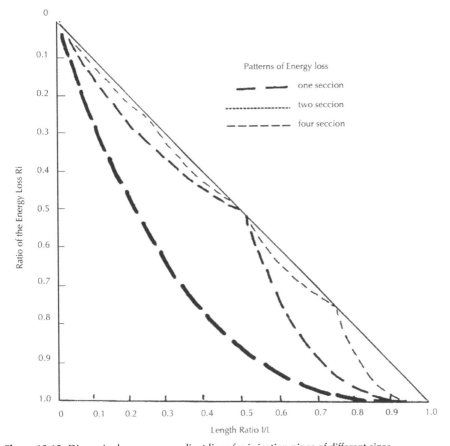

**Figure 12.12.** Dimensionless energy gradient lines for irrigation pipes of different sizes.

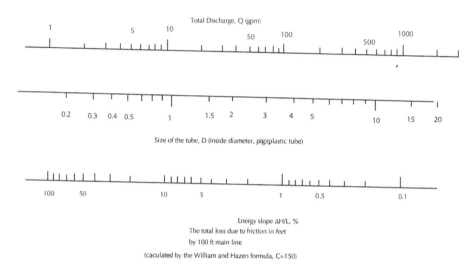

**Figure 12.13.** Nomograph for the design of main and secondary lines in F.P.S. units (for multiple sections with varying sizes).

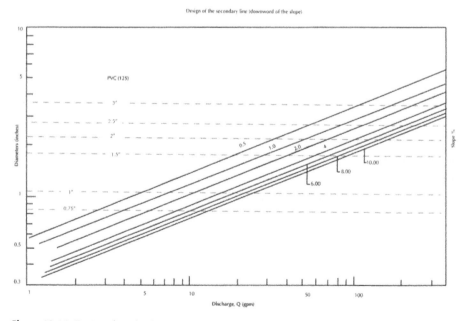

**Figure 12.14.** Design chart for the secondary line for slopes greater than 0.5% [F.P.S. units].

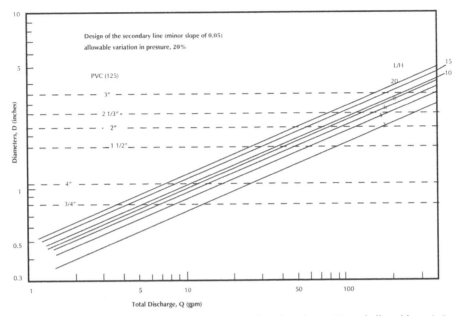

**Figure 12.15.** Design chart for the secondary line for slope less than 0.5% and allowable variation of pressure of 20% in F.P.S. units.

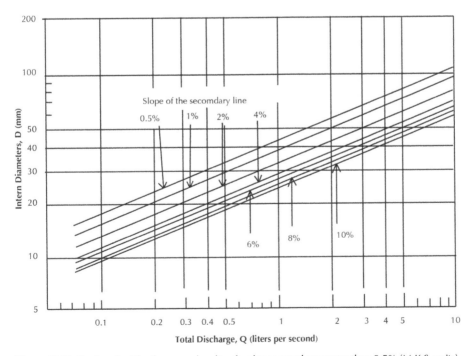

**Figure 12.16.** Design chart for the secondary line for slopes equal or greater than 0.5% (M.K.S. units).

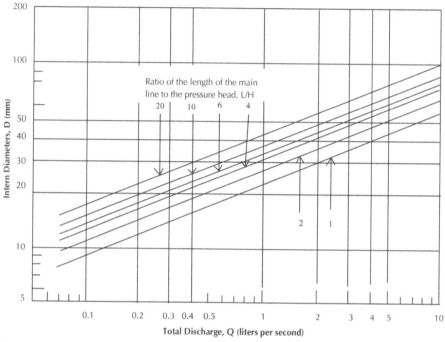

**Figure 12.17.** Design chart for the secondary line for slope less than 0.5% and allowable variation in pressure of 10% (M.K.S. units).

## Design of Main Line

The design of the main line is not a problem. The main line is designed based on total discharge, the length and the allowable energy loss according to equation (1) in Table 12.1. When the main line supplies water to many secondary lines (or to many sub-fields), the total discharge in each section of main line diminishes along the length of the main line. The size of the main line for different sections will depend on the shape of the energy gradient line for the main line. The optimum solution can be obtained using a computer simulation. The concept of straight energy gradient line was developed [4] for the design of the main line, which simplifies the design procedure. Given that the design procedure is simple, it can be used to design alternate arrangements of the main lines in the field. The design procedure is:

**Step 1:** Plot the profiles (slopes) of the main line and the required pressure head for the drip irrigation according to Figure 12.18 for English units and Figure 12.19 for metric units.

**Step 2:** Plot an energy gradient line from the available operational pressure head to the profile of required pressure (Figures 12.18 and 12.19), so that for any point along the main line, the gradient of energy line agrees with the profile of required pressure.

**Step 3:** Determine the slope of the straight energy gradient line: $\Delta H/L$.

**Step 4:** Determine the discharge required for each section of the main line.

**Step 5:** Design the size of the main line using Figure 12.20 or 12.21 based on the slope of energy (determined in the step 3) and the total discharge (determined in the step 4) for each section of the main line.

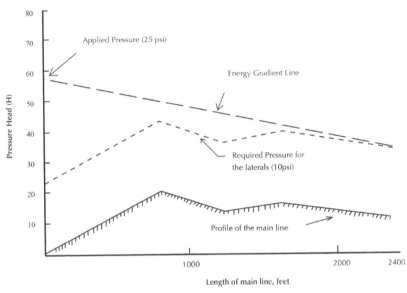

**Figure 12.18.** Profile of energy gradient line and profile of the main line (F.P.S. units), pressure head in feet.

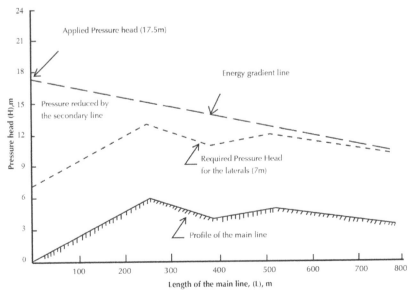

**Figure 12.19.** Profile of energy gradient line and profile of the main line (M.K.S. units).

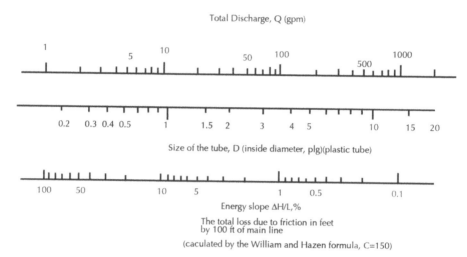

**Figure 12.20.** Nomograph for the design of main and secondary lines (F.P.S. units).

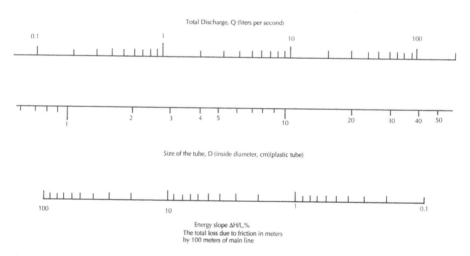

**Figure 12.21.** Nomograph for the design of main and secondary lines (M.K.S. units).

## DESIGN EXAMPLES

### Design of Lateral Line on Uniform Slope

*Design example number 1*

The operational pressure of the lateral line is 6.5 psi (or 15 feet of head); the length of the lateral line is 300 feet; the total discharge is two gpm; the slope of the lateral line is of 2% (downward of the slope); and the size of the lateral line is 0.5 inches (I.D. = 0.625 inches). Review the acceptability of the design. One can use Figures 12.1 and 12.3 as follows:

1. Calculate L/H = 300/15 = 20
2. From Figure 12.1 (or Figure 12.3) in the third quadrant, move vertically from L/H = 20 toward the line of total discharge (Q = 2 gpm) in the second quadrant; then establish a horizontal line toward the first quadrant.
3. Move horizontally from L/H = 20 in the third quadrant to the line of slope of 2% in the fourth quadrant; then establish a vertical line toward the first quadrant.
4. The point of intersection of two lines in the first quadrant shows a CU, $C_u$ = 97% and variation in the dripper flow = 13%. The design is acceptable.

## Design example number 2

The length of the lateral line in a vegetable field is 150 feet and the slope of the line is 1% (downward of the slope). The dripper spacing is one feet. The dripper flow is one gph at an operational pressure of 15 psi. Design the size of the lateral lines.

Length of the lateral line, L = 150 ft

Operational pressure, H = 15 psi =15/2.3 = 34 ft of head.

Number of drippers = 150

Total discharge, Q = 150 gph for 150 drippers = 2.5 gpm.

## Design procedure:

**Step 1:** Calculate L/H = 150/34 = 4.4

**Step 2:** In Figure 12.7 and for the desirable uniformity (Zone A), determine the value of $\Delta$ H/L = 6.

**Step 3:** In Figure 12.9, and for total discharge Q = 2.5 gpm and $\Delta$ H/L = 6, determine the minimum size of the lateral line = 0.5 inches (I.D.).

## Design example number 2A

Assuming the same conditions indicated in the example 4.1.2, design the size of the lateral line, for the following data:

Length of the lateral line        = 50 meters

Operational pressure head        = 10 meters

Number of drippers        = 600 1ph = 600/3600 = 0.167 lps

## Design procedure:

**Step 1:** Determine L/H = 50/10 = 5

**Step 2:** In Figure 12.5, move vertically from L/H to the total discharge of the line Q = 0.167; then establish a horizontal line toward the first quadrant.

**Step 3:** Move horizontally from L/H = 5 to the line of the slope of 1% in the fourth quadrant; then establish a vertical line toward the first quadrant.

**Step 4:** The line of intersection is in the acceptable region. Therefore, the lateral line of 12 mm is used in the design.

## Design example number 3

The length of the lateral line in a vegetables field is 100 meters; the slope of the lateral line is 1% toward underneath of the side. The dripper flow is four lph at an operational pressure head of 10 meters. Design the size of the lateral line with the following data:

Length of the lateral line    = 100 meters
Operational pressure H        = 10 meters
Number of drippers            = 300
Total discharge Q             = 4 x 300 = 1200 lph = 1200/3600= 0.334 lps

### Design procedure:

**Step 1:** Determine L/H = 100/10 = 10

**Step 2:** In Figure 12.8, move horizontally from L/H = 10 toward the line of 1% of slope in the fourth quadrant. From this point, establish a vertical line toward the first quadrant.

**Step 3:** Establish a point along this line in the first quadrant, in the upper margin of the desirable region. From this point, establish a horizontal line toward the second quadrant.

**Step 4:** Establish a vertical line in the second quadrant from L/H = 10, so as to intersect the horizontal line of the step three.

**Step 5:** For this point, determine $\Delta H/L$ = 3.5 in the second quadrant.

**Step 6:** In Figure 12.10, for total discharge Q = 0.334 lps and $\Delta H/L$ = 3.5: Determine the minimum size of the lateral line = 19 mm (or use size greater than 19 mm).

## Design of the Lateral Lines on Non Uniform Slope

## Design example number 4

A lateral line of 0.5 inches (16 mm) diameter and 400 feet (122 m) of length is laid on a non uniform slope. The non uniform slope can be expressed as: 0–100 feet (0–30 m) = 3% of slope downward; 100–200 feet (30–60 m) = 2% of slope downward; 200–300 feet (60–90 m) = 0% of slope downward; 300–400 feet (90–120 m) = 3% of slope downward. The operational pressure head for the drip irrigation is 34 feet (or 10 m) and the total discharge for the lateral line is 2 gpm (0.13 lps). Suppose that the total energy loss by friction is 5 feet (1.5 m). Review the variation in pressure with the following data:

H = 34 ft (10 m).

L = 400 ft (120 m).

$\Delta H$ = 5 ft (1.5 m) calculated by equation (2) in Table 12.1.

### Design procedure:

Determine: L/H = 11.76 for British system or 12 for metric system.

$\Delta H/H$ = 0.147 for British system or 0.15 for metric system.

Non uniform slopes: Plot 1/L vs. $\Delta H'_i$ /L in the first quadrant (Figure 12.22).

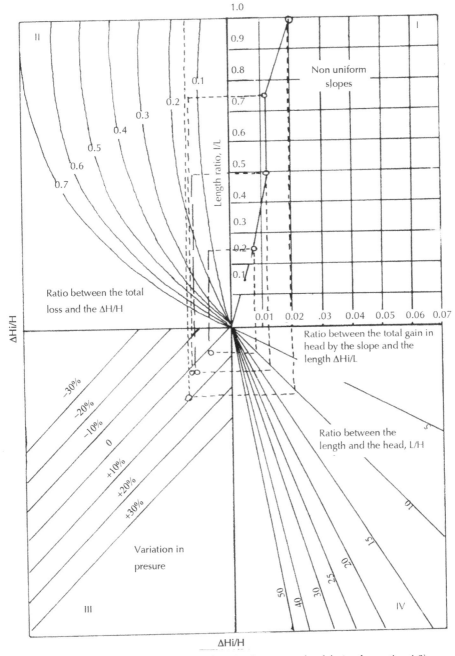

**Figure 12.22.** Design chart for non uniform slope (with an example of design for section 4.2).

| 1/L | H$_i$ (m) | $\Delta$H'$_i$ (m) | $\Delta$H'$_i$ /L |
|------|-----------|---------------------|---------------------|
| 0.25 | 3 | 0.9 | 0.0075 |
| 0.50 | 5 | 1.5 | 0.0125 |
| 0.75 | 5 | 1.5 | 0.0125 |
| 1.00 | 8 | 2.4 | 0.0200 |

Continue the procedure given in the previous section to revise the variation in pressure for four points and for a length ratio of 0.25, 0.50, 0.75, and 1.00, respectively.

The results are given in Figure 12.22. It has been found in third quadrant that the variation in pressure along the lateral lines is less than 10%. The design is acceptable.

## Design of Lateral Line with Different Sizes of Pipes

### Design example number 5

In a papayas field, the length of lateral line is 1000 feet and is laid on a slope 5% downward. The papayas are planted in a zigzag pattern on both sides of the lateral line. The plant spacing is 5 feet. A micro tube of polyethylene is connected to the lateral line. Each spaghetti can deliver a flow of two gph at an operational pressure of 10 psi. Design the size of the lateral line, when different sizes of tubing of the lateral line can be used.

### Design procedure:

Total Discharge, Q = [(1000 x 2)/5] x [2]/60 = 13.33 gpm

1. Slope of the lateral line = 5%
2. If the lateral lines line is divided into 10 sections, the size of the lateral line can be determined from Figure 12.13.

Size of lateral line on a slope of 5% downward (4.3.1):

| Section | Discharge per section (gpm) | Size of lateral line (in) |
|---------|------------------------------|----------------------------|
| 1 | 12.60 | 1.25 |
| 2 | 11.30 | 1.25 |
| 3 | 9.98 | 1.00 |
| 4 | 8.50 | 1.00 |
| 5 | 7.30 | 1.00 |
| 6 | 6.00 | 1.00 |
| 7 | 4.66 | 0.75 |
| 8 | 3.33 | 0.75 |
| 9 | 2.00 | 0.50 |
| 10 | 0.67 | 0.50 |

## Design example number 6

Size of lateral line on a non uniform slope (4.3.2)

Design the size of the lateral line. If the lateral line is laid on a non uniform slope in example 5 (downward the slope) as given below:

0–200 ft = 5%

200–400 ft = 3%

400–600 ft = 1%

600–800 ft = 3%

800–1000 ft = 5%

## The design procedure:

1. Total discharge, Q = 13.33 gpm
2. Slope of the lateral line = non uniform
3. If the lateral lines line is divided into 10 sections, the size of the lateral lines can be determined from Figure 12.13. The results are given below:

| Section | Discharge per section (gpm) | Slope (%) | Size of lateral line (in) |
|---------|------------------------------|-----------|----------------------------|
| 1 | 12.6 | 5 | 1.25 |
| 2 | 10.3 | 5 | 1.25 |
| 3 | 9.9 | 3 | 1.25 |
| 4 | 8.5 | 3 | 1.25 |
| 5 | 7.3 | 1 | 1.25 |
| 6 | 6.0 | 1 | 1.25 |
| 7 | 4.7 | 3 | 1.00 |
| 8 | 3.3 | 3 | 0.75 |
| 9 | 2.0 | 3 | 0.75 |
| 10 | 0.7 | 5 | 0.50 |

## Design of Secondary Line Using an Individual Size

## Design example number 7

The rectangular sugarcane field has an area of one acre. The length of the secondary line is 100 feet and has a slope of zero. The secondary line supplies 15 gpm, at an operational pressure head of 34 feet. Design the size of secondary line.

## Design procedure:

1. Total discharge = 15 gpm
2. Ratio of length of the secondary line and the operational pressure head, L/H = 100/34 =3
3. Slope of the secondary line = 0
4. In Figure 12.7, $\Delta H/L$ is found to be 10% (variation in pressure of 20% for region A).

5. In the Figure 12.9, the size of the line = 1 inch
6. Simplified design chart (Figure 12.15) can also be used.

### Design example number 8

In example 7, the secondary line is on a 5% slope (downward of the slope). Design the size of the secondary line.

### Design procedure:

1. Total discharge Q = 15 gpm
2. Ratio of length of the secondary line and the operational pressure head, L/H = 100/34= 3.3
3. Slope of the secondary line: 5% downward.
4. From the Figure 12.14, the size of the secondary line is approximated as one inch.

### Design example number 9

A secondary line of 10 meters of length is installed in a vegetable field. The spacing between the lateral lines is one meter. These lateral lines are connected to the secondary line. The secondary line is laid on a zero slope. Design the size of secondary line with the following data:

Length of the secondary line, L = 10 m
Operational pressure head, H = 10 m
Total discharge for each lateral line, q = 0.167 lps

### Design procedure:

1. Total discharge, Q = 0.167 x 10 = 1.67 lps
2. Ratio of length of the secondary line and the pressure head, L/H = 10/10 = 1.0
3. Slope of the secondary line = 0
4. From Figure 12.17, the size of the secondary line is approximated as 25 mm.

## Design of Secondary Lines of Different Sizes

### Design example number 10

In a sugarcane field, the secondary line is used for a rectangular subfield of two acres. The secondary line has a length of 300 feet and is laid on a slope 3% (downward the slope). Design the size of secondary line, when the engineer can design for different sizes.

### Design procedure:

a. Total discharge = 40 gpm
b. Slope = 3% uniform (downward)
c. If the line secondary is divided into 10 sections, the size of each section of the secondary line can be determined from Figure 12.13.

| Section | Mean Discharge (gpm) | Size of Secondary Line (inches) |
|---------|----------------------|----------------------------------|
| 1 | 38 | 2 |
| 2 | 34 | 2 |
| 3 | 30 | 2 |
| 4 | 26 | 1.5 |
| 5 | 22 | 1.5 |
| 6 | 18 | 1.5 |
| 7 | 14 | 1.5 |
| 8 | 10 | 1.25 |
| 9 | 6 | 1 |
| 10 | 2 | 3/4 |

## Design of Main Line

### Design example number 11

A drip irrigation system is designed for a papaya field of 50 acres (= 20 hectares). The field is rectangular and is divided into subplots of one acre (0.4 hectares). Each subplot is irrigated by a secondary line. The main line is located in the center of the field with 25 acres on each side. Each subplot is approximately 435 feet in length (130 m) and 100 feet wide (30 m). Each section of main line is 100 feet in length. The design capacity is 30 gpm (2 lps) for each subplot. There are 24 sections in total. At end of each section, there is an exit to supply 60 gpm (4 lps) for irrigating on both sides. The slopes of the main line are shown in Figure 12.18 or 12.19. The pressure of water required for the lateral line is 10 psi (pressure head of 17.5 m). Design the main line.

### Design Procedure:

1. Plot slopes (profiles) of the main line as shown in Figure 12.18 or 12.19.
2. Plot the required pressure (10 psi or 17.5 m) along the main line as shown in Figure 12.18 or 12.19: 23 feet above the ground file.
3. Determine the energy slope. From Figures 12.18 and 12.19, the energy slope is determined as 1%. The size of main line can be determined from Figure 12.20 or 12.21, utilizing 1% of the energy gradient line. The results are shown in Table 12.2.

Table 12.2. Sizes of the main line as determined from Figure 12.20 or 12.21.

| Section | British Units | | Metric Units | |
|---------|-----------------|-------------------|------------------|---------------|
| | Discharge (gpm) | Diameter (inches) | Discharge (lps) | Diameter cm) |
| 0 | 1500 | — | 48 | 25 |
| 1 | 1440 | 10 | 46 | 25 |
| 2 | 1380 | 10 | 44 | 25 |
| 3 | 1320 | 10 | 42 | 20 |

**Table 12.2.** *(Continued)*

| Section | British Units | | Metric Units | |
|---|---|---|---|---|
| | Discharge (gpm) | Diameter (inches) | Discharge (lps) | Diameter cm) |
| 4 | 1260 | 8 | 40 | 20 |
| 5 | 1200 | 8 | 38 | 20 |
| 6 | 1140 | 8 | 36 | 20 |
| 7 | 1080 | 8 | 34 | 20 |
| 8 | 1020 | 8 | 32 | 20 |
| 9 | 960 | 8 | 30 | 20 |
| 10 | 900 | 8 | 28 | 20 |
| 11 | 840 | 8 | 26 | 20 |
| 12 | 780 | 8 | 24 | 20 |
| 13 | 720 | 8 | 22 | 20 |
| 14 | 660 | 8 | 20 | 15 |
| 15 | 600 | 6 | 18 | 15 |
| 16 | 540 | 6 | 16 | 15 |
| 17 | 480 | 6 | 14 | 15 |
| 18 | 420 | 6 | 12 | 15 |
| 19 | 360 | 6 | 10 | 15 |
| 20 | 300 | 5 | 8 | 12.5 |
| 21 | 240 | 5 | 6 | 12.5 |
| 22 | 180 | 4 | 4 | 10 |
| 23 | 120 | 4 | 2 | 10 |
| 24 | 60 | 3 | | |

There is an outlet at the entrance of the section one to irrigate the plots on both sides of section one.

## SUMMARY

This chapter discusses basic concepts of hydraulics needed to develop the design charts for the lateral and secondary lines of a drip irrigation system. It provides a thorough explanation of the chart and design procedures for lateral and secondary lines, on uniform and non uniform slopes. It also explains the design procedure for lateral and secondary lines of different sizes and the design of the main line. These procedures are carefully outlined. To enhance the understanding of the design procedure for each case, the design examples are presented.

**KEYWORDS**

- Blasius formula
- Coefficient of uniformity
- Control system
- Drip irrigation system
- Dripper
- Energy gradient line
- GPM, British units of volumetric flow, gpm
- Hydraulic gradient
- Lateral line
- LPH, Metric units of volumetric flow
- LPS, Metric units of volumetric flow
- Main line
- Orifice
- Pipe or tube
- Polyethylene (PE)
- Secondary line
- Smooth surface
- Spaghetti

**BIBLIOGRAPHY**

Refer to bibliography section 12.1 for the literature cited in this chapter.

# Chapter 13

## Design of Lateral Lines

Victor A. Gillespie, Allan L. Phillips, and I. Pai Wu

### INTRODUCTION

In a drip irrigation system, a major design criteria is a minimization of the discharge (or emitter flow) variation along a drip irrigation line, either a lateral or a submain. The discharge variation can be kept within acceptable limits in laterals or submains of a fixed diameter by designing a proper length for a given operating pressure. The discharge (or emitter flow) variation is controlled by the pressure variation along the

[1]Gillespie, Victor A., Allan L. Phillips, and I. Pai Wu, 1979. Drip irrigation design equations. *Journal of the Irrigation and Drainage Division*, ASCE, 105 (IR3): 247–257. Proc. Paper # 14819.

line which results from the combined effect of friction drop and line slopes. When the kinetic energy is considered to be small and neglected in a drip irrigation line, the pressure variation will be simply a linear combination of the friction drop and energy gain or loss due to slopes as suggested by Wu and Gitlin [6] and Howell and Hiler [2].

A lateral length (or submain) can be designed by using a step by step calculation using a computer. The computer program can be used to simulate different situations to develop design charts as shown by Wu and Fangmeier [5]. Simplified design procedures were developed [3, 7] by using a general shape of the energy gradient line and line slopes. Design charts for lateral line design were introduced by Wu and Gitlin [7], however trial and error techniques are required in the design procedure. The purpose of this chapter is to derive mathematical expressions for lateral lines (or submains) which will simplify design techniques.

The derivations apply to different types of uniform slope conditions, where there is no change in land slope along the length of the emitting line. The derived equations relate design length to the total pressure head, and are in a form which facilitates the use of computerized design methods. The calculations can be done by a digital computer, or these may be solved using a pocket calculator so that the design engineer can do a design right in the field. The graphical solutions can also be developed. Although basically still a trial and error technique, the adaptability of these design equations to computerized solutions should represent a significant advance in drip irrigation system design.

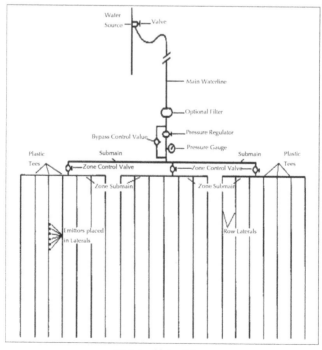

Design of lateral lines

## SOME BASIC EQUATIONS

One commonly used friction drop equation for pipe flow is the Williams–Hazen equation [4].

$$H_f = \frac{K_1 * V^{1.852} * L}{C^{1.852} * D^{1.167}} \qquad (1)$$

where:

K$_1$ = Constant equal to 3.023 for British units and 0.0837 for SI units;
H$_f$ = Friction drop, in feet (meters);
V = Mean velocity, in feet per second (meters per second);
L = The pipe length, feet (meters);
D = Inside diameter, in feet (meters);
C = A roughness coefficient.

Equation (1) can be modified to use a total discharge as follows:

$$H_f = \frac{K_2 * Q^{1.852} * L}{C^{1.852} * D^{4.871}} \qquad (2)$$

where:

K$_2$ = Constant equal to 10.45 for British units and 2.264 × 10$^7$ for SI units;
Q = Expressed as gallons per minute (liters per second);
D = Inside diameter, inches (centimeters).

Equation (2) calculates the friction drop using total discharge. For lateral line or submain, the discharge in the line decrease with respect to the length of the line. The total friction drop at the end of the line can be calculated by applying a correction factor of, [1/2.852], indicated by Wu and Gitlin [8]. The total energy drop due to friction at the end of a lateral line or submain can be expressed as:

$$\Delta H = \frac{K_3 * Q^{1.852} * L}{C^{1.852} * D^{4.871}} \qquad (3)$$

where:

K$_3$ = Constant equal to 3.6642 for British units and 7.94 × 10$^6$ for SI units;
ΔH = The total friction drop at the end of a lateral line or submain, in feet (meters).

Assuming the emitter flow q is uniform or is designed with a certain variation, one can rearrange equation (3) into:

$$\Delta H = \frac{K_3 * q^{1.852} * L^{2.852}}{C^{1.852} * S_p^{1.852} * D^{4.871}} \qquad (4)$$

where: q = Average emitter flow, in gallons per minute (liter per second);
S$_p$ = Emitter spacing, in feet (meter).

In a drip irrigation design, the terms q, C, S$_p$, and D are usually known, therefore:

$$\Delta H = K*L^{2.852}, \text{ where } K = \frac{K_3 * q^{1.852}}{C^{1.852} * S_p^{1.852} * D^{4.871}} = \text{Constant} \tag{5}$$

The total friction drop divided by total length is a dimensionless term S [=$\Delta H/L$]. Equation (5) can be expressed as:

$$S = K*L^{1.852} \tag{6}$$

The total friction drop shown in equation (5) is the total friction drop over the full length of the line. The friction drop along the line can be determined from a dimensionless energy gradient line as derived by Wu and Gitlin [8]. It can be expressed as follows:

$$\Delta Hp = [1 - \{1 - P/L\}^{2.852}]* \Delta H \tag{7}$$

where:

$\Delta H_p$ = The total friction drop at a distance P, from the inlet.

When a lateral line or submain is laid on uniform slopes, the total energy gain (down slope situation) or loss (up slope situation) due to change in elevation can be expressed as:

$$\Delta H' = S_o *L \tag{8}$$

where:

$\Delta H'$ = The total energy gain or loss due to uniform slope at the end of the line, in feet (meters);

$S_o$ = The line slope.

The energy gain or loss at a point along the line due to uniform slopes can be shown as:

$$\Delta H'_p = S_o *p \tag{9}$$
$$\Delta H'_p = P/L * \Delta H' \tag{10}$$

where: $\Delta H'_p$ = The energy gain or loss due to slopes at a length P measured from the inlet;

$S_o$ = The land slope;

$\Delta H'$ = The energy gain or loss due to slope over the total length of the line.

## PRESSURE PROFILES

The pressure head profiles along the lateral or submain can be determined from the inlet pressure, friction drop, and energy change due to slopes. It can be expressed as:

$$H_p = H - \Delta H_p \pm \Delta H'_p \tag{11}$$

where: H = The inlet pressure or operating pressure expressed as pressure head, in feet (meters), the plus (+) sign means down slope and the minus (−) sign means up slope.

Substituting equations (7) and (10) into equation (11), we have:

$$H_p = H - [1 - \{1 - P/L\}^{2.852}] \Delta H \pm P/L * \Delta H' \tag{12}$$

The equation (12) describes pressure profiles along a lateral line or submain. The shape of profiles will depend on the inlet pressure (initial pressure), total friction drip, and total energy change by slopes. There are five typical pressure profiles as shown in Figure 13.1 and these can be explained as follows:

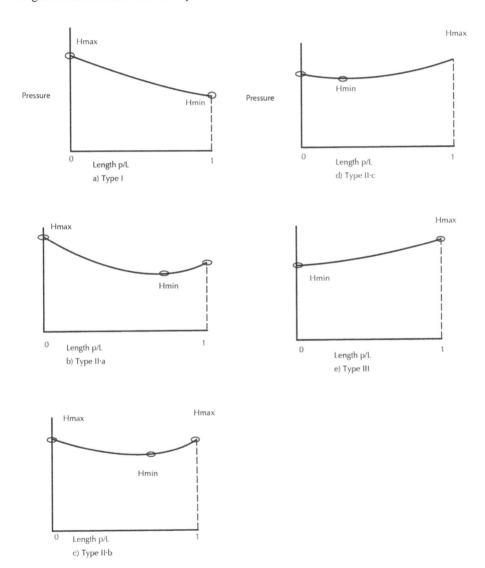

a) Type I

d) Type II·c

b) Type II·a

e) Type III

c) Type II·b

## Profile Type I

This occurs when the lateral line (or submain) is on zero or uphill slope. Energy is lost by both elevation change due to up slope and friction. The pressure decrease with respect to the length of the line and the maximum pressure, $H_{max}$ is at the inlet and minimum pressure, $H_{min}$ is at the downstream end of the line.

## Profile Type II

### Type a
This occurs when the lateral line (or submain) is on down slope situation, where a gain of energy by slopes at downstream points is greater than the energy drop by friction but the pressure at the end of the line is still less than the inlet pressure. The maximum pressure, $H_{max}$ is at the inlet and a minimum pressure is located somewhere along the line.

### Type b
This is similar to Type II-a but the profile is such that the end pressure is equal to the inlet pressure. The maximum pressure, $H_{max}$ is at the inlet and the end of the line. The minimum pressure, $H_{min}$ is located somewhere near the middle section of the line.

### Type c
This occurs when the line slope is even steeper so the pressure at the end of line is higher than the inlet pressure. In this condition, the maximum pressure, $H_{max}$ is at the downstream end of the line and the minimum pressure is located somewhere along the line.

## Profile Type III
This occurs when the lateral line (or submain) is on steep down slope conditions where the energy gain by slopes is larger than the friction drop for all sections along the line. In this condition, the maximum pressure is at the downstream end of the line and minimum pressure is at the inlet. The location of the minimum pressure along the pressure profile II-a-b-c, can be determined by differentiating equation (12) with respect to the length P and setting the derivative equal to zero.

$$\{2.852*(1 - P/L)^{1.852} * \Delta H/L\} - \Delta H'/L = 0 \qquad (13)$$

If the term $\Delta H/L$, the ratio of total friction drop to length, is set as energy slope S, equation (13) becomes:

$$\{2.852*(1 - P/L)^{1.852} * S\} - S_o = 0 \qquad (14)$$

Simplifying:

$$P/L = 1 - [0.3506 \, S_o/S]^{0.54} \qquad (15)$$

Equation (15) shows the location of the point of minimum pressure when both $S_o$ and S are known.

## DESIGN EQUATIONS
Since the five pressure profiles are smooth curves as shown in Figure 13.1, pressure variation can be used as design criteria. The pressure variation is defined as:

$$H_{var} = \frac{H_{max} - H_{min}}{H_{max}} \qquad (16)$$

where: $H_{var}$ = Pressure variation. The maximum and minimum pressure along the line are $H_{max}$ and $H_{min}$, respectively.

The emitter flow variation can also be expressed as follows:

$$q_{var} = \frac{q_{max} - q_{min}}{q_{max}} \tag{17}$$

where: $q_{max}$ = Maximum emitter flow along the emitting line;

$q_{min}$ = Minimum emitter flows along the emitting line produced by $H_{max}$ and $H_{min}$.

For the orifice type of emitter flow, the relationship between $q_{var}$ and $H_{var}$ is given by equation (18):

$$H_{var} = 1 - [1 - q_{var}]^2 \tag{18}$$

Using pressure variation as design criteria, the design equations for each pressure profile type are derived as follows:

## Profile Type I

The inlet pressure is the maximum pressure and the minimum pressure is at the end of the line:

$$H_{min} = H - (\Delta H + \Delta H') \tag{19}$$

The pressure variation can be expressed as:

$$H_{var} = \frac{H - [H - (\Delta H + \Delta H')]}{H} \tag{20}$$

$$H_{var} = \frac{\Delta H + \Delta H'}{H} \tag{21}$$

Both sides of equation (21) can be multiplied by H/L, to obtain:

$$\frac{H_{var}}{L} * H = \frac{\Delta H}{L} + \frac{\Delta H'}{L} \tag{22}$$

$$H_{var} * H = (S + S_o)L \tag{23}$$

$$L = \frac{H_{var} * H}{S + S_o} \tag{24}$$

The values of $H_{var}$ and H are selected by the designer; $S_o$ can usually be obtained from field measurements; S and L are unknown. If S is as a function of L [equation (6)], it is possible to substitute equation (6) for S and derive a computational form of equation (24) that contains only one unknown variable, L. It is the same equation as given by Howell and Hiler [2].

$$L = \frac{H_{var} * H}{K * L^{1.852} + S_o} \tag{25}$$

## Profile Type II

### Type a

The inlet pressure is the maximum pressure and the minimum pressure is somewhere along the line. The line slope is downhill and there is energy gain due to slope. The pressure variation can be expressed as:

$$H_{var} = \frac{H - [H + (\Delta H'_p + \Delta H_p)]}{H} \tag{26}$$

$$H_{var} = \frac{\Delta H_p - \Delta H'_p}{H} \tag{27}$$

$$\frac{H_{var} * H}{L} = \frac{\Delta H_p}{L} - \frac{\Delta H'_p}{L} \tag{28}$$

Substituting equations (7) and (10) into (28) and simplifying, we obtain:

$$L = \frac{H_{var} * H}{[1 - \{1 - P/L\}^{2.852}] * S - \frac{P * S_0}{L}} \tag{29}$$

Substituting equation (15) into (29) and simplifying, we obtain:

$$L = \frac{H_{var} * H}{S + S_o[0.3687 * (S_o / S)^{0.54} - 1]} \tag{30}$$

The computerized form of equation (30) is obtained by substituting the equation (6) into (30):

$$L = \frac{H_{var} * H}{K * L^{1.852} + S_0[0.3687 * (S_o / K * L^{1.852})^{0.54} - 1]} \tag{31}$$

### Type b

This is similar to Type II-a, the only difference is that S and $S_o$ are equal as defined by equations (6) and (8). It is therefore possible to substitute $S_o$ for S in both equations (30) and (31). Equation (30) can be shown as:

$$L = \frac{H_{var} * H}{S + S_o[0.3687 * (S_o / S)^{0.54} - 1]} \tag{32}$$

Simplifying:

$$L = \frac{H_{var} * H}{0.3687 * S_o} \tag{33}$$

### Type c

The maximum pressure is located at the downstream end of line and the minimum pressure is somewhere along the line. The pressure variation can be expressed as:

$$H_{var} = \frac{[H + (\Delta H' - \Delta H)] - [H + (\Delta H_p - \Delta H'_p)]}{H + (\Delta H' - \Delta H)} \quad (34)$$

$$H_{var} = \frac{\Delta H' - \Delta H + \Delta H_p - \Delta H'_p)]}{H + (\Delta H' - \Delta H)} \quad (35)$$

$$H_{var} H = [\Delta H'/L - \Delta H/L] + \Delta H_p/L - \Delta H'_p/L] - H_{var}/L*[\Delta H'/L - \Delta H/L] \quad (36)$$

Substituting equations (7), (10), (15) and simplifying, we obtain:

$$L = \frac{H_{var} * H}{S_o[0.3687(S_o/S)^{0.54}] - [(S_o - S) * H_{var}]} \quad (37)$$

$$L = \frac{H_{var} * H}{S_o[0.3687 * (S_o/KL^{1.852})^{0.54}] - H_{var} * [S_o - KL^{1.852}]} \quad (38)$$

## Profile Type III

The derivation of the Type III profile is simpler than the other down slope situations because there is no minimum point along the pressure profile. The value $H_{min}$ is H at the head of the emitting line, and $H_{max}$ is at the end of the emitting line. The pressure variation can be expressed as:

$$H_{var} = \frac{H + (\Delta H' - \Delta H) - H}{H + (\Delta H' - \Delta H)} \quad (39)$$

$$H_{var} H + H_{var} * (\Delta H' - \Delta H) = (\Delta H' - \Delta H) \quad (40)$$

$$\frac{H_{var} * H}{L} = (S_o - S) - H_{var} * (S_o - S) \quad (41)$$

$$L = \frac{H_{var} H}{(S_o - S) * (1 - H_{var})} \quad (42)$$

The design length can be expressed as:

$$L = \frac{H_{var} H}{(S_o - KL^{1.852}) * (1 - H_{var})} \quad (43)$$

## CRITERIA FOR THE SELECTION OF THE APPROPRIATE DESIGN EQUATION

The criteria for selecting which of the five design equations to use for a given land slope and flow situations is dependent on the relationship between S and $S_o$. The criteria for the Type I profile is simplest, equation (25) is used when there is zero slope or for uphill slopes. The criteria for choosing which of the four down slope design equations to use are based on the magnitude of S and $S_o$ and on equation (15). The Type II-a profile is characterized by S being greater than $S_o$:

$$S > S_o; S/S_o > 1; \frac{KL^{1.852}}{S_o} > 1 \quad (44)$$

The profile Type II-b is characterized because S is equal to $S_o$:

$$S = S_o; S/S_o = 1; \frac{KL^{1.852}}{S_o} = 1 \qquad (45)$$

The profiles II-c and Type III are characterized, because S is smaller than $S_o$:

$$S < S_o; S/S_o < 1; \frac{KL^{1.852}}{S_o} < 1 \qquad (46)$$

If the land slope and flow conditions satisfy this inequality, it is possible to use equation (15) to determine which design equation to use for the Type II-c pressure profile. The minimum point occurs at P/L greater than zero and less than one. This occurs if the following inequality holds true:

$$0 < \left(0.3506\frac{S_o}{S}\right)^{0.54} < 1 \qquad (47)$$

$$0 < \left(0.3506\frac{S_0}{S}\right) < 1 \qquad (48)$$

$$0 < \frac{S_o}{S} < 2.852 \qquad (49)$$

$$0 < \frac{S_o}{KL^{1.852}} < 2.852 \qquad (50)$$

Conversely, if the land slope and flow situations is such that $S < S_o$, but $S_o/S \geq 2.852$, then it is appropriate to use the Type III profile design equation. When the design procedure begins, S is usually unknown, and it is not possible to solve the down slope criteria equations until L is either known or calculated. If S is not known, it is necessary to intuitively select one of the four down slope design equations and to solve it for a value of S. This value can then be tested using the appropriate criteria equation. If the criteria is not satisfied, it is necessary to solve for another value of S using another of the downhill design equations. In most down slope situations, it is appropriate to first use the Type II-a profile design equation. Generally the designer will try to design a lateral line to have a gentle down slope condition which usually will result in a Type II-a, b, or c profile.

## DESIGN EXAMPLES

In the developed design equations, the design length cannot be solved directly. One can use a calculator and use a trial and error method to determine the length, or use Newton's method of approximation iteratively to determine the length using a computer program. Two design examples are shown as follows:

### Example 1
A lateral line is on a 1% uphill slope. The following data are given, and it is necessary to determine the maximum L for the land slope and flow conditions using Type I profile

design equation: Emitter spacing, $S_p$ = 2.0ft (0.61m); Emitter diameter, d = 0.010 in (0.254 mm); Design emitter flow, q = 0.0047 gal/min (2.1 x $10^{-5}$ L/s); Lateral line diameter, D = 0.56 in (14.2 mm); Inlet pressure, H = 10.4 ft (3.17 m); Pressure variation magnitude, $H_{var}$ = 0.19; Land slope uphill, $S_o$ = 0.010; and Roughness coefficient, C = 150. The equation (5) is written as:

$$K = \frac{7.94x10^6 q^{1.852}}{C^{1.852} * S_p^{1.852} * D^{4.871}}$$
(51a)

$$K = \frac{7.94x10^6 (2.1x10^{-5})^{1.852}}{(150)^{1.852} * (0.61)^{1.852} * (1.42)^{4.871}}$$
(51b)

$$K = 7.82 \times 10^{-8}$$
(51c)

From the equation (25):

$$L = \frac{Hvar * H}{K*(L^{1.852})+So} = \frac{(0.19)(3.17)}{(7.82x10^{-8})(L^{1.852})+(0.010)}$$
(52a)

$$L = 178 \text{ feet (54.27 meters)}$$
(52b)

We can also obtain a graphical solution solving H for various L and tracing a graph to determine "L." The particular type of line can extend approximately to 180 feet (55 m) for an up hill slope of 1%, before $H_{var}$ will exceed 19% that corresponds to a $q_{var}$ of 10%.

## Example 2

A lateral line is on a 1.5% downhill slope. The first design equation used is that for the Type II-a profile. Emitter spacing, $S_p$ = 8 ft (2.4 m); Emitter diameter, d = 0.019 in (0.48 mm); Design emitter flow, q = 0.026 gal/min (1.14 × $10^{-4}$ L/s); Emitting line diameter, D = 0.612 in (15.5 mm); inlet pressure, H = 28.37 ft (8.65 m); Pressure variation magnitude, $H_{var}$ = 0.19; Land slope downhill, $S_o$ = 0.015; and Roughness coefficient, C = 137. Using these values, we obtain:

$$K = 7.98 \times 10^6 (1.14 \times 10^{-14})^{1.852}$$
(53a)

$$K = 9.93 \times 10^{-7}$$
(53b)

From equation (31):

$$L = 9.93 \times 10^{-7} L^{1.852} + 0.015[0.3687(0.015/9.93 \times 10^{-7} L^{1.852})^{0.54} -1]$$
(54a)

$$L = 659 \text{ ft (201 meters)}$$
(54b)

The solution can also be obtained, graphing L and H as shown in Example 1. The answer need to be verified to determine, if the right equation was used.

$$\frac{S}{S_o} = \frac{9.93x10^{-7}(201)^{1.852}}{0.015} = 1.22$$
(55)

## SUMMARY

Five pressure profiles are presented and considered. These represent design conditions which result from a lateral line (or submain) laid on uniform slopes. Procedures are developed to identify pressure profiles by land slope and total friction drop at the end of the line. Equations for designing lateral length (or submain) based on a given criteria, pressure variation, are derived. These equations cannot be solved directly, but solutions can be obtained using trial and error technique on a pocket calculator or by using Newton's method of approximation in a computer program. The developed mathematical equations can be useful in the future development of computerized drip irrigation system design.

## KEYWORDS

- **Discharge variation**
- **Friction drop equation**
- **Pressure head profiles**

## BIBLIOGRAPHY

Refers to bibliography Section 13.1 for the literature cited.

## LIST OF SYMBOLS

The following symbols are used in this chapter:

| | |
|---|---|
| C | Roughness coefficient. |
| D | Diameter of lateral or sub principal line, in feet (meters). |
| D | Diameter of emitters, in inches (millimeters). |
| H | Total pressure head or operating pressure head, in feet (meters). |
| $H_f$ | Frictional head loss, feet (meters). |
| $H_p$ | Pressure charge at distance p, from the initial point (inlet), in feet (meters). |
| $H_{max}$ | Maximum pressure head, in feet (meters). |
| $H_{min}$ | Minimum pressure head, in feet (meters). |
| $H_{var}$ | Pressure head variation, in feet (meters). |
| K | Coefficient (a constant value). |
| $K_1,K_2,K_3$ | Coefficients for system of units. |
| L | Total length of the lateral or submain line, in feet (meters). |
| P | Length of the lateral line measured from the header (inlet), in feet (meters). |
| Q | Total discharge at the inlet section of the lateral or sub principal line, in gallons per minute (liters per second). |
| $Q_e$ | Discharge of one emitter, in gallons per minute (liters per second). |
| $q_{max}$ | Maximum flow of an emitter, in gallons per minute (liters per second). |
| $q_{min}$ | Minimum flow of an emitter, in gallons per minute (liters per second). |
| $q_{var}$ | Variation in flow of an emitter, in gallons per minute (liters per second). |
| S | Slope of energy line defined as $\Delta H/L$. |
| $S_o$ | Slope of the lateral or sub principal line. |

| $S_p$ | Distance between the emitters, in feet (meters). |
|---|---|
| V | Average velocity, in feet per second (meters per second). |
| $\Delta H$ | Total frictional head loss at a side of the lateral or sub principal line, in feet (meters). |
| $\Delta H'$ | Gain or total loss of energy head due to uniform slope at a side of the lateral or sub principal line, in feet (meters). |
| $\Delta H_p$ | Total friction energy head at a distance "p" from the initial point, in feet (meters). |
| $\Delta H'_p$ | Total gain of energy head due to uniform slope at a distance "p" from the initial point, in feet (meters). |

# Chapter 14

## Evaluation of the Uniformity Coefficients

Vincent F. Bralts

Associate Dean, College of Engineering, Purdue State University

### INTRODUCTION

The uniformity of irrigation application is one of the most difficult factors to evaluate. The initial cost, the operational cost, and the plant response are related with the uniformity of water application. A considerable effort has been directed to this problem in the design and management of the irrigation system. This chapter presents a simple method to evaluate uniformity of water application in drip irrigation. This method does not require mathematical equations and sophisticated equipments.

In an irrigation system, there is a direct relationship between the uniformity of water application and the initial cost. The pressure decreases as the water flows through the pipe lines due to loss by friction. This results in reduction of water application rates at the farthest sections of the irrigation system. The water is distributed more uniformly and the loss by friction is reduced if the pipe lines have a large diameter. Since

[1]This chapter is modified and translated from "Goyal, Megh R., y Luis E. Rivera, 1990. Evaluación de la uniformidad de riego por goteo. Capitulo XV en: *Manejo de riego por goteo* editado por Megh R. Goyal, paginas 415–431. Río piedras, PR: Servicio de Extensión Agrícola, UPRM." For more details, one may contact by E-mail: goyalmegh@gmail.com

the pipe lines of larger diameter are more expensive, therefore a system with high uniformity is more expensive compared to poorly designed system with low uniformity. Before purchasing the system, the buyer should evaluate the cost of the system, its capabilities, and uniformity.

The operational cost of an irrigation system are directly associated with uniformity of water application. In many cases the water is applied uniformly when the system is operated at high pressures. This practice goes against one of the advantages of drip irrigation in relation to savings in energy consumption. The operational cost of a system with smaller diameter pipes is higher than an appropriate design. This is due to the fact that the small size pipes need more time to apply the desired quantity of water at the farthest end of the lateral.

The crop production efficiency is also related with the uniformity. In general, it is difficult to determine the loss in efficiency by low uniformity, because the efficiency is affected by many factors. In general, the efficiency losses are due to the fact that some plants do not receive the adequate amount of water while others receive in excess. Excessive applications of water may wash away the nutrients that are accessible to the plants.

To fulfill the objectives of the drip irrigation, the system must be designed to apply the water uniformly within the economical limits. This way, each plant in the field will receive the same amount of water. This facilitates the operator to adjust the quantity of water applications according to the crop requirements.

## FACTORS THAT REDUCE THE UNIFORMITY OF WATER APPLICATION

The following factors can interact to reduce the uniformity of water application:
1. Defective irrigation pump.
2. Broken or twisted distribution lines.
3. Obstruction of drippers and/or filters by the physical, biological, and chemical agents.
4. Corrosion of some parts in the irrigation system.
5. Obstructed or defective valves.
6. Inadequate design.

## PROCEDURE FOR THE EVALUATION OF UNIFORMITY

A simple procedure was developed to evaluate the uniformity of water application in a drip irrigation system. This procedure can be used by farmers, designers, and sales persons:
1. This method can be used by a potential customer to evaluate a system before acquiring it. In addition system can be evaluated to determinate if it complies with minimum requirements, before the final payment is made to the seller.
2. This method can be used by a designer or seller to determine if the system was designed and installed properly. The system components can also be evaluated.
3. The irrigation operator can use it to detect variations in the uniformity of water application. The operator can also detect problems due to the obstructed drippers

and filters. The lack of uniformity of the water application due to changes in hydraulic characteristics of the drippers and other components of the system can be detected. Therefore, the defective parts can be repaired and replaced.

## DEFINITION OF UNIFORMITY

The uniformity (U) of water application is defined in statistics by the following equation:

$$U = 100 \times [1.0-V] \tag{1}$$

where, U = Uniformity or the emitter discharge rate, intervals between 0 and 100%,

V= Coefficient of variation

The coefficient of variation (V) is a variation in flow of each dripper compared to average flow rate of all drippers. The uniformity is expressed in relative terms so that it does not depend on the magnitude of flow of drippers. Instead, it depends on the variation between the flow of an individual dripper and average flow.

A uniformity of 100% in equation (1) corresponds to a coefficient of variation of zero. This indicates a perfect uniformity, therefore there is no variation in the flow among the drippers. Uniformities of 100%, 90%, 80%, 70%, and 60% corresponds to coefficient of variation of 0.0, 0.1, 0.2, 0.3, and 0.4. This uniformity can be classified as:

| Classification | Statistical Uniformity | Emission Uniformity |
| --- | --- | --- |
| Excellent | For U = 100–95% | 100–94% |
| Good | For U = 90–85% | 87–81% |
| Fair | For U = 80–75% | 75–68% |
| Poor | For U = 70–65% | 62–56% |
| Not Acceptable | For U < 60% | <50% |

## STANDARD FOR UNIFORMITY OF WATER APPLICATION

The American Society of Agricultural and Biological Engineers (ASABE) has developed a standard for the uniformity of water application in drip irrigation. This standard establishes minimum acceptable uniformity for the design of a drip irrigation system. The standards for uniformity are presented in Figures 14.1 and 14.2. The intervals in Figures 14.1 and 14.2 represent the efficient economical values of uniformity. Table 14.1 shows acceptable intervals. Design for a uniformity level less than the design value will result in a reduction in the irrigation efficiency; and cause loss of water and fertilizer due to poor uniformity of water application. Design based on high values of uniformity will increase the initial cost.

An irregular topography of the land affects the design and uniformity of water application. It will result in a high cost of the system. In soils with an irregular topography, allow a smaller uniformity to compensate for the initial and operational costs of the system.

The planting distance for a crop also affects the desired uniformity of the water application. The uniformity should be higher in crops with larger planting distance when one or two drippers per plant are used to apply water to a plant. In narrow planting and narrow dripper spacing, the variation in the flow per dripper reduces.

Each plant may have two or more drippers and this way the effects of random variation on the dripper emission rate are much less. However, this does not eliminate the effects of continuous reduction along the laterals due to loss by friction. Therefore, allowable reductions in uniformity for crops with narrow planting distance are smaller (10%) compared to the crops with wider planting distance.

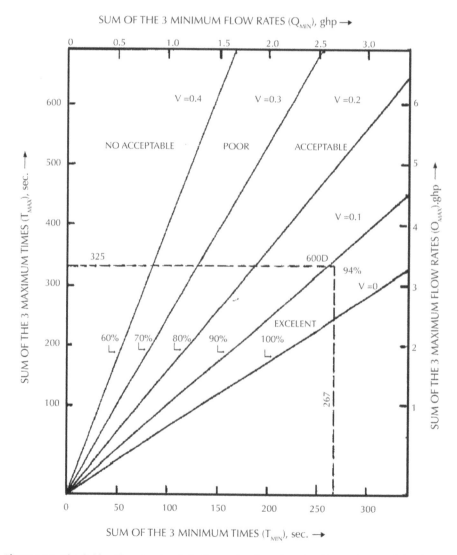

**Figure 14.1.** The field uniformity of an irrigation system based on the dripper times and the dripper flow rate, with an example in this chapter.

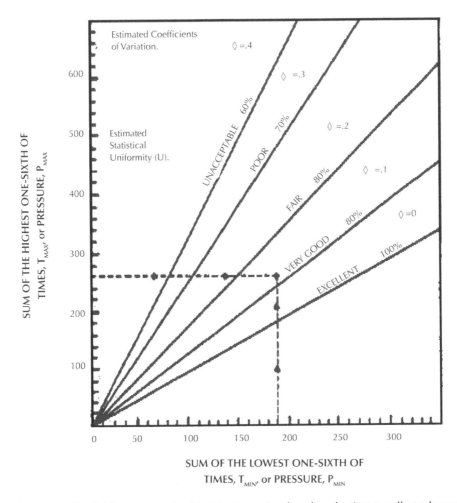

**Figure 14.2.** The field uniformity of a drip irrigation system based on the time to collect a known quantity of water or based on pressure for hydraulic uniformity.

**Table 14.1.** Acceptable intervals of uniformity in a drip irrigation system.

| Type of dripper | Slope | Uniformity interval, % |
|---|---|---|
| Point Source: located in planting distance > 3.9 m. | Level* | 90–95 |
| | Inclined** | 85–90 |
| Point Source: located in planting distance < 3.9 m. | Level* | 85–90 |
| | Inclined** | 80–90 |
| Drippers inserted in the lines for annual row crops. | Level* | 80–90 |
| | Inclined** | 75–85 |

* Level = Slope less that 2%.
** Inclined = Slope greater than 2%.

## EVALUATION PROCEDURE

The evaluation procedure utilizes a known size container to determine the uniformity of water application in drip irrigation. The time required to fill the container is used to calculate the flow rate and the uniformity. This required time can be measured by a stop watch.

One must take at least 18 samples (one sample per dripper) by recording the time to fill the container. One can take more than 18 samples if it is necessary. The selected drippers must be a representative of the area [some at the beginning, others in the middle and others at the end of the lateral line]. Some samples should also be taken at highest elevations of the field and at lowest elevations. The representative drippers should be recorded along with location for analysis. The sum of the three highest observations is denominated as maximum time ($T_{max}$). The sum of the three lowest observations is denominated as minimum time ($T_{min}$). $T_{max}$ and $T_{min}$ are used to determine the uniformity of water application using Figure 14.1. Example: In Table 14.2, the three highest observations are 107, 110, and 108 seconds. The sum of these three observations = 325 seconds = $T_{max}$. The three lowest observations are 89, 87, and 91 seconds. The sum of these lowest observations = 267 seconds = $T_{min}$. The vertical line, T max and the horizontal line for $T_{min}$ intersects at a point to the uniformity using Figure 14.1. The interpolation between the uniformity lines for this particular point gives us a uniformity of 94%. This is interpreted as an excellent uniformity. If the data are representative of the field, then it can be concluded that the system is well designed and well constructed.

**Table 14.2.** Time required to fill the container in a given field, [example of the field data.

| | | |
|---|---|---|
| 89 sec. (smaller) | 97 sec. | 110 sec. (higher) |
| 104 sec. | 107 sec.(higher) | 93 sec. |
| 92 sec. | 100 sec. | 103 sec. |
| 96 sec. | 94 sec. | 108 sec. (higher) |
| 100 sec. | 98 sec. | 91 sec. (smaller) |
| 99 sec. | 102 sec. | 87 sec. (smaller) |

T max. = 107 + 110 + 108 = 325 seconds

T min. = 89 + 91 + 87 = 267 seconds

## CONFIDENCE INTERVALS

The example in section 6.0 is only for 18 drippers in a larger area. Therefore, the results may not be entirely accurate. The only way to determine the exact uniformity is to measure the flow of each dripper in the field. From statistical point of view, it can be proved that the values in Figures 14.1 and 14.2 are precise. The Table 14.3 gives us confidence limit of 95% for the uniformity values presented in Figure 14.1. Interpolations in Table 14.3 indicate that the confidence limit for a uniformity coefficient of 94% is U ± 1.7% for 18 measurements in the field. If we get a uniformity of 94% from Figure 14.1, the true uniformity of the field is 94 ± 1.7%. Therefore uniformity varies from 92.3 to 95.7%. This interval of 95% of confidence indicates that the value of the field uniformity should be within a confidence interval (92.3% a 95.7%), 95 times out of 100, if the sampling procedure is repeated. In Table 14.3, the confidence limit increases as the uniformity decreases. It implies that at low uniformity, the results are less accurate.

For example if the field uniformity was only 60%, the Table 14.3 shows a confidence limit of 60% ± 13.3%.

**Table 14.3.** Confidence limits for field uniformity (U).

| Field uniformity | 18 drippers Confidence limit | | 36 drippers Confidence limit | | 72 drippers Confidence limit | |
|---|---|---|---|---|---|---|
| | N Sum* | % | N Sum | % | N Sum | % |
| 100% | 3 | U ± 0.0 | 6 | U ± 0.6% | 12 | U ± 0.0% |
| 90% | 3 | U ± 2.9 | 6 | U ± 2.0% | 12 | U ± 1.4% |
| 80% | 3 | U ± 5.8 | 6 | U ± 4.0% | 12 | U ± 2.8% |
| 70% | 3 | U ± 9.4 | 6 | U ± 6.5% | 12 | U ± 4.5% |
| 60% | 3 | U ± 13.3 | 6 | U ± 9.2% | 12 | U ± 6.5% |

*N Sum = 1/6 part of the total measured drippers. This is a number of samples that will be added to calculate $T_{max}$ and $T_{min}$.
A level of 95% shows that the field uniformity should be between the confidence interval of the exact uniformity of the field, 95 times out of 100, if the procedure is repeated.

Therefore the true field uniformity has a confidence interval of 46.7–73.3%. There is wider range, because we took only 18 drippers in the whole field. The variability between the drippers is larger as shown by low uniformity. With a random selection of the drippers, there is a higher chance for a representative data.

In Table 14.3, the confidence limits are given for trials of 18, 36, and 72 drippers. The certainty for the results can be increased if more samples are taken. This way confidence interval can be reduced.

## FLOW RATE MEASUREMENTS

The time to fill the container will be infinite if the dripper is obstructed completely. Therefore, one cannot use Figure 14.1 directly. In this case, we shall add the three highest flow rates and the three lowest flow rates. Now we shall use flow rate units in Figure 14.1 to calculate the field uniformity.

Now the flow rate measurements will be more difficult to take and to calculate. It requires the use of a calibrated container to measure the volume of water in a given period of time. Then the flow rate is calculated.

## SUMMARY: THE PROCEDURE FOR FIELD EVALUATION

In this chapter, the procedure to evaluate the uniformity coefficient for a trickle irrigation system is presented. The uniformity of water application is affected by the degree of clogging, accuracy of the design, and periodic maintenance of the system. Nomograph for the determination of uniformity is presented. The procedure involves taking water samples in a known time from the representative drippers. The three highest and lowest values are summed to give $T_{max}$ and $T_{min}$ The evaluation procedure is summarized below:

1. Allow the system to operate at design operational pressure for enough time to remove all the air from the lines.
2. Measure the required time to fill up the containers in each of the 18 drippers. Be sure that the drippers represent all parts of the field.
3. Calculate $T_{max}$ adding the three highest times, or 1/6 of the total number of the required drippers to fill the container.
4. Calculate $T_{min}$ using the three lowest times, or 1/6 of the total number of the required drippers to fill the container.
5. Using Figure 14.1, determine the field uniformity for a point where lines of $T_{max}$ and $T_{min}$ intersect. Interpolate to calculate the uniformity if it is necessary.
6. If the uniformity of the field is lower or the confidence limit is higher, it is convenient take more data or repeats the procedure to make sure that the system is not poorly designed.

This method is particularly advantageous for use in the field due to a limited number of required data and the simplicity of the procedure. To facilitate the compilation of data, one may use the data sheet in section 12.0. Nomograph for statistical uniformity is shown in section 13.0.

## TROUBLE SHOOTING

| Causes | Remedies |
|---|---|
| **Uniformity 60%** | |
| 1. Few samples were taken. | Take more than 18 drippers as a sample. |
| 2. Clogging in the filters, lines or drippers. | Clean the filters flush the lines and drippers with acid. |
| | Clean or replace the clogged drippers. |
| **Considerable difference in T max and T min.** | |
| 3. It is possible that the drippers lines are obstruct or broken. | Repair the broken lines and clean the obstructed lines. |
| **Loss of pressure due to excess friction in the lines.** | |
| 4. Pipes are of small diameter than the design values. | Revise the design and use correct size of pipes. |

## KEYWORDS

- **Coefficient of variation**
- **Irregular topography**
- **Planting distance**
- **Uniformity**

## BIBLIOGRAPHY

Refer to bibliography Section 14.1 for the literature cited.

## EVALUATION OF UNIFORMITY: DATA SHEET

Name of evaluator: _____ Date of evaluation: Month _____Day _____
Year_____
Name of farmer: _____ Direction: _____

### Description of the trickle system:

High pressure / low pressure
Size of pump _____KW or _____ HP
Size of the farm _____ acres
Size of the filter mesh _____
Area of block where sample are taken _____ acres

### Procedure:

1. Turn on the system to eliminate the air from the lines.
2. Measure the time (seconds) to fill the container in each of the 18 drippers.
3. Calculate the maximum time, sum three highest times.
4. Calculate the minimum time, sum three lowest times.
5. Determine the field uniformity (Figure 14.1).
6. If the field uniformity is low or the confidence interval is high, take more samples.

| Dripper # | Time (Seconds) | Dripper # | Time (Seconds) |
|---|---|---|---|
| 1. | | 10. | |
| 2. | | 11. | |
| 3. | | 12. | |
| 4. | | 13. | |
| 5. | | 14. | |
| 6. | | 15. | |
| 7. | | 16. | |
| 8. | | 17. | |
| 9. | | 18. | |

Sum of the three highest times = _____ + _____ + _____ = _____ T max.

Sum of the three lowest times = _____ + _____ + _____ = _____ T min.

Using Figure 14.1, Uniformity = _____ % [where $T_{max}$ and $T_{min}$ intersect]

**Observations and recommendations:**

_____
_____
_____
_____
_____
_____
_____

## NOMOGRAPH FOR STATISTICAL UNIFORMITY

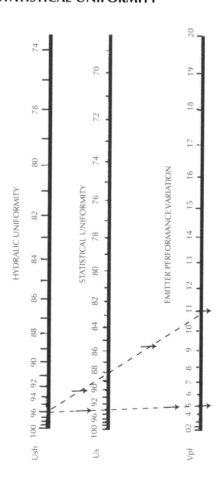

# Chapter 15

## Use of Soil Mulch

### INTRODUCTION

The use of soil mulch is one of the traditional techniques that have been used for centuries in the intensive production of crops. The practice consists in placing mulch over the soil surface to create a favorable microclimate among soil-water-plant. The mulch has a significant effect on soil climate and the microclimate surrounding the plant. It affects the thermodynamic environment, the moisture, the erosion, the physical soil structure, the incidence of pests and diseases, the crop yield, and the crop growth.

Any type of material can be used as mulch (soil cover). In practice, however some are more appropriate than others for agricultural use. In the past, natural and synthetic mulches have been used and evaluated, such as: Paper products, glass fabrics, bituminous emulsions, metal fabrics of aluminum and polyethylene laminates, crop organic waste, and so forth.

The selection of the mulch material depends mainly on the availability, the cost, the efficiency of use, and the purpose of use. For this reason, the soil plastic mulch (polyethylene plastic) is a commonly used soil cover in the intensive production of vegetables, throughout the world.

### ADVANTAGES OF THE USE OF SOIL MULCH

1. The use of soil mulch in combination with drip irrigation technology helps to increase significantly the commercial yield of various crops. It improves the crop quality.
2. The silver coated plastic mulch increases the photosynthesis of some crops. Therefore, the growth is accelerated causing an early flowering and fruit formation.

3. The mulches of opaque color reduce the germination and growth of weeds. This reduces the competition with the crop and the labor requirement for inter-cultivation.

4. The direct water loss due to evaporation is greatly reduced, which then contributes to a uniform soil moisture in the root zone.

5. It protects the soil from the direct impact of rain drops, contributing to the conservation of soil structure. It also reduces erosion.

6. It has an effect on the soil climate and the microclimate surrounding the plant, thus affecting thermodynamic environment.

7. The use of bright or reflecting mulches reduce the incidence of certain insects.

8. Soil mulch serves as a barrier to certain soil pathogens. It helps to maintain the fruits free of dirt, thus requiring less attention in the classification, packing, and processing phase.

9. The use of soil mulch is a complement to the drip irrigation system. This simplifies the localization and management of soil fertilizers, soil fumigants, and reduces the losses due to leaching to deeper layers.

Installation of plastic mulch.

Laying of plastic mulch for vegetable production.

## DISADVANTAGES OF THE USE OF SOIL MULCH

The soil mulch, as many other agricultural techniques, cannot adapt to all crops, places, and specific targets. The major disadvantages are as follows:

1. Expensive: Its use is recommended in high value and rentable crops that are adaptable to the mechanization.
2. Most of the plastics mulches do not decompose easily. Therefore these must be removed out of the field at the end of crop season. However, biodegradable mulches can solve this problem.
3. Installation and removal of mulch increase the labor requirements.
4. It complicates the banded application of solid fertilizers. The fertilizer must be applied before the installation of mulch. However, the soluble fertilizers can be applied through the irrigation water.

Despite many disadvantages, there are a series of conditions that justify the use of soil mulch:

1. The lack of water in arid and semiarid regions around the world requires intelligent use of irrigation water by increasing the application efficiency.
2. The necessity of increasing demand for food requires efficient methods of increasing the crop yield with limited resources.

3. There is a great demand for high quality products, which requires that the crop does not come in contact with soil.

4. The high cost of fertilizers requires an effective use.

5. The necessity of reducing the incidence of insects. Less use of pesticides to control pests and diseases which reduce the contamination of our environment.

## RESEARCH ADVANCES IN THE USE OF SOIL MULCH

The Agricultural Experimental Stations throughout the world have evaluated the effects and viability of the soil mulch on the crop production under limited natural resources. Most of this technology has been directly transferred and adapted from one place to another around the world. The micro irrigation system has been adapted in conjunction with soil mulch. However, many of these research findings and experiences in these places are not absolutely applicable to all situations and needs, because the data are obtained under variable conditions of climate and soil.

### Effects on the Crop Yield

Experiments realized at the Agricultural Experimental Substation of Fortuna in Juana Díaz—Puerto Rico demonstrated that the use of silver coated black plastic in combination with the drip irrigation system increased significantly the yield of various vegetables, in comparison with non-soil mulch treatments (Table 15.1). It was found that the use of silver coated black plastic significantly increased the yield of sweet peppers compared to treatments using transparent plastic, white plastic, or black plastic. These findings are in agreement with the ones obtained by other investigators around the world. In scientific terms, the increase in crop yield is due to a series of physical, climatic, and biological effects that are favored by the mulch. In practical terms, this increase in crop yield means more income for the farmer.

**Table 15.1.** Labor input requirements and average yield of various drip irrigated vegetable crops under soil mulch.

| Crop | Labor Requirements Man-hours/ha | | Average Yield ton/ha[1] | |
|---|---|---|---|---|
| | With Mulch[2] | Without Mulch | With Mulch[2] | Without Mulch |
| Cantaloupe | 1413 | 1119 | 16.8 | 12.9 |
| Cucumber | 1561 | 1386 | 41.3 | 39.5 |
| Green beans | 1247[3] | 1247[3] | 3.5 | 2.8 |
| Pepper | 1964 | 1900 | 36.5 | 24.2 |
| Tomato | 2529 | 2467 | 56.4 | 47.5 |
| Watermelon | 1495 | 1205 | 25.8 | 16.2 |

[1]Average yield.
[2]Installation of silver coated black plastic with manual operations.
[3]A single pass and it was sown in the cucumber field, once the cucumber cultivation was over.

## Labor Input Requirements

The use of mulch increases the labor input because of installation, maintenance, and removal operations, as shown in Table 15.1. In practical terms this means that the farmer have to incur additional expenses when using the soil mulch. Different studies have demonstrated that the use of soil mulch reduces the labor requirements for weeding to ≤4%, depending on the crop and season of the year. Without mulch, the weeding requires 13–27% of the total labor input requirements. This reduction, in addition to other benefits, can justify the use of soil mulch. The use of soil mulch with manual operations requires 15–28% of the total labor requirements. These values can be significantly reduced by mechanizing of the installation and mulch disposal operations.

## Conservation of Soil Moisture

The soil mulch reduces water loss due to evaporation. Moreover, it inhibits the weed growth and reduces the water use by weeds. This contributes to a more efficient irrigation management. It was observed that the plastic mulch treatments were more efficient in conserving and maintaining uniform soil moisture in the root zone, in comparison with the organic mulch. Both types of mulch were more efficient than non-mulch. Maintaining more uniform moisture storage and increasing the water use efficiency will imply less frequent irrigations, and savings in water and energy. On short and long term basis, this means net in the pocket and more profit for the farmer, under well irrigated conditions.

## Soil Temperature

The effects of different soil mulch treatments—plastic mulch (transparent, white, black, and silver), organic mulch and control (without mulch)—were evaluated in Puerto Rico on the soil temperature at four depths (0, 7.5, 15.0, 22.5 cm) during each time of the day (morning, noon, and evening) and two seasons (summer and winter) in drip irrigated sweet peppers. It was found that the use of transparent plastic mulch causes a rise in the soil temperature higher than the other treatments at all depths and time of the day in winter and summer. This is because the transparent mulch allows passage of most of the sun radiations received on the soil surface, causing a soil heating by absorption. Generally, the inferior side of the plastic is covered of condensed water drops in the morning before the sun rises. The water creates a barrier that allows the entrance of infrared radiations of short wave, but large wave radiations cannot escape out. This is the reason that most of the heat that will be irradiated to space will be retained in soil. Furthermore, the surface evaporation decreases considerably conserving the evaporation energy. This increment in the heat flow contributes in the rising of soil temperature. In Israel, the phenomenon "conservatory effect" has been used as a method for "sterilizing" and reducing the incidence of certain pests and diseases in the soil. In Puerto Rico, the rise in temperature generated by the transparent mulch is not sufficient to carry out this practice. The temperature in all plastic mulch treatments was higher than the ones in the control, during the morning at all depths, in winter and summer. This is attributed to the heat absorption by drops of condensed water under the mulch and to the reduction in the heat loss by evaporation. The interaction of these physical and biological effects can cause an early appearance of the fruit and an increase in the crop yield.

The soil temperatures in reflecting plastic mulch treatments (white and silver) were lower than the ones in the control, during noon and evening at all depths in both crop seasons. This effect is attributed to soil temperature which varies depending on the reflection, absorption, or transmission when receiving the solar radiation. The reflecting mulches transmit to the soil just a few of the radiations received. In general terms, significant differences in temperature in various treatments were not observed for very cloudy or rainy days. It was attributed to the low intensity of soil radiation received by the soil surface. It could be observed that the variations in temperature were more pronounced during the first 30 days of the growing cycle.

### Integral Weed Control

The integral weeding is of great importance in vegetable crops using soil mulch. In pepper and tomato production, the highest net profit was obtained using plastic mulch combined with weeding or the direct application of some herbicide. The use of plastic mulch in combination of agricultural practices is recommendable. Nevertheless, the use of an additional chemical agents or weeding (manual or mechanical) will depend greatly on the economic considerations in each case. The opaque mulches avoid the germination of weeds. The transparent mulches stimulate the growth of weeds since the light can pass through the soil. The *Cyperus rotundus L.* weed could not be controlled using soil mulches because during its germination the buds perforated the mulch.

### SUMMARY

For centuries the use of soil mulch has been a traditional technique for the crop production. Its selection depends mainly on the availability, cost, efficiency of use, and purpose of use. The commonly used mulches are paper products, glass and metal fabrics, bituminous emulsions, and organic matter. This chapter discusses technology of plasticulture, advantages and disadvantages of soil mulch, and research experiments on effects of plastic mulch on vegetable production. The use of soil mulch increases the labor input because of the installation, maintenance, and removal operations of plasticulture. The soil mulch reduces water loss due to evaporation. Maintaining more uniform moisture storage and increasing the water use efficiency will imply less frequent irrigations, and savings in water and energy. The integral weed control is of great importance in vegetable crops using soil mulch. The use of plastic mulch in combination with agricultural practices is recommendable. Opaque mulches avoid the germination of weeds and the transparent mulches stimulate the growth of weeds.

### KEYWORDS

- **Opaque mulches**
- **Polyethylene plastic**
- **Root zone**
- **Soil mulch**

## BIBLIOGRAPHY
Refer to Bibliography Section 15.1 for the literature cited.

Eggplant field with nut sedge emerging through the plastic mulch.

# Chapter 16

## Viability Studies

[1]This chapter is modified and translated from "Goyal, Megh R. y Eladio A. Gonzalez, 1990. Estudios de Viabilidad. Capitulo XVI en: *Manejo de Riego por Goteo.* Editado por Megh R. Goyal, páginas 433–446. Rio Piedras, P.R: Servicio de Extensión Agrícola, UPRM." For more details one may contact by E-mail: goyalmegh@gmail.com

## INTRODUCTON

Usually professional assessment is required before investing in a drip irrigation system. A consulting firm, with engineers and agronomists, can make a viability study based on the information relevant to the case. This information is valuable for a large scale as well as a small scale business. Some basic considerations for the viability study are:

1.  Water and land resources.
2.  Conditions of the agricultural business.
3.  Pattern to follow with the new irrigation system.

4. Design and costs with present alternatives for the drip irrigation system.

5. Financial analysis about the development of the drip irrigation system.

Other factors, that can be included in the study for the development of an enterprise a drip irrigation systems, are:

1. Availability of the resources (plants and materials) in the country.

2. Management and installation.

3. Training of administrator, farmers, and workers.

4. Financing for the installation, marketing, and other services for the farmer.

5. Development of modes of transportation and communication media within the project and external market.

The drip irrigation system with all its components is not permanent in the field because it needs maintenance and some specialized care. In fact, the viability studies should include recommendations and service related to the equipment and materials for different irrigation systems. The project studies must be considered at the farm. Estimates should be conducted to compare the benefits due to new practices in the development and techniques of an irrigation system. The objective of this chapter is to show how to develop guidelines for the viability studies for the drip irrigation system. All the information can be compiled using Section 11.

## WATER, LAND, AND SOIL RESOURCES

### Soil Analysis and Topographical Studies

The soil analysis is necessary to identify the soil chemistry, soil texture, soil structure, drainage characteristics, and the agricultural potential of an irrigation system. The soils are classified according to the physical and chemical characteristics. The drainage characteristics include evaluation of the soil structure, the soil permeability, soil infiltration, deep percolation, and the horizontal/vertical horizons of the soil.

The agricultural potential is a function of the soil type and its characteristics as limiting aspects such as soil salinity and land topography. The studies may be done with aerial photography or through coordination with soil conservation specialists. The topographical maps are normally used for the design of an irrigation system. These maps include necessary information such as roads, water channels, and general structures. The degree of precision in the studies should agree with the economical necessities of the program. For this reason, it is essential to make a thorough study of the topographical maps to look for better alternatives while implementing the irrigation system. The topographical maps of the distribution of different type of soils and the agricultural potential in the area may available from the Soil Conservation Service or other agricultural agency. The specialist, who will develop the irrigation system, needs clear and detailed information of the specific area of the project.

### Water Resource

Water is a limiting resource in many regions of the world. It should be used efficiently, because besides being important to agriculture, water enables the development of hydroelectric, industrial and residential projects. To estimate the quantity and quality of water for an irrigation system, hydrologic analysis is made. It is recommended to study the historic file of the water source to evaluate the viability. State water authority, the Department of Natural Resources, and Geological Survey of a country have useful information regarding the water resources that are destined to agricultural use. If there is no information on the quality of the water resources, samples may be taken and sent to a laboratory for chemical and physical analysis. The water from the subterranean wells is generally of good quality. It is necessary and important to determine the quantity of available water for the crop.

## CLIMATE

The climate is an important component to establish an irrigation system. Many places have weather stations to provide long term data. Agricultural Experimental Stations and the local airports have weather stations to measure rainfall, temperature, solar radiation, wind, relative humidity, and class A pan evaporation. The farmers may use this information and instruments such as tensiometers to determine the amount of water that should be applied to the crop. It is very important to know crop water requirements for the design an irrigation system, to avoid unnecessary expenses and losses.

The water requirements can be estimated from the meteorological data. The studies with specific crops in specific areas have shown relationship between the climatic data and the crop yield. Throughout the world, the irrigation needs for various crops have been estimated using temperature, rainfall, solar radiation, wind, and humidity data with several evapotranspiration models. However, the experimental data in the field is not available for most places (Lysimeter studies). While designing the irrigation systems, the engineer must be careful accepting the irrigation requirements from another region. An error of 20% in the estimation of water use may change the economical analysis, especially if water cost is high.

## CROP TYPE

Following are considerations while proposing an irrigation system:
1. Agronomic practices (Distance between plants and rows).
2. Plant height.
3. Root depth.
4. Growth stages (germination, growth, vegetative, fruit formation, and maturity).
5. Incidence of pests and diseases.
6. Areas irrigated by infiltration, sprinkler, and trickle irrigation systems.

## AGRICULTURAL SOCIOLOGY

### Agricultural Practices, Transportation, and Marketing
The Department of Agriculture of each state of a country has statistical information about the agricultural situation. We should select the most relevant and reliable information. One may consider the following factors:
1. Number of farms and size of holdings.

2. Agricultural methods and techniques in use.

3. Irrigated and non irrigated acreage.

4. Types of agricultural machinery in use.

5. Crop yield per hectare.

6. Total crop production and costs.

7. Availability of agricultural labor force.

The consultation is essential for consideration of the economical benefits. For proposals of irrigation systems and crop patterns, one must demonstrate that:

1. The soils and climate are adequate.

2. The rotation of the crops is efficient.

3. The necessary water is available.

4. There are markets capable of absorbing the increase in production at a rentable cost.

5. The farmers have necessary experience to cultivate with success.

6. There exist an appropriate training techniques and facilities.

7. The credit is adequate.

The increase in production can only benefit, if it can reach the consumer. The actual marketing systems and transportation must be studied to explore, if the marketing facilities exist to accommodate the increase in crop production. It should be possible to evaluate the rentability of the project by predicting: The price of a product in the market, the transportation cost, and the farm prices related to the volume of production of various crops with the new technology. The benefits of some crops are very sensitive to the costs of transportation and marketing expenses.

These viability studies can help the farmers to decide what crops to cultivate and which irrigation system to establish.

## Social Changes

The modern irrigation technology may impose many challenges among the agricultural community and society. These changes may result in new influences and may affect over the lifestyle and the attitude of the people. Although the implications of this program cannot be measured precisely yet some of the changes can be predicted. One should consider factors such as: Agricultural experience, skills in irrigation methods and attitude toward changes, the life style, and social structure. These factors are imperative to provide adequate information and education to the farmers so that they can adapt to the new irrigation practices. For this purpose, a scientific study on the social change is necessary. Agricultural agencies can provide the information about the advantages of the irrigation system. If a farmer is convinced that a higher efficiency implies lower costs and better crop yield, then the mental barriers are overcome.

## Traditional Rights: Land and Water

The irrigation design has some limitations that are imposed by the traditional rights and actual legislation. For example, problems may arise frequently with the water supply through traditional rights, which are defended tenaciously by the owners. These rights may interfere with the use of modern practices. On the other hand, when a commercial bank extends financing for an irrigation project, the loan may be subjected to legislative changes and to local laws, which are not favorable to the project. When the government is the owner of the most part of the irrigated land, the status of the farmer is reduced to a worker. But if the farmer owns the land, he has all the rights to develop the irrigation system. However, they might need cooperation from government agencies, educational institutions, and private agricultural agencies.

## ORGANIZATION AND MANAGEMENT

The managers of the irrigation system must have thorough knowledge about the project in order to justify their functions. The viability study insists in hiring of authorized personnel to make a better performance of an irrigation system.

### Installation and Operation

The installation and implementation of an irrigation system should be well controlled and coordinated. It is necessary to hire experienced personnel to install and establish an irrigation system. The installation of an irrigation system should be done in a specified time to avoid delays and economic losses. The project should be efficiently managed to achieve the best installation. There should be a team of agricultural engineers, agronomists, and technicians with experience. The specialized personnel should be competent to:

1. Advise the farmers on cropping practices and cultivation.
2. Design and supervise the work in the field.
3. Control the flow of water according to the crop requirements.
4. Initiate new irrigation practices.

### Agricultural Extension Service

The success of the irrigation system depends on the speed of the learning process and the receptivity of the farmers. A program of the Agricultural Extension Service should include:

1. Demonstration farms where the farmers can see the results of methods and practices with the irrigation system.
2. Training centers to offer short courses to the youths and courses designed for the specific needs of the farmers.
3. A center for the applied research of a specific problem in the agricultural region.
4. Field advisors supported by specialists, laboratory equipment to analyze the soils, plants, and crop diseases.
5. Marketing cooperatives.

## ENGINEERING

The engineering section for the viability study includes investigation of the available resources, the field condition, the legal aspects, and all design aspects. Proposals are made about the water resources, its distribution and associated structures.

### Field Conditions

The exploration study includes the study of the subsoil, soil type, and the conditions that affect the design of an irrigation system. Later soil and rock samples are taken through perforations. Following considerations are important:

1. Three dimensional soil samples.
2. The soil texture and compaction characteristics of the soil.
3. Subsurface water resources.
4. Field capacity and permanent wilting percentage of a soil.

The cost of the research is a fraction of the total project cost, but it is necessary for success of a drip irrigation system. The study focuses mostly on the soil texture, soil fertility, soil structure, soil water, and soil compaction for agricultural purpose. If there is no previous study of the region where the irrigation system is to be installed, it should be compared with nearby areas. But if there exists some type of research, it should be used to make better decisions in the design, evaluation, and installation of the irrigation system. The viability study determines the agricultural and economical development. The mechanics of water distribution should be studied so that the viability of the agricultural development can be used to evaluate the cost-benefit ratio.

### Equipments and Materials

Before completing the design of an irrigation system, one must consider material specifications and the available equipments in the market and must look for the local distributors. The study must include import restrictions and the delays in the orders. A list of all the materials with specifications and the necessary equipment should be made. Also, an examination of the construction industry should be conducted for the availability of workers, contractors, engineers, and other personnel.

### Water Quantity

It is necessary to know the crop water requirements and evapotranspiration needs for the region. The engineer must estimate the irrigation requirements for each crop and each section of the field. He should also include water needs for other purposes on the farm.

## ECONOMIC AND FINANCIAL ASPECTS

### Economic Aspects to Complete the Project

An extensive project, if planned carefully, can be successfully financed for many years. The money can be invested as necessary without investing all at once, which allows a margin for the contingencies at the appropriate moments. A bad financial planning can cause the project to fail.

**Costs and Benefits**

Financing a large project is complicated for its ramifications. A project should look for its own maintenance and subsistence. The implementation of the irrigation system is different because the same is installed to improve the crop production and increase the income. Therefore, there is an indirect relationship between the maintenance and the subsistence of the system. The benefits of the irrigation system are social and economical; some can be direct or indirect, some can be accessible and other intangibles. In general, the farmers should pay the water at a reasonable price in order to be able to cover the operational cost. This way, the drip irrigation system is economical and is efficient in the water use. All the information compiled in the viability study, gives us an idea of the cost of installation and development of the irrigation system. The cost study is used to estimate and compare the expenses that are incurred by the farmer to buy machinery, materials, and labor cost should be included. The benefits of the project are based on the crop production and its value in the market. The operational and management costs are deduced out from the total value to calculate the net benefit. Also, one can estimate the production and the efficiency of the system to achieve greater operational benefits.

**SUMMARY**

The drip irrigation system is not a permanent system in the field because its components need maintenance and some specialized care. Therefore, the viability studies should include recommendations regarding the equipment, materials, and the management of the irrigation system. In the viability studies, one should compare the benefits that can include the new practices and the development of methods and techniques. All the information compiled in the viability study is a generalized idea of what are the development costs of an irrigation system. The specialized personnel needs clear and detailed information of the specific area, where the system will be installed. The topographical maps of the area can be supplied by the local agricultural agency, office of Soil Conservation Service. The conclusions and results of the viability study for the drip irrigation system must be included in the report.

**KEYWORDS**

- **Agricultural potential**
- **Drainage**
- **Exploration study**
- **Topographical maps**

**BIBLIOGRAPHY**

Refer to bibliography Section 16.1 for the literature cited.

## APPENDIX I: DATA SHEET FOR A VIABILITY STUDY

1.  Date_____
    Name of farmer_____
    Name of farm _____
    City_____ Region _____
    Slope and topographic configuration (include map) _____

2.  **Possession**
    Owner _____ Area (hectares) _____
    Leased _____ Area (hectares) _____
    Cost of lease ($/ha): _____
    Lease term _____

3.  **Crop System**
    Crop_____
    Variety_____
    Total area _____ ha, Drip irrigated _____ ha
    Date of planting (month, day, year) _____, _____, _____
    Date of a harvest (month, day, year) _____, _____, _____
    Distance between rows_____ cm. Distance between plants _____ cm.
    Seeds per hole _____ Observations _____

4.  **Drip/Trickle or Micro Irrigation system**
    a) Water Supply
    Source: Lake/well/river/potable water _____
    Where it discharges _____
    Available quantity _____ liters per hour
    Variation in quantity _____ to _____ liters per hour
    Total amount of suspended solids in the water (ppm) _____
    Total amount of dissolved solids in the water (ppm) _____
    Specific substances in suspension and in the solution _____
    Observations _____
    b) Energy Source _____
    Electrical Yes/No Mechanical Yes/No
    Phase Single/double/triple Fuel type _____
    Voltage 110/230/460 V Engine capacity _____ KW.
    c) Pump tation
    Engine _____
    Make and model_____
    Voltage 110/230/460 V Cycles _____ HZ

Capacity _____ RPM _____

Date of purchase _____ Cost _____

Distributor _____ Pump _____

Make and model _____ Type _____

Voltage 110/230/460 V        Cycles _____ HZ

Capacity (L/hr) _____ RPM _____

Date of purchase _____ Cost ($) _____

Distributer _____

Additional pump (If any) _____

Make and model _____ Type _____

Capacity _____ liters per hour Bhp _____ KW.

Pressure head _____ m or (psi)

Cost ($) _____ Date of purchase _____

Distributer _____

d)  Filter

i. Primary Filter

Make and model_____ Type _____

Make mesh (size) _____ Mesh (type) _____

Cost ($) _____ Date of purchase _____

Capacity _____ liters per hour

Distributor _____

Number of times the filter is flushed

_____Weekly            _____Monthly

_____Quarterly            _____Half yearly

_____Yearly

How to clean the filter _____

ii. Secondary Filter

Make and model_____ Type _____

Mesh (size) _____ Mesh (type) _____

Cost ($) _____ Date of purchase _____

Capacity _____ liters per hour

Distributor _____ Cost _____

Amount of time the filter is flushed

_____Weekly            _____Monthly

_____Quarterly            _____Half yearly

_____Yearly

How to clean the filter _____

e)  Injector for Chemical Compounds

Make and model _____

Cost ($) _____ Date of purchase _____

Distributor _____

Application frequency (weekly, monthly, etc.) _____

Accessories _____

f)  Other Accessories of the Main System

Part _____ Size _____

Quantity _____ Price/Unit _____

Total _____ Observation _____

g)  Main Line

Material _____ Type _____

Surface _____, Subsurface          Depth _____
cm.

Number of irrigation outlets _____ Diameter _____ cm.

Length _____ m, Price per unit _____ $/m Total Cost _____

Date of purchase _____ Usage life _____ years (personal
opinion)

Distributor _____ Observaciones _____

h)  Sub Main or Secondary Line

Part _____ Quantity _____

Diameter _____ cm. Price per unit _____ $/m

Length _____ m Total _____

Date of purchase _____ Distributor _____

Observations _____
_____
_____

i)  Lateral Lines (or Drip Lines)

Crop _____

Type _____

Diameter _____ cm

Usage life (years) _____

Length _____ m

Price per unit _____ $/m

Total cost _____ $

Date of purchase _____

Distributor _____

Observations_____

j)  Dripper/Emitter/Micro Jets

Crop _____ Lines (number) _____

Type (make) _____ Length _____ m

Distance between the drippers _____ cm.    Total _____ m

Dripper discharge _____ liters/hour    Usage life (years) _____

Connection type _____ Total cost, # _____

Distributor _____Date of purchase _____

Observations _____

k) Other field accessories

Part _____ Quantity _____

Size _____ Price per unit _____

Unity _____    Total cost _____

## 5. Manual Labor

|  | Hours | $/hr | Total ($) |
|---|---|---|---|
| a. Installation of Pump | _____ | _____ | _____ |
| b. Installation of Primary and Secondary Line | _____ | _____ | _____ |
| c. Installation of Laterals | _____ | _____ | _____ |
| d. Other Work | _____ | _____ | _____ |

## 6. Financing

Source _____ Interest rate _____%

Purpose _____ Term _____ years

Loan amount _____ Monthly payment _____ $/monthly

## 7. General Observations (Who installed, designed the system, etc.)_____

_____

_____

# Glossary of Technical Terms

Agriculture is a dynamic field that has evolved throughout history and is an integral part of our food chain that sustains the mankind. Agriculture is subdivided into several specialties such as: Planting, irrigation, fertilization, growth and development of plants, diseases and pests, crop physiology, soil science, agronomy, agricultural engineering, economics, and so forth. Each one of these disciplines includes a series of terms and technical words. This chapter defines a series of commonly used terms in drip, micro, or trickle irrigation. This will help the farmers, agronomists, soil scientists, engineers, technicians, and other people related to agriculture to broaden their knowledge in drip irrigation.

**ABSORPTION**—refers to a process in which a substance penetrates and is absorbed by another substance. For example, the penetration of water into soil or the entrance of gases, water, nutrients, and other substances into the plant.

**ACTIVE ROOT ZONE**—is a root zone where the absorption of nutrients and the water movement is active.

**ADAPTER**—section of a pipe utilized to change the size of a tube. Using the adapter, we can join pipes with or without threads or inserted threads. There are male, female, and inserted adapters.

**ADHESION**—the increase of concentration of molecules or ions above a surface, including interchangeable cations and anions in soil particles.

**ADSORPTION**—the attraction of ions or composites to the soil surface of a solid.

**AERODYNAMIC TRANSFERENCE PHENOMENA**—Heat transferred by a surface that evaporates and is dragged by means of turbulence convection. The turbulence is created by the wind on the plant cover.

**AGGREGATE**—the arrangement of smaller or bigger size soil particles to form the aggregates.

**ALBEDO**—rate of sun radiations received at the soil surface and the reflected solar radiation.

**ALGAECIDE**—chemical agent to kill or to inhibit the growth of algae.

**AMMONIFICATION**—formation of composites of ammonia.

**ANNUAL PLANTS**—plants that grow and complete their life cycle in a perennial year or before.

**ANTITRANSPIRANTS**—composites or materials that are applied to the plant to reduce transpiration. These antitranspirants are not detrimental for the plant physiological processes.

**AREA, DESIGN**—where a drip irrigation system will be installed.

**AREA, SURFACE**—area of a soil, usually expressed in square meters per gram.

**ASSETS**—in economical terms as work or capital asset. It refers to an irrigation system that offers a productive service to business during a one year period.

**ATMOSPHERE**—is a pressure unit. One atmosphere is equal to 1.013 centimeter of a mercury column, equal to 14.71 psi, equal to 1 bar or equal to 101.3 KPa.

**AUTOMATION OF DRIP IRRIGATION SYSTEM**—(an automatic or semiautomatic system) provides a number of operational possibilities. This consists of a broad range of controls for a complete automatic system: water meter, pressure gage, solenoid valve, hydraulic valve, timer, and so forth. The system closes as soon as a guaranteed amount of water has been applied.

**AUTOMATION, NON SEQUENTIAL SYSTEM**—includes hydraulic or electrical valves, which independently operate in terms of the amount of water to be replaced and the irrigation frequency. Each unit can supply amount of volume of water and can open at same time due to a predetermined program by use of a solenoid valve. The Control panel contains electrical circuits to operate the pump or the main valve, to add fertilizer and to measure the soil moisture so that the irrigation system can fulfill the needs of a crop.

**AUTOMATION, PERIODICALLY ELECTRICALLY OPERATED**—It releases a desired amount of water regulated by a volumeter (solenoid valve). The duration of operation is determined by a timer and is activated by solenoid valves. Each unit has its own switch. The periodic irrigation automatically happens with a closed unit and the successive units are operated in sequence. The power unit can be A.C. or D.C.

**AUTOTROPHIC BACTERIA**—are bacterias that synthesize or produce their own food.

**AVAILABLE FREE CHLORINE**—is an excess of chlorine (hypochloric acid, HOCl) that remains in the irrigation water, after the reaction has been completed with inorganic compounds, organic substances, and the metals.

**BACKFLOW PREVENTION DEVICE**—a device that prevents back flow of contaminated water.

**BACTERICIDE**—a chemical agent to prevent the bacterial growth.

**BAR**—is equal to one atmosphere at level of the sea. It corresponds approximately to 101.3 KPa and 760 mm Hg.

**BIOCIDE**—prevents growth of micro or macroorganisms in the irrigation water.

**BIOLOGICAL AGENT**—growth of slime, algae and, other microorganisms that can obstruct or clog the laterals and drippers (emitters).

**BIOLOGICAL FACTORS (AGENTS)**—growth of microorganisms or macroorganisms that may obstruct the irrigation system.

**BLANEY-GRIDDLE METHOD**—it is used to calculate the evapotranspiration in a system based on the correlation between the monthly average of water use and the temperature, the percentage of daily sunshine hours and the crop coefficient.

**BUBBLER**—a water emission device that tends to bubble water directly onto the ground or that throws water a short distance.

**CALIBRATION**—adjustment of a chemical injection equipment so as to apply a desired quantity of chemical compound.

**CAPILLARY CONDUCTIVITY**—allow the flow of water through soil capillary pores in response to a potential gradient.

**CAPILLARY MECHANISM**—The capillary phenomena is due to: the force of attraction of the water by solids in the walls of the pores through which it flows and the surface tension of the water, that resists in any form except a smooth surface in the liquid-air interphase.

**CAPILLARY PORES**—are small pores, which maintain the soil moisture at a tension 60 cm or more. It is a fraction of the soil volume that is not occupied by soil particles.

**CAPILLARY SPACE**—is a micropore of the soil, or is a total volume of small pores that retain water in the soil at tension greater than 60 cm of water.

**CENTIBAR**—one hundredth part of a bar. The tensiometer is usually calibrated in centibars.

**CENTRIFUGAL FILTER**—Separates sand. Hydrocyclone or centrifugal filters remove suspended particles of a specific gravity greater than of water. These filters are ineffective to remove most of organic solids.

**CHEMICAL AGENT**—see chemical factors.

**CHEMICAL FACTORS (AGENTS)**—during the fertigation, pestigation, chloration, chemigation, chemical precipitates may be formed to cause obstruction at different locations of the drip irrigation system.

**CHEMICAL WATER TREATMENT**—chemical treatment of the water to make it acceptable for use in micro irrigation systems. This may include the use of acids, floculants, fungicides, and bactericides to help prevent emitter plugging or for pH adjustment.

**CHEMIGATION**—is a application of chemicals through the irrigation system.

**CHLORATION OR CHLORINTION**—is an application of chlorine through the irrigation system to control the growth of algae, bacteria, and other microorganisms in the irrigation lines and emitters.

**CHLOROSIS**—is one of the common symptoms of mineral deficiency of chlorine. It appears as a green color or yellowing of the green parts of a plant, particularly the leaves.

**CLAY**—soil particles whose diameter is smaller than 0.002 mm.

**CLOUDY**—the atmosphere loaded with clouds.

**COEFFICIENT OF VARIATION**—a statistical measure of the relative dispersion for an independent variable as defined in ASABE Engineering Practice.

**COEFFICIENT OF VARIATION**—it compares emitter flow rate of each dripper with the average flow rate of drippers.

**COEFFICIENT OF VARIATION OF THE MANUFACTURER**—describes the variation in pressure in the pressure controller (regulating) valve in relation to the operating pressure of the drippers.

**COEFFICIENT, PERMANENT WILTING**—is a percentage of soil moisture at which the plants have wilted permanently at a tension of 15 atm. The plants do not recover.

**COEFFICIENT, REFLECTION (ALBEDO)**—ratio between the amount of solar radiation received by the soil surface and the reflected sun radiation.

**COEFFICIENT, CROP**—is a ratio between crop evapotranspiration (ET-cultivation) and the potential evapotranspiration under optimal conditions of growth.

**COEFFICIENT, HYGROSCOPIC**—is a soil moisture percentage, after it has been dried to the atmospheric conditions or after the soil has reached a balance with a relative humidity of the atmosphere.

**COEFFICIENT, PAN**—ratio between the potential evapotranspiration and class A pan evaporation.

**COEFFICIENT, WILTING**—soil moisture content at which the plants have wilted. It corresponds to the upper limit of the usable moisture by plants.

**COEFFICIENTS OF SOIL WATER**—are the limits or margins between the 3 levels of water in the soil: gravitational, capillary, or hygroscopic.

**COMPOUNDS, INSOLUBLE**—do not dissolve in the water.

**COMPOUNDS, SOLUBLE**—dissolve completely in water.

**COMPOUNDS, SYSTEMATIC**—are composites that can travel through epidermis of the leaves, stems and roots or seeds, getting up itself to the sap, which then transfers these to all parts of the plant.

**CONDUCTIVITY, ELECTRICAL (CONDUCTANCE)**—the property of a substance to transfer an electrical charge. It is reciprocal of an electrical resistance.

**CONDUCTIVITY OF SOIL**—see permeability of the soil.

**CONDUCTIVITY, ELECTRICAL, OR SATURATION EXTRACT, (ECC)**—is an amount of salts in the soil solution when the soil is saturated of water at current or normal conditions.

**CONFIDENCE LIMITS**—a statistical statement which relates the probability that the true value of a variable falls within the described interval.

**CONTROL AND WATER TREATMENT STATION**—the control station may include facilities for water measurement, filtration, treatment, addition of amendments, pressure control, and timing of application.

**CONTROL SYSTEM**—group of tubes, accessories, and controls that regulates the water to the lateral lines.

**CONTROL VALVE**—a device used to control the flow of water.

**CONVECTION**—heat transfer from one place to another by the particle movement of a hot liquid gas.

**CORROSIVE**—substance is disturbed or eroded by the action of chemical agents.

**COUPLING**—a section of pipe with female or male threads or without threads. These are used to connect the pipes of different lengths.

**CROP REFERENCE**—is used to develop and to calibrate methods to calculate reference evapotranspiration. The reference crops are alfalfa and grasses.

**CROP WATER REQUIREMENTS**—Depth of water for the evapotranspiration (ET crop) required by a crop or diversified pattern of crop during a given period (mm/day).

**CROP, CONSUMPTIVE USE**—total amount of water that must be replaced in soil during the crop growth period. (see evapotranspiration).

**CROPPING PATTERN**—sequence of different crops that are seeded in a given order.

**DENITRIFICATION**—is a process by which the organic nitrates, nitrites, or soil deposits are reduced to nitrogen oxides by the action of bacteria.

**DEPRECIATION COST**—is a systematized method to find the value of the irrigation system based on its cost: and is a method as to prorate the cost for the active service life.

**DESIGN AREA**—the specific land area which is to be irrigated by the micro irrigation system.

**DESIGN EMISSION UNIFORMITY**—an estimate of the uniformity of emitter discharge rates throughout the system, as described in ASAE Engineering Practice EP40: Design and Installation of Micro irrigation Systems.

**DETERIORATION**—is a reduction in the value of an object due to the elements of nature.

**DIAPHRAGM EMITTER**—emitter that contain a stretched membrane with a small opening. When particles plug the opening, pressure builds stretching the membrane until the particle is forced out. The membrane and opening then return to normal size.

**DIFFUSION**—is a net movement of molecules of a substance from a region of high molecular activity to one of low activity as result of a random molecule movement.

**DILUTION RATE**—relation between the chemical agent concentration to be applied and the amount of water that flows through the irrigation system.

**DISCHARGE RATE OF DRIPPERS**—amount of water from drippers at a given pressure that is applied near the plant per unit time.

**DISSOLVE**—is a formation of a homogenous mixture between a liquid and a solid of which first usually is in greater quantity.

**DOSAGE**—amount used of a product.

**DRAINAGE**—is a process by which the superficial or underground water is removed from a soil in an area.

**DRIP OR DROP**—is formed by water movement through small orrifices.

**DRIP IRRIGATION**—a method of irrigation for slow application of water under low pressure through tube openings or attached devices just above, at or below the soil surface.

**DRIP LINE**—the circle that can be made on the soil around a tree below the tips of the outermost branches of the tree.

**DRIPPER, INTERNAL SPIRAL TYPE**—these are based on the principle of long trajectory. The capillarity effect is produced by means of the union of a pipe extended in an inner cylinder. This is a spiral of a circular or rectangular transverse section. The water enters the terminal and leaves by the side due to small force. The design provides drippers with long trajectory with the integration of components to function uniformly.

**DRIPPER, SPIRAL TYPE**—Capillary of long trajectory, made of polypropylene, in spirals normally hardened to maintain shape permanent. This dripper is inserted on the lateral and the water leaves the dripper drop by drop.

**EFFICIENCY, APPLICATION**—the percentage of the total water applied that is actually stored in the root zone.

**ELECTRICAL CONDUCTIVITY, MAXIMUM (ECmax)**—is a limit of the concentration of salts in the saturated soil (ECC) at which the growth will become paralyzed (null production). Units are mmhos/cm.

**ELEVATION, MAXIMUM**—difference in elevation between the water source and the most distant location in the field.

**EMISSION POINT**—point where the water is discharged from lateral line through emitters.

**EMITTER (DRIPPER)**—a dispensing device in a micro-irrigation system that regulates the flow of water.

**EMITTER (DRIPPER) SPACING**—is a space between two drippers or emitters. It depends on the type of crop and the emitter flow rates.

**EMITTER DISCHARGE EXPONENT**—the emitter discharge exponent; $x$, as described by the equation $q = kh^x$, which characterizes the type of emitter. For example, an $x$ value of 0.5 is common for orifice type emitters, whereas the $x$ value for a pressure compensating emitter would range from 0 to 0.5.

**EMITTER DISCHARGE RATE**—     the instantaneous discharge rate at a given operating pressure from an individual point-source emitter or from a unit length of line-source emitter, expressed as a volume per unit of time.

**EMITTER MANUFACTURER'S COEFFICIENT OF VARIATION**—a statistical term used to describe the variation n discharge rate for a sample of new emitters when operated at a constant temperature and at the emitter design operating pressure.

**EMITTER PERFORMANCE COEFFICIENT OF VARIATION**—statistical term used to describe the variation in emitter discharge due to the combined effects of emitter manufacturer's variation, emitter wear, and emitter plugging.

**EMITTER**—the device used to control the discharge from a lateral line at discrete or continuous points.

**EMULSION FORMULA**—is a liquid formulation of a given pesticide that can form emulsion when mixed with water.

**ENERGY BALANCE**—is a physical system that regulates the microclimate in the vicinity of the plant. Incoming solar radiation at the soil surface influences the energy balance.

**EROSION**—eroding down of the soil surface, loosening and movement of the soil or rocks, caused by runoff, wind, ice, or another geologic agent, including processes as the gravitational drags.

**EVAPORATION**—water loss by physical processes. It converts the liquid phase to gas phase.

**EVAPORATION, CLASS A PAN**—is a common instrument to measure evaporation the water from an open surface. The pan is constructed of galvanized iron sheet with diameter of 125 cm and 25 cm of depth. The pan is painted white.

**EVAPOTRANSPIRATION RATE**—is the total water used by the plant and the inevitable losses. It includes the amount of water extracted by the plants from the soil, transpired through the leaves, and the water evaporated directly from soil surface.

**EVAPOTRANSPIRATION**—the combined loss of water by evaporation from the soil surface and transpiration from plants.

**EVAPOTRANSPIRATION, REFERENCE**—is rate of evapotranspiration from an extensive surface of 8–15 cm of height from the soil with grass or alfalfa, uniform in height, actively growing, completely covering the soil, and without limitations of water.

**EVAPOTRANSPIRATION, POTENTIAL**–the evapotranspiration of a soil area covered with vegetation without limitations of water.

**FERTIGATION**–is an application of fertilizers through the irrigation system.

**FERTILIZER**—is a chemical compound that is added to soil to increase the crop yield.

**FERTILIZER, ACID**—the salts of acidic compounds (as the ammonia and urea) help to acidify the soil; and these have an acidification affect.

**FERTILIZER, BASIC**—compound that has an alkaline effect on the soil. Some examples are potassium nitrate, sodium nitrate, calcium nitrate, and calcium cynamide.

**FERTILIZER, NEUTRAL**—exerts a minimum residual effect on pH of the soil. It is added to resist the tendency of acidification of the nitrogen fertilizers.

**FILTER**—a canister device with a screen of a specified mesh to catch particles large enough to clog emitters.

**FILTER, DISK OR RING**—it is like a mesh filter. It consists of rings or disks (with grooves) to retain sand particles and suspended solids.

**FILTER, DOUBLE MESH**—consists of two sieves, which can remove suspended particles of different sizes.

**FILTER, MESH (SCREEN)**—consists of a mesh of metal, plastic, or synthetic fabric. The meshes are classified according to the number of squared holes in one square inch of a screen. The number of mesh increases according to the size of the squared. A sieve of 75 for microns (mesh 200) will have squares with each side of 0.075 mm (0.0029 inches). Drip irrigation we use sieves of 150 or 75 microns (mesh of 100–200).

**FILTER, SAND**—consists of fine gravel and sand of selected sizes in a tank. It can eliminate amounts of suspended solids from the irrigation water.

**FILTER, VORTEX (PRIMARY)**—is a hydrocyclone filter to remove suspended solids from the water. The solids heavier than water are thrown outside at the periphery of a filter. Due to the gravitational force, the solid particles descend to a lower portion of the cone. Then, these are expelled out of the filter.

**FILTER, VORTEX (SECUNDARY)**—the secondary vortex elevates the treated water upwards and allows it to escape through the superior flow.

**FILTRATION SYSTEM**—the assembly of components used to remove suspended solids from irrigation water. This may include both pressure and gravity-type devices and such specific units as settling basins or reservoirs, screens, media beds, and centrifugal units.

**FITTINGS**—the array of coupling and closure devices used to construct a drip system including connectors, tees, elbows, goof plugs, and end caps. Fittings may be of several types including compression, barbed, and locking

**FIXED COST**—In economic analysis, this is a fixed cost and is not variable during the service life of any equipment. It involves initial capital investment.

**FLOODED**—soil saturated with water.

**FLOW**—the rate or amount of water that moves through pipes in a given period of time. Flow is expressed in gph (gallons per hour), for micro-irrigation (drip) devices. It is expressed in gpm (gallons per minute), for high-pressure sprinkler systems.

**FLOW, INVERSE (BACK FLOW)**—is a water flow that is in the direction opposite to the direction of a normal flow. The back flow is allowed for flushing of the filters.

**FLOW, NORMAL**—Amount or volume of water that flows through the lines to maintain uniform application of water.

**FLUSHING (CLEANING)**—to separate suspended solids from the irrigation water to avoid clogging.

**FRACTION OF EXHAUSTION OF SOIL MOISTURE**—Fraction of the water available in the soil that can be obtained by the crop, allowing an evapotranspiration without limits and the crop growth.

**FREE ENERGY**—is a available energy of a given body (without change in temperature) to do the work.

**FRICTION**—negative force exerted by the tubes and the components of the system and that prevents the free water movement through system. It is also called drag.

**FRICTION LOSS OF PIPE**—for a pipe of a given length and a particular material, the amount of pressure loss due to the friction depends on the diameter of tube, flow rate, viscosity, and density of water.

**FUNGIGATION**—application of fungicides through the irrigation system.

**FUNGUS**—talofite plant that does not contain chlorophyll. These are microscopic organisms and some are pathogens.

**GAGE, PRESSURE**—device that indicates existing pressure in the irrigation system. Absolute pressure = atmospheric pressure + gage pressure.

**GAGE, VACUUM (SUCTION)**—important part of a tensiometer that records soil moisture suction values in the soil.

**GERMINATION**—renewal of the growth of a seed after a period of dormancy in the presence of soil moisture.

**GOOF PLUGS**—insertable caps to plug holes in mainline and microtubes where drip devices have been removed or are not needed.

**GRAVITATIONAL FORCE**—natural force that cause objects to move towards the center of earth.

**GROWTH HARMONE**—regulates the plant growth.

**GUTTATION**—is a water loss in form of drops caused by the root pressure.

**GYPSUM BLOCKS**—are tips that are installed at the desired depth to determine the soil tension. A pair of electrodes is encapsulated in a gypsum block for measuring the electrical resistance of soil.

**HARVEST INTENSITY**—is a percentage of total area programmed for a given period of an irrigated crop.

**HERBICIDE**—substances to destroy the unwanted weeds that grow along with growth and development of a crop.

**HERBICIDE, SELECTIVE**—herbicide to kill weeds of a certain specie only.

**HERBIGATION**—is an application of herbicides through the irrigation system.

**HYDRAULIC GRADIENT**—is the difference in the height of the water on and under the soil column. The volume of water that moves downward in the soil column will depend on the force, and the hydraulic conductivity of the soil.

**HYDROPHILLIC QUALITY**—where a substance can be absorbed or be dissolved in water.

**HOPKIN'S LAW OF BIOCLIMATOLOGY**—it establishes that there is a delay in the flowering as the number of degrees of latitude or altitude is increased.

**HYDRAULIC CONDUCTIVITY OF SOIL, K**—ratio of water flow through a cross-sectional area of the soil, for a unit of hydraulic gradient. It is called permeability or transmissibility.

**HYDRAULIC CONTACT**—that makes one substance with the water or other liquids at rest or movement.

**HYDRAULIC DESIGN COEFFICIENT OF VARIATION**—a statistical term used to describe the variation in hydraulic pressure in a sub main unit or throughout a micro irrigation system. Care should be taken not to confuse this term with the emitter discharge coefficient of variation due to hydraulics.

**HYDRAULIC EQUILIBRIUM**—is a balance reached between the water within stem of a tensiometer and the soil moisture.

**INCORPORATE**—to mix some compound or chemical agent with soil.

**INDUSTRIAL REVOLUTION**—it began around 1760. The invention of machines like engines were developed. It contributed to the advance of irrigation practices.

**INFILTRATION**—the downward water movement in the soil. Penetration of a liquid through solid pores.

**INFILTRATION OR PERCOLATION**—instantaneous rate of water that moves downward in a soil.

**INITIAL DEVELOPMENT STAGE**—is time (for a given crop) during germination or initial growth, when the soil covered by the plant growth is less than 10%.

**INJECTION SYSTEM FOR CHEMICAL PRODUCTS**—the system for the application of chemical agents through the irrigation system.

**INJECTION, APPLICATION IN THE SUCTION LINE**—a hose or tube can be connected on the suction side of a pump to inject chemicals. Another hose or tube is connected on the line to replace water and to mix in a tank. This method should not be used with dangerous substances, since the possibility of contamination of the water source is possible. A foot valve or safety valve at the end of suction line can avoid contamination.

**INJECTION, PRESSURE DIFFERENTIAL**—this method is easy to operate. There is a tank under pressure, it is interconnected to the discharge line at two points, one serves as water entrance and the second is an exit of the water with chemical agents. A pressure difference is created by use of a ball valve in the main line. The pressure difference across this valve is sufficient to cause a water flow through the tank. The water with chemical agents mixes with the water in the main. The injection rate is recalibrated.

**INJECTION, PRINCIPLE OF VENTURI**—it injects chemical agents in the irrigation line. The principle of operation includes a pressure reduction at the venture. The difference in pressure created across the Venturi is sufficient to cause suction of the chemical solutions from a reservoir.

**INJECTION, USE OF PRESSURE PUMP**—rotating diaphragm or piston type pump can be used to force the chemical to flow from the tank towards the irrigation line. The internal parts of the pump must be corrosion resistant. This method is precise enough to measure the quantity of chemical agent that is injected into to the irrigation system.

**INJECTOR**—to inject the gas or liquid into the irrigation system.

**INOXIDABLE**—that cannot be oxidized. The term is related to metals and alloys that are resistant to oxidation.

**INSECTIGATION**—is an application of insecticides through the irrigation system.

**INTEMPERIZATION**—are edaphic and biotic processes that transform rocks into soil particles.

**INTEREST**—is a rate charged by commercial bank on the loan. The interest rate depends on the number of payments, quantity, and terms of loan.

**INTERPOLATION**—obtain a desired value using a reference value but next to it.

**INTERVAL (RANGE) OF OPERATION OF THE TENSIOMETER**—range of readings of a tensiometer from 0 to 80 centibars when the instrument is operating satisfactorily and under normal wet conditions of a soil.

**INTERVAL, CONFIDENCE**—confidence or safety margin for a result that is not exact. This value is used in the uniformity of the water application.

**INTERVAL, IRRIGATION**—Time between the initial or successive applications of irrigation in the same plot (field).

**INTERVAL, RE-ENTRY**—during the pesticide application, an operator must wait to enter into the sprayed field with out any risks and without any protective clothing.

**IRRIGATION**—an artificial application of soil moisture that is essential for the plant growth. It supplements the rainfall when it is not sufficient.

**IRRIGATION DEFICIT**—the percentage of the soil volume in the root zone which is not wetted to field capacity or above (after irrigation) divided by the total potential root zone of the crop.

**IRRIGATION DEPTH**—is amount of water that is applied to supply the moisture to the roots zone at field capacity.

**IRRIGATION DURATION**—time interval between the beginning and successive applications of irrigation in a given area.

**IRRIGATION EFFICIENCY (WATER USE EFFICIENCY)**—ratio of the amount of water used by the crop to the water applied.

**IRRIGATION FREQUENCY**—time that passes from irrigation to the other irrigation. It is a number that implies how many times a crop will be irrigated.

**IRRIGATION PRACTICES**—are those that provide water to agricultural land to compensate for the water shortage. These include a dynamic set of circumstances: micro climatic of the crop, soil characteristics, water resource, and design restrictions.

**IRRIGATION REQUIREMENTS**—amount of water necessary to compensate the evapotranspiration needs of a crop; excluding the contributed by the precipitation, the soil water storage, the operational losses, and flushing requirements.

**IRRIGATION REQUIREMENTS, SEASONAL**—total amount of water necessary for the normal plant growth during its growing cycle.

**IRRIGATION, DRIP**—is a highly controlled irrigation method. Here the water is deposited directly in root zone. The water is received slowly at a desired pressure, to maintain soil moisture at a desired level to allow optimum plant growth. Also it is known as "trickle," "tension," "capillary," "daily." "high frequency," and "continuous moisture." The distribution of the water is by means of the use of secondary lines and emitters.

**IRRIGATION, DURATION**—is a period of time during which the irrigation system is in operation.

**IRRIGATION, FLOOD**—water is flooded over a predetermined area.

**IRRIGATION, INFILTRATION**—the water from the river is taken to the field by leveled irrigation furrows.

**IRRIGATION, OVERHEAD**—is a pressure irrigation that uses perforated pipes or the pipes with sprinklers to supply the water similar to a rainfall.

**IRRIGATION, SPRINKLER**—is based on the principle that the water is released at pressure through portable irrigation system: Light weight, lines of irrigation of fast connection or united with mounted sprinklers at intervals. An ample range of sprinklers is available to operate under varied pressures, spaces, sizes, and flow. This irrigation system simulates the rainfall, adaptable to several agricultural conditions.

**LATE GROWTH STAGE**—is time between the end of the average stage of vegetative growth and the harvest or maturity.

**LATERAL**—the water delivery pipeline that supplies water to emitters from the main or from a manifold pipeline or pipelines.

**LIFE, SERVICE OR USEFUL**—period after which the equipment is rendered useless.

**LINE, DISCHARGE**—It is part of an irrigation pipe that takes the water from the pump to the principal line or main.

**LINE OF DISTRIBUTION**—irrigation line that takes the water from the main line to the lateral one (or drip line).

**LINE, LATERAL**—is a part of irrigation piping that takes water from the secondary line/lines to the drippers.

**LINE, MAIN**—it is a part of irrigation pipe that takes the water from the pump house to the secondary line or to distribution.

**LINE, SECONDARY**—it is a part of irrigation piping that takes water from the main line to the lateral lines.

**LINE, SUCTION**—It is line on the downstream side of a pump. It takes water from the source to the pump.

**LIXIVIATION**—washing of salts in the soil by percolation process.

**LYSIMETER**—section of a representative block of a land and is surrounded by soil. It measures the water flow through a soil surface cover.

**MAIN AND SUB MAIN**—the water delivery pipelines that supply water from the control station to the manifolds.

**MAIN LINE**—tubing used in the drip system and is sometimes called lateral line. It is a soft polyethylene material.

**MAINTENANCE AND SERVICE**—revision and periodic care of the irrigation system.

**MANIFOLD**—the water delivery pipeline that supplies water from the main to the laterals. Pressure regulation is often applied at the inlet to the manifold.

**MANOMETER**—measure pressure difference.

**MASS FLOW**—this theory maintains that a high pressure of turgor in the cells of the leaves soon produces the flow of the sugar solution and other substances through plasmodium to the adjacent cells and through the crib tubes towards the adjacent cells and soon through the crib tubes towards the root cells.

**MAXIMUM NUMBER OF SUN-SHINE HOURS**—number of hours of sunshine during a period of 24 hours.

**MECHANISM OF DIODE VORTEX**—an Israeli designer proposed the unique variation of the orifice concept. This mechanism is a simple device that produces resistance to the flow creating a vortex in a circular chamber. The water enters tangentially at the circumference of the chamber and causes fluid turn around in the chamber at a high speed. This causes a swirling action.

**METABOLIC INHIBITOR**—chemical inhibitor of the metabolic reactions of the catalyzed by enzymes.

**METHYL BROMIDE**—is a soil disinfectant in the form of a gas that is applied through the drip irrigation system.

**MICROIRRIGATION**—the frequent application of small quantities of water directly above or below the soil surface; usually as discrete drops, continuous drops, tiny streams, or microspray; through emitters or applicators placed along a water delivery line. Microirrigation encompasses a number of methods or concepts; such as drip, subsurface, bubbler, and microspray irrigation.

**MICROIRRIGATION SYSTEM**—the physical components required to apply water through microirrigation. System components that may be required include the pumping station, control and water treatment station, main and sub main lines, manifold lines, lateral lines, filtration, emitters, valves, fittings, and other necessary items.

**MICROPORES**—are smallest pores of the soil.

**MICROSPRAY**—a low pressure sprayer device generally placed on a stake that is designed to wet soil with a fan or jet of water.

**MICROTUBES (SPAGHETTI TUBES)**—the 1/4 inch flexible pipe used to link emitters or sprayers to the mainline. Plastic stakes are frequently used to hold the tubes and the dispensing device attached to the end in place.

**MISTER**—a device that delivers fog-like droplets of water often for cooling purposes.

**MOISTURE EQUIVALENT**—is a percentage of retained weight of water; saturated previously by a soil mass of 1 cm in thickness after it has been subjected to a centrifugal force of 1000 times the gravitational force for 30 minutes.

**MOISTURE PERCENTAGE**—is a ratio of the volume of soil moisture at field capacity in the root zone, to the total the potential root zone.

**NEMATODE**—is a multicellular, generally microscopic organism with a main physiological and circulatory systems. In general these are thin worms, are cylindrical, extended, some segmented externally with differentiations in the head and tail.

**NEMEGATION**—is an application of nematicides through the irrigation system.

**NET IRRIGATION WATER REQUIREMENT**—quantity of water that is required for crop production, exclusive of effective precipitation.

**NEUTRON DISPERSOR OR SCATTERER**—is an equipment based on the registry of reduction of neutrons installed in an accessible tube in the soil. It consists of a neutron emitter source and a scale. A method to determine soil moisture content and the soil density by means of calibration curve.

**NIPPLE**—section of a pipe with male threads on both ends.

**NITROGEN**—is a main element of the plant growth and development of leaves. Most of nitrogen is absorbed during the vegetative growth period.

**NUTRIENT**—any element that is absorbed by the plant.

**OASIS EFFECT**—effect of dry zones that surround the microclimate of relatively small soil extensions, where an air mass that move towards an irrigated area can provide sensible heat.

**OBSTRUCTION**—loss in pressure of the system due to particles or suspended solids are accumulated in lines, filters or drippers. The obstructions are a most serious problem in the drip irrigation system. Clogging can cause serious losses

**ONION SHAPE**—pattern of moisture caused by the drip irrigation. It simulates a half onion.

**ORGANIC MATTER**—the residual of plants and animals in several states of decomposition, alive organisms of the soil and substances synthesized by these organisms.

**OSMATIC MECHANISM**—is a pressure exerted on the living bodies due to unequal concentration of salts on both sides of a cellular wall or a membrane. The water will move from a low concentration of salts through the membrane towards the high concentration of salts, and therefore, exerts additional pressure on the side of membranes.

**OSMOSIS**—Diffusion through a semi permeable membrane.

**PANEL, FIELD**—In computerized irrigation system with a series of field controls the system manually or by remote control. In the completely automated system, the signals are transmitted to individual points in the field; the field panel also collects information (value of water pressure, etc) and transmits to the principal panel.

**PANEL, CENTRAL CONTROL**—the central panel allows a complete control of all the field operations, sending instructions to the field panels, and obtaining continuous data on the operation of the irrigation system. Irrigation programmer, a unit for the transmission information consists of a unit, units standard for the control of water in lateral and a unit for danger signs.

**PARASITE**—it is an organism that obtains its food while living in or on the body of other animals or plants.

**PARTICLES, SUSPENDED**—particles found suspended in water. These consist of soil particles of several sizes (sand, silt, and clay), soil materials washed in open channels and taken through the pumping units, eroded accumulated material in the deposits, and particles eroded from the surface of pipes and fittings.

**PATHOGEN**—organism that cause plant diseases.

**PEAK EVAPOTRANSPIRATION**—the maximum rate of daily evapotranspiration as determined from the average of the seven highest consecutive days expected to occur.

**PEAK IRRIGATION WATER REQUIREMENT**—the water quantity needed to meet the peak evapotranspiration.

**PERCENTAGE, WETTING**—the percentage of the soil volume in the root zone which is wetted to field capacity or above (after irrigation) divided by the total potential root zone of the crop.

**PERCOLATION**—the downward movement of the water through the soil (see filtration).

**PERIOD, GROWTH**—the time between the day of sowing, transplanting, and harvesting for a given crop.

**PERMANENT WILTING PERCENTAGE**—soil moisture at 15 atmosphere of tension. At this limit, water is not available to plants causing wilting.

**pF**—logarithm of the soil moisture tension in centimeters of a column.

**pH**—logarithm of the reciprocal of the hydrogen ion concentration. It is expressed in terms of pH of the solution. For example, a pH of <7 is acid and from 7 to 12 is alkaline.

**PHOTOSYNTHESIS**—is a production of nutritional substances from carbon dioxide, water, and minerals in the presence of a solar energy and the chlorophyll in green plants. It is opposite of respiration.

**PHSYCROMETER**—instrument to measure the air humidity.

**PHYSICAL AGENT**—accumulation of sand, clay, and organic matter that can obstruct lines and drippers.

**PLANT POPULATION**—number of plants per unit of area of a crop.

**PLASMOLYSIS**—is a separation of the cytoplasm of the cellular wall as a result of the osmotic water movement. The plasmolysis increases the concentration of salts. The osmotic movement of the water occurs from a cell towards a concentrated soil solution.

**PLASTIC RESIN**—is an adapted material to mold emitters (emitting or jets). It must be strong, resistant to the solar radiation and black color to avoid the growth of algae. The finished product must maintain its original dimensions under climatic variations, and must not alter the flow characteristics of an emitter.

**PLOW LAYER (COMPACT SOIL LAYER)**—is a plow layer or hard soil that does not permit penetration and growth of the roots and affects water infiltration through this layer.

**PLUG** - an accessory that is threaded or is without threads. It is placed at the end of a pipe to stop flow of water.

**POLYETHYLENE**—is an ethylene of the cured thermoplastic resin group ($-CH2$ $CH2-$). It is very resistant to chemical agents and any degree of flexibility or rigidity can be given during the manufacturing process. Each manufacturing step adds chemical agent to create desired properties.

**POLYMER**—an organic molecule composed of many similar molecules, for example, the starch is composed of glucose units.

**POLYPROPYLENE**—it is plastic resin. It is used frequently to manufacture drippers. This plastic is a propylene polymer, a substance which is similar to polyethylene.

**POLYVINYL CHLORIDE, PVC**—is a vinyl chloride polymer. It is a thermoplastic resin ($-CH_2CH_1-$) resistant to ambient conditions. If it is exposed to sun light for longer periods, it will lose the plastic properties gradually. Therefore, the PVC tubes should not be left on the soil surface but should be buried or installed below the soil surface.

**POROSITY**—a fraction of soil space that is not occupied by soil particles.

**POROUS CERAMIC TIP**—consists of porous membrane cup with a vast number of small pores. It is closed at one end and other end is connected to tensiometer. It is installed in the root zone at the desired depth, to measure tension.

**POROUS SPACE**—is the total space that is not occupied by soil particles in the total volume of soil.

**POTASSIUM**—the plant need high quantities of potassium that is supplied by the natural reserves in the soil or chemical or chemical fertilizers.

**POTENTIAL OF THE SOIL MOISTURE, TOTAL**—is a sum of all the contributing forces that act on the soil moisture. $Pt = Pg + Po +$

**POTENTIAL, SOIL WATER**—is a result of absorption and capillary forces. The net effect of these two forces is to reduce the free energy of the soil water. It affects the soil moisture retention during the soil water movement.

**POTENTIAL (m)**—is an energy to take a unit weight of the object from a datum line to a desired location. Units are force x distance. It can also be measured in terms of potential head (P=γh).

**POTENTIAL, CAPILLARY**—is a work that is required to move a unit mass of water against the capillary forces from the surface of the water to a specified location.

**POTENTIAL, GRAVITATIONAL**—is due to a gravity force on the free soil water.

**POTENTIAL, OSMATIC (SOIL)**—it is attributed to the salts dissolved in the soil solutions. The dissolved substances may be or may not be ionic, but the net effect to reduce the free energy of the water. The osmotic potential has little effect the movement of the soils water. It may affect the water absorbed by the plant roots. The membrane of the roots, which transmits water more water vapor movement because the water vapor pressure reduces the presence of undisolved salts.

**POTENTIAL, PRESSURE**—includes the effect that may increase or decrease the pressure in the free energy of the soil water.

**PRECIPITATION**—total amount of precipitation (rainfall, drizzle, snow, ice, hail, fog, and condensation), expressed in depth of water that could cover a horizontal plane in absence of runoffs infiltration or evapotranspiration (mm/día).

**PRECIPITATION OF CHEMICAL COMPOUNDS**—deposit of solid compounds in the lines and emitters as a result of chemical reactions or chemigation.

**PREEMERGENT**—a chemical agent (fertilizers or herbicides or pesticides) that is applied before the emergency of the weeds or crops.

**PRESSURE**—the force propelling water through pipes. Common household static (non-flowing) pressure is 50–70 psi (pounds per square inch). Irrigation systems operate under dynamic (flowing) water pressure that is reduced with elevation gain and friction loss through rubbing on the sides of pipes. Long lengths of pipe generally result in low pressure at the ends of the run. Divide a large irrigation zone into smaller ones, to minimize frictional losses.

**PRESSURE, VAPOR**—pressure exerted by the water vapor contained in the air (millibar).

**PRESSURE CONTROLER**—is used in pressure irrigation systems such as sprinkler or trickle. It regulates pressure in the system. It is also called pressure regulator valve or pressure-relief valve.

**PRESSURE LOSS**—reduction in pressure from a pump to any point in the field.

**PRESSURE MEMBRANE**—is a permeable membrane to water and less permeable to the wet gases: through which the water can escape from the soil sample in response to a pressure gradient.

**PRESSURE REGULATOR**—a device that reduces incoming water pressure for low-pressure drip systems. Typical household water pressure is 50–60 psi while drip systems are designed to operate at around 20–30 psi depending on the manufacturer.

**PRESSURE, SATURATED VAPOR**—is a vapor pressure when the air is saturated at given temperature (millibar).

**PRESSURE, GAGE**—measures the pressure across the filters and the pressure in the secondary lines or at any location.

**PRESSURE, ROOT**—pressure developed in the roots as a result of the osmosis. It is a mechanism to fill the guttation and exudation from the cut branches.

**PRESSURE, TURGOR**—pressure exerted in the liquid by the walls of the plant cells.

**PRESSURE, COMPENSATING EMITTER**—an emitter designed to maintain a constant output (flow) over a wide range of operating pressures and elevations.

**PRESSURE, SENSITIVE EMITTER**—an emitter that releases more water at the higher pressures and less at lower pressures found with long mainlines or terrain changes.

**PRINCIPAL AMOUNT**—total sum of money that is borrowed for the purchase of the irrigation system.

**PROPYLENE**—is a flammable gaseous hydrocarbon, usually obtained from petroleum and propane.

**PUMP HOUSE**—is a small house that has permanent installation of pump, filtration system, controls, and other regulators of the irrigation system.

**PUMP**—is a motorized unit that pumps water from a water source through the irrigation pipes. It helps to provide a desired emitter.

**PUMP, VACUUM (MANUALLY OPERATED)**—is used to remove the air from the stem of a tensiometer.

**PUMP, CENTRIFUGAL**—consists of impellers. It can be vertical or horizontal.

**PUMP, FERTIGATION**—is used to inject the fertilizer solution into the irrigation system. These can be of two types: (a) The pump that is operated by an external power and (b) The proportional pump that is operated by the pressure of the irrigation water.

**PUMP, SUBMERSIBLE TURBINE**—is connected directly to an electrical motor. The pumping unit has analogous characteristics to the classic pumps for deep wells with the same pressures and capacities. The high flow rates at desired pressures are possible. These types of pumps are used for wells whose depth can exceed 400 meters. Pumps with are submersible motor are available.

**PUMPING STATION**—the pump or pumps that provides water and pressure to the system, together with all necessary equipments such as base, pump, screens, valves, motor controls, motor protection devices, fences, shelters, and backflow prevention device.

**QUALITY OF IRRIGATION WATER**—refers to the purity of the water to the obstruction problems in the lateral and emitter.

**RADIATION, EXTRATERRESTRIAL**—amount of received radiation horizontally at the top of atmosphere.

**RADIATION, NET**—balance between all the radiation of long wave that enters and leaves (Rn = Rns + Rnt): it is a measurement of the energy available at the soil surface.

**RADIATION, NET (LONG WAVE)**—balance between the radiation of long wave that leaves and enters. (Negative)

**RADIATION, NET SOLAR**—difference between the radiation of short wave received at the soil surface and the reflected radiation by the soil (surface of water or crop).

**RADIATION, SOLAR**—amount of radiation of short wave received in a horizontal plane at the soil surface.

**RADIOACTIVE SOURE**—is a compound or an element that emits radioactive rays.

**RAINFALL**—is the main source of soil moisture during the dry periods (see precipitation).

**RAINFALL, EFFECTIVE**—usable rainfall in agreement with the water requirements of the crop. This excludes deep percolation, run-off, and interception.

**RATOONS**—are plants that bloom repeatedly conserving all and freshness and greenish. All the meristem does not enter the reproductive phase at same time.

**REDUCER**—an accessory to reduce the size of exit terminal of a pipe. Both or combination of both ends can be threaded or not.

**RELATIVE HUMIDITY**—amount of real water vapor present in the air. It is an amount of water vapor that the air can retain at a given temperature.

**RESIDUAL VALUE**—is a junk value after which the equipment is not useful.

**RESPIRATION**—is sugar consumption by the plants in the presence of oxygen and moisture. It is opposite to oxidation.

**RISK RANGE**—one talks about the variable amounts of water provided to the soil: the discharge of the drippers, the amount of water applied during an irrigation interval. It is possible to predict the effect of each one of these factors, because the soil properties affect these factors.

**ROOT ZONE**—the root zone of a plant is defined as the soil depth (m) from which the plant extracts moisture.

**RUNOFF**—is an excess water that runs on the soil surface. This is a principal contributor to the soil erosion.

**SALINITY**—excess of soluble salts in the soil solution with sufficient concentration to harm the plants and to reduce the crop yield.

**SALT INDEX**—is defined as the degree of increase in the osmatic pressure of the soil solution due to a fertilizer, in comparison with the osmatic pressure created by an equivalent amount of $NaNO_3$. With an increase in the osmatic pressure of the soil solution, the ability of the plant to absorb reduces and the rate or plant growth and crop yield are reduced.

**SALTS**—amount of soluble salts in the soil, expressed in terms of percentage or parts by million, and so forth.

**SAND**—soil particles whose diameter is from 0.05 to 1 mm.

**SATURATION**—condition of the soil when all the capillary spaces are full of water.

**SILT**—it is a soil particle whose diameter varies from 0.002 to 0.05 mm.

**SLOPE EFFECT**—is a variation in the hydrostatic pressure in the main, submains, and laterals of irrigation system to a change in elevation in ground surface.

**SOIL**—a natural substance of the intemperization of minerals and the decomposition of organic matter that covers the soil surface, in which plants grow.

**SOIL, APPARENT DENSITY**—is a ratio mass of soil to volume of soil particles including the pore space of the sample. It is expressed in grams per cubic centimeters.

**SOIL COVERAGE, COMPLETE**—when the soil is 100% covered by crop or vegetative growth.

**SOIL COVERAGE, EFFECTIVE**—is a percentage of covered soil offered by the crop when the evapotranspiration is near its maximum value.

**SOIL FLUSHING REQUIREMENTS**—fraction of the irrigation water that enters the soil and that flows indeed to traverse and beyond of the roots zone by salinity. It is the minimum amount of water necessary for the control of soil salinity.

**SOIL MOISTURE**—see water content of soil.

**SOIL PERMEABILITY**—a process with which the gases, liquids or plant roots penetrate or pass through a soil or

**SOIL PROFILE**—it is a vertical section of the soil.

**SOIL SAMPLER (AUGER)**—is used for taking soil samples from a desired depth. It is necessary for the installation of tensiometer.

**SOIL STRUCTURE**—is an arrangement of soil particles to form aggregates, which are in a variety of recognized forms: large, small, and so forth.

**SOIL TEXTURE**—is a percentage of sand, silt, clay in a given soil class: (a) Sand: mineral particles with diameter of 0.05–1.0 millimeter; (b) The silt: mineral particles with diameter of 0.002—0.05 millimeter; (c) Clay: mineral particles smaller of diameter of <0.002 millimeter.

**SOIL, ALKALINE**—soil class that has high percentage of clay an excessive degree of saturation with free sodium, and with an non appreciable amount of soluble salts.

**SOIL, CATION EXCHANGE CAPACITY**—is a total amount of cations that a soil can adhere by cation exchange, usually expressed in milliequivalents in 100 grams of soil. The measurement of the cationic capacity depends on the method employed.

**SOIL, CLAYEY**—is applied to any class of ground that contains a high percent of clay. Soil with high fine particle content, particularly clay, is difficult to plow.

**SOIL, FIELD CAPACITY**—amount of water retained by the soil after abundant irrigation or heavy rainfall; and when the rate of downward water movement has ceased; and the gravitational water has been eliminated. It is water content in the soil at a tension of 0.33 bar.

**SOIL, INFILTRATION CAPACITY**—is a speed at which the water moves through the soil.

**SOIL, LOW WATER TENSION OF THE**—in drip irrigation, the soil moisture content does not exceed the field capacity. Deficit implies that the matrix tension does not exceed 30–50 bars. Low moisture tension is difficult to maintain with any other irrigation system.

**SOIL PARTICLES**—soil particles of diameter between from 0.20 to 0.005 microns.

**SOIL, SANDY**—soil class with high percentage of sand and is easy to cultivate.

**SO--IL, SILTY**—soil class with 1–21% clay, 28–50% silt, and <52% of sand.

**SOLAR EFFECT**—is determined by calculating the theoretical length of the day during the vegetative period of different crops.

**SOLUBILITY**—property with which a solid is dissolve in a liquid. The solubility can depend on the dissolvent, the dissolved substance and temperature.

**SOLUBLE**—the relative capacity of a solid compound to dissolve in a given liquid.

**SPECIFIC GRAVITY OF THE SOIL**—ratio of weight of dry soil to soil volume.

**STATION, CONTROL**—can include facilities to measure and to regulate the water flow, operation, filtration, chemigation, pressure, and other irrigation instruments.

**STATION, SELECTOR**—it is an indicating point that is used for two purposes (i) A visual reference that demonstrates to the operator which field station is being irrigated during an automatic cycle of irrigation; (ii) Manual selector that is manually operated to irrigate a desire field.

**STATION, PUMP**—is a permanent small house that has a pump, engine or motor, filtration system, chemical injection equipment, main irrigation controller and all necessary fittings, valves, and accessories.

**STATION, TENSIOMETER**—is a representative location in the field where the tensiometer is installed to obtain suction readings.

**STATISTICAL UNIFORMITY**—an estimate of the uniformity of emitter discharge rates throughout an existing microirrigation system.

**STOMATE**—is a small opening between the epidermic cells specialized in epidermis of the leaves and herbaceous stems, through which the exchange of gases occurs.

**SUBMAIN UNIT**—independently controlled irrigation unit usually covering from 1 to 5 ha and including a submain manifold lateral lines and emitters.

**SUNSHINE DAY LIGHT HOURS**—number of daily sunshine hours. It is also defined as the duration or burns on a hydrograph of "Campbell Stokes" recorder.

**SUPPLY DEMAND**—water replaced independently to satisfy the necessity of irrigation needs.

**SUPPLY FACTOR**—the rate between the possible maximum supply and the real supply in cubic meters per second.

**SUPPLY REQUIREMENT**—the rate of the maximum requirement by daily provision during the period of maximum water use.

**SUPPLY, MAXIMUM IRRIGATION**—average water requirement during the highest crop water use, based on cropping patter, crop rotation, and climate.

**SYSTEMIC**—are compounds with a property to pass through epidermis of the leaves, stems, roots or seeds; and can go up it selves to the sap, which them transports it to all the parts of plant.

**TEMPERATURE, DEW POINT**—temperature at which the air is cooled off and is saturated of water and at which the water vapor begins to condense.

**TEMPERATURE, WET BULB**—temperature given by a thermometer whose bulb is surrounded by a fine cotton piece. This allows reducing the temperature by means of loss of latent heat due to evaporation.

**TENSIOMETER**—measures the soil suction caused by soil capillary pores. It consists of a porous ceramic tip that is introduced in the soil at a desired depth and is hermetically connected to a tube and a vacuum gage that is graduated in centibars (0–100 cbars).

**TENSIOMETER GRAPH**—is a graph of the tensiometer readings. This can help to indicate when to irrigate and the duration of irrigation.

**TENSIOMETER, DEAERATION**—is a process to remove the accumulated air in tensiometer. A manual vacuum pump or a polyethylene tube of 0.125 centimeter (1/8 inch) of diameter can be used for this purpose.

**TENSION**—see tension of soil moisture.

**TENSION, (SUCTION) SOIL MOISTURE**—suction with which the moisture is retained in the soil. The tension is caused by forces with which the water is attached to soil particles.

**TENSION, OSMATIC**—retention of water by soil particles in the soil solution.

**TERM, LOAN**—time for which the loan is approved by the bank to purchase the irrigation system.

**TEST D.P.D**—calorimetric method to measure total chlorine and free chlorine available in the water.

**TEXTURE**—in clayey soils, the clay predominates. In coarse texture, sand predominates.

**THORNTHWAITE METHODO**—Thornthwaite developed an empirical equation based on the monthly average temperature to calculate evapotranspiration.

**TIME, MAXIMUM**—highest value of all observations to take a known quantity of water from drippers. This value is used to evaluate uniformity of application of water.

**TIME, MINIMUM**—lowest value of all observations to take a known amount of water from the dripper.

**TOXIC**—substance that can cause poisoning.

**TRANSPIRATION**—loss of water from the surface of a plant that is regulated by physical processes.

**TREATMENT, ACID**—to prevent partial obstruction of drippers because of salts. The irrigation system can be washed with some acids that are not detrimental to plant growth.

**TREATMENT, HOT WATER**—water at 80 centigrade is allowed through the irrigation system to avoid damage by freezing.

**TUBE WALL (DOUBLE WALL OR BIWALL)**—are drip lines that consist of two chambers. The inner chamber maintains a high pressure (50–150 cbs) in a diameter of approximately 1.25 cm. The outer chamber contains orifices in the wall of a tube at given intervals. The water leaves the tube at low pressure through the orifices to wet the soil.

**TURBINE PUMP FOR A DEEP WELL**—has an impeller with radial vanes. It may be suspended vertically inside the unloading pipe of a well. The impeller can be centrifugal, axial, or of an intermediate type depending on the desired pressure for a given flow rate.

**TURBULENT FLOW EMITTER**—emitters with a series of channels that force water to flow faster not allowing particles to settle out and plug the emitter.

**UNIFORMITY OF WATER APPLICATION**—to evaluate a drip irrigation system with respect to the rate of application of water by the emitters.

**UNLEVELED**—change in elevation of the land may cause gain or loss in the pressure. A change of elevation of 2.3 feet is equivalent to change in the pressure of 2.3 psi.

**VALVE, AIR RELIEF (VACUUM BREAKER)**—removes air in the irrigation line. It must be installed at the highest location in the field.

**VALVE, CHECK (SAFETY)**—allows the flow of the water only in the forward direction. It avoids contamination.

**VALVE, FLUSHING**—is at the end of the drip lines to clean the system when taking out the dirty water out of lateral lines.

**VALVE, FOOT**—it stops back flow of water to the water resource. It is made of bronze, galvanized iron, steel, PVC, polyethylene, and so forth.

**VALVE, GATE (BALL)**—allows certain amount of water through the pipe.

**VALVE, ONE WAY**—see check valve.

**VALVE, SOLENOID**—controls the amount of water in the submain or lateral by an electrical current.

**VALVE, VOLUMETRIC**—measures volume of water that is applied to a field.

**VAPOR PRESSURE GRADIENT**—is a difference between the free air vapor content on the crop and the content of vapor in the evaporative surface.

**VENTILATION**—is a process by which there is exchange of air and other gases of the soil. The index of ventilation of the soil depends on the size and number of soil pores and the soil moisture.

**WATER AMENDMENTS**—in addition to fertilizers, pesticides, micro nutrients, and other water additives to improve the crop yield, water amendments are used for chemical treatment of the water to reduce the obstruction of drippers.

**WATER AVAILABLE**—amount of water in the soil that is available to the plants. It is a soil moisture content between the field capacity and the point of permanent wilting.

**WATER BALANCE IN THE SOIL**—sum of all moisture entering into the soil system and water loss, in a given period of time, mm/period.

**WATER CONTENT OF THE SOIL**—is a depth of water contained in the soil. It is a ratio of the mass of water to the mass of dry, or a ratio of the volume of water to a unit of volume of soil, sample. It can be expressed on the basis of wet or dry weight of a soil sample. It is called soil moisture content.

**WATER SOURCE**—it supplies amount of water needed for irrigation system. It can be a well, lake, river, reservoir tank, or a natural fall of water, and so forth.

**WATER STORAGE IN SOIL**—quantity of stored water in a root zone due to rainfall, snow, or irrigation applications. This meets crop water requirements in subsequent periods, partially or completely.

**WATER, CAPILLARY**—water that is retained by capillary pores of the soil. It is an amount of water that is retained in the soil between the field capacity and the hygroscopic tension.

**WATER, GRAVITATIONAL**—water that is retained in the soil between the point of saturation and the field capacity. This water can be drained easily from the soil due to a gravitational force.

**WATER, HYGROSCOPIC**—absorbed water in the vapor form due to suction forces that attract it to the surface of soil particles.

**WATER, NON-AVAILABLE**—water retained in the soil and that is not available to the plants.

**WEEDS**—are unwanted plants in the vicinity of a given crop.

**WILTING PERCENTAGE, PERMANENT**—is a seasonal wilting shown by the plants due to the deficiency of water to compensate the loss by transpiration. This is called incipient wilting.

**WIND VELOCITY**—speed of wind at a height of two meters from the soil surface in an environment without barriers.

**ZONE**—the section of an irrigation system that can be operated at one time.

# Appendices

## APPENDIX A
## CONVERSION SI AND NON SI UNITS

| To convert the column 1 in the Column 2, Multiply by | Column 1 Unit SI | Column 2 Unit Non-SI | To convert the column 2 in the column 1 Multiply by |
|---|---|---|---|
| **LINEAR** | | | |
| 0.621 | kilometer, km ($10^3$m) | miles, mi | 1.609 |
| 1.094 | meter, m | yard, yd | 0.914 |
| 3.28 | meter, m | feet, ft | 0.304 |
| $3.94 \times 10^{-2}$ | millimeter, mm ($10^{-3}$) | inch, in | 25.4 |
| **SQUARES** | | | |
| 2.47 | hectare, he | acre | 0.405 |
| 2.47 | square kilometer, km$^2$ | acre | $4.05 \times 10^{-3}$ |
| 0.386 | square kilometer, km$^2$ | square mile, mi$^2$ | 2.590 |
| $2.47 \times 10^{-4}$ | square meter, m$^2$ | acre | $4.05 \times 10^{-3}$ |
| 10.76 | square meter, m$^2$ | square feet, ft$^2$ | $9.29 \times 10^{-2}$ |
| $1.55 \times 10^{-3}$ | mm$^2$ | square inch, in$^2$ | 645 |
| **CUBICS** | | | |
| $9.73 \times 10^{-3}$ | cubic meter, m$^3$ | inch-acre | 102.8 |
| 35.3 | cubic meter, m$^3$ | cubic-feet, ft$^3$ | $2.83 \times 10^{-2}$ |
| $6.10 \times 10^4$ | cubic meter, m$^3$ | cubic inch, in$^3$ | $1.64 \times 10^{-5}$ |
| $2.84 \times 10^{-2}$ | liter, L ($10^{-3}$ m$^3$) | bushel, bu | 35.24 |
| 1.057 | liter, L | liquid quarts, qt | 0.946 |
| $3.53 \times 10^{-2}$ | liter, L | cubic feet, ft$^3$ | 28.3 |
| 0.265 | liter, L | gallon | 3.78 |
| 33.78 | liter, L | fluid ounce, oz | $2.96 \times 10^{-2}$ |
| 2.11 | liter, L | fluid dot, dt | 0.473 |

## WEIGHT

| | | | |
|---|---|---|---|
| $2.20 \times 10^{-3}$ | gram, g ($10^{-3}$ kg) | pound, | 454 |
| $3.52 \times 10^{-2}$ | gram, g ($10^{-3}$ kg) | ounce, oz | 28.4 |
| 2.205 | kilogram, kg | pound, lb | 0.454 |
| $10^{-2}$ | kilogram, kg | quintal (metric), q | 100 |
| $1.10 \times 10^{-3}$ | kilogram, kg | ton (2000 lbs), ton | 907 |
| 1.102 | mega gram, mg | ton (US), ton | 0.907 |
| 1.102 | metric ton, t | ton (US), ton | 0.907 |

## YIELD AND RATE

| | | | |
|---|---|---|---|
| 0.893 | kilogram per hectare | pound per acre | 1.12 |
| $7.77 \times 10^{-2}$ | kilogram per cubic meter | pound per fanega | 12.87 |
| $1.49 \times 10^{-2}$ | kilogram per hectare | pound per acre, 60 lb | 67.19 |
| $1.59 \times 10^{-2}$ | kilogram per hectare | pound per acre, 56 lb | 62.71 |
| $1.86 \times 10^{-2}$ | kilogram per hectare | pound per acre, 48 lb | 53.75 |
| 0.107 | liter per hectare | galloon per acre | 9.35 |
| 893 | ton per hectare | pound per acre | $1.12 \times 10^{-3}$ |
| 893 | mega gram per hectare | pound per acre | $1.12 \times 10^{-3}$ |
| 0.446 | ton per hectare | ton (2000 lb) per acre | 2.24 |
| 2.24 | meter per second | mile per hour | 0.447 |

## SPECIFIC SURFACE

| | | | |
|---|---|---|---|
| 10 | square meter per kilogram | square centimeter per gram | 0.1 |
| $10^3$ | square meter per kilogram | square millimeter per gram | $10^{-3}$ |

| To convert the column 1 in the Column 2, Multiply by | Column 1 Unit SI | Column 2 Unit Non-SI | To convert the column 2 in the column 1 Multiply by |
|---|---|---|---|

## PRESSURE

| | | | |
|---|---|---|---|
| 9.90 | megapascal, MPa | atmosphere | 0.101 |
| 10 | megapascal | bar | 0.1 |
| 1.0 | megagram per cubic meter | gram per cubic centimeter | 1.00 |
| $2.09 \times 10^{-2}$ | pascal, Pa | pound per square feet | 47.9 |
| $1.45 \times 10^{-4}$ | pascal, Pa | pound per square inch | $6.90 \times 10^3$ |

## TEMPERATURE

| 1.00 (K − 273) | Kelvin, K | centigrade, °C | 1.00 (C + 273) |
| (1.8 C + 32) | centigrade, °C | Fahrenheit, °F | (F − 32)/1.8 |

## ENERGY

| $9.52 \times 10^{-4}$ | Joule J | BTU | $1.05 \times 10^3$ |
| 0.239 | Joule, J | calories, cal | 4.19 |
| 0.735 | Joule, J | feet–pound | 1.36 |
| $2.387 \times 10^5$ | Joule per square meter | calories per square centimeter | $4.19 \times 10^4$ |
| $10^5$ | Newton, N | dynes | $10^{-5}$ |

## WATER REQUIREMENTS

| $9.73 \times 10^{-3}$ | cubic meter | inch acre | 102.8 |
| $9.81 \times 10^{-3}$ | cubic meter per hour | cubic feet per second | 101.9 |
| 4.40 | cubic meter per hour | galloon (US) per minute | 0.227 |
| 8.11 | hectare-meter | acre-feet | 0.123 |
| 97.28 | hectare-meter | acre-inch | $1.03 \times 10^{-2}$ |
| $8.1 \times 10^{-2}$ | hectare centimeter | acre-feet | 12.33 |

## CONCENTRATION

| 1 | centimol per kilogram | milliequivalents per 100 grams | 1 |
| 0.1 | gram per kilogram | percents | 10 |
| 1 | milligram per kilogram | parts per million | 1 |

## NUTRIENTS FOR PLANTS

| 2.29 | P | $P_2O_5$ | 0.437 |
| 1.20 | K | $K_2O$ | 0.830 |
| 1.39 | Ca | CaO | 0.715 |
| 1.66 | Mg | MgO | 0.602 |

## NUTRIENT EQUIVALENTS

| Column A | Column B | Conversion A to B | Equivalent B to A |
| --- | --- | --- | --- |
| N | $NH_3$ | 1.216 | 0.822 |
| | $NO_3$ | 4.429 | 0.226 |
| | $KNO_3$ | 7.221 | 0.1385 |

| Column A | Column B | Conversion A to B | Equivalent B to A |
|---|---|---|---|
| | $Ca(NO_3)_2$ | 5.861 | 0.171 |
| | $(NH_4)_2SO_4$ | 4.721 | 0.212 |
| | $NH_4NO_3$ | 5.718 | 0.175 |
| | $(NH_4)_2HPO_4$ | 4.718 | 0.212 |
| P | $P_2O_5$ | 2.292 | 0.436 |
| | $PO_4$ | 3.066 | 0.326 |
| | $KH_2PO_4$ | 4.394 | 0.228 |
| | $(NH_4)_2HPO_4$ | 4.255 | 0.235 |
| | $H_3PO_4$ | 3.164 | 0.316 |
| K | $K_2O$ | 1.205 | 0.83 |
| | $KNO_3$ | 2.586 | 0.387 |
| | $KH_2PO_4$ | 3.481 | 0.287 |
| | Kcl | 1.907 | 0.524 |
| | $K_2SO_4$ | 2.229 | 0.449 |
| Ca | CaO | 1.399 | 0.715 |
| | $Ca(NO_3)_2$ | 4.094 | 0.244 |
| | $CaCl_2\ 6H_2O$ | 5.467 | 0.183 |
| | $CaSO_4\ 2H_2O$ | 4.296 | 0.233 |
| Mg | MgO | 1.658 | 0.603 |
| | $MgSO_4\ 7H_2O$ | 1.014 | 0.0986 |
| S | $H_2SO_4$ | 3.059 | 0.327 |
| | $(NH_4)_2SO_4$ | 4.124 | 0.2425 |
| | $K_2SO_4$ | 5.437 | 0.184 |
| | $MgSO_4\ 7H_2O$ | 7.689 | 0.13 |
| | $CaSO_4\ 2H_2O$ | 5.371 | 0.186 |

# APPENDIX B
# PIPE AND CONDUIT FLOW

PIPE AND CONDUIT FLOW

For sudden enlargements and sudden contractions the equivalent length is in meters or feet of pipe of the smaller diameter, d. The dashed line shows the determination of the equivalent length of a 6-in, standard elbow.

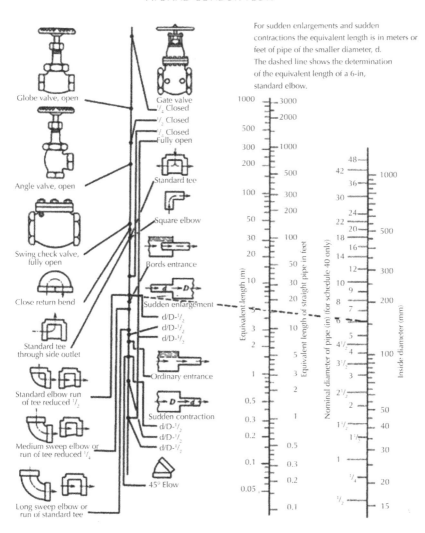

Globe valve, open

Angle valve, open

Swing check valve, fully open

Close return bend

Standard tee through side outlet

Standard elbow run of tee reduced $\frac{1}{2}$

Medium sweep elbow or run of tee reduced $\frac{1}{4}$

Long sweep elbow or run of standard tee

Gate valve
$\frac{3}{4}$ Closed
$\frac{1}{2}$ Closed
$\frac{1}{4}$ Closed
Fully open

Standard tee

Square elbow

Bords entrance

Sudden enlargement
d/D-$\frac{1}{4}$
d/D-$\frac{1}{2}$
d/D-$\frac{3}{4}$

Ordinary entrance

Sudden contraction
d/D-$\frac{1}{4}$
d/D-$\frac{1}{2}$
d/D-$\frac{3}{4}$

45° Elow

Equivalent length (m)

Equivalent length of straight pipe in feet

Nominal diameter of pipe (in) (for schedule 40 only)

Inside diameter (mm)

## APPENDIX C

## PERCENTAGE OF DAILY SUNSHINE HOURS: FOR NORTH AND SOUTH HEMISPHERES

| Latitude | Jan | Feb | Mar | April | May | June | July | Aug | Sept | Oct | Nov | Dec |
|---|---|---|---|---|---|---|---|---|---|---|---|---|
| | | | | | *NORTH* | | | | | | | |
| 0 | 8.50 | 7.66 | 8.49 | 8.21 | 8.50 | 8.22 | 8.50 | 8.49 | 8.21 | 8.50 | 8.22 | 8.50 |
| 5 | 8.32 | 7.57 | 8.47 | 3.29 | 8.65 | 8.41 | 8.67 | 8.60 | 8.23 | 8.42 | 8.07 | 8.30 |
| 10 | 8.13 | 7.47 | 8.45 | 8.37 | 8.81 | 8.60 | 8.86 | 8.71 | 8.25 | 8.34 | 7.91 | 8.10 |
| 15 | 7.94 | 7.36 | 8.43 | 8.44 | 8.98 | 8.80 | 9.05 | 8.83 | 8.28 | 8.20 | 7.75 | 7.88 |
| 20 | 7.74 | 7.25 | 8.41 | 8.52 | 9.15 | 9.00 | 9.25 | 8.96 | 8.30 | 8.18 | 7.58 | 7.66 |
| 25 | 7.53 | 7.14 | 8.39 | 8.61 | 9.33 | 9.23 | 9.45 | 9.09 | 8.32 | 8.09 | 7.40 | 7.52 |
| 30 | 7.30 | 7.03 | 8.38 | 8.71 | 9.53 | 9.49 | 9.67 | 9.22 | 8.33 | 7.99 | 7.19 | 7.15 |
| 32 | 7.20 | 6.97 | 8.37 | 8.76 | 9.62 | 9.59 | 9.77 | 9.27 | 8.34 | 7.95 | 7.11 | 7.05 |
| 34 | 7.10 | 6.91 | 8.36 | 8.80 | 9.72 | 9.70 | 9.88 | 9.33 | 8.36 | 7.90 | 7.02 | 6.92 |
| 36 | 6.99 | 6.85 | 8.35 | 8.85 | 9.82 | 9.82 | 9.99 | 9.40 | 8.37 | 7.85 | 6.92 | 6.79 |
| 38 | 6.87 | 6.79 | 8.34 | 8.90 | 9.92 | 9.95 | 10.10 | 9.47 | 3.38 | 7.80 | 6.82 | 6.66 |
| 40 | 6.76 | 6.72 | 8.33 | 8.95 | 10.02 | 10.08 | 10.22 | 9.54 | 8.39 | 7.75 | 6.72 | 7.52 |
| 42 | 6.63 | 6.65 | 8.31 | 9.00 | 10.14 | 10.22 | 10.35 | 9.62 | 8.40 | 7.69 | 6.62 | 6.37 |
| 44 | 6.49 | 6.58 | 8.30 | 9.06 | 10.26 | 10.38 | 10.49 | 9.70 | 8.41 | 7.63 | 6.49 | 6.21 |
| 46 | 6.34 | 6.50 | 8.29 | 9.12 | 10.39 | 10.54 | 10.64 | 9.79 | 8.42 | 7.57 | 6.36 | 6.04 |
| 48 | 6.17 | 6.41 | 8.27 | 9.18 | 10.53 | 10.71 | 10.80 | 9.89 | 8.44 | 7.51 | 6.23 | 5.86 |
| 50 | 5.98 | 6.30 | 8.24 | 9.24 | 10.68 | 10.91 | 10.99 | 10.00 | 835 | 7.45 | 6.10 | 5.64 |
| 52 | 5.77 | 6.19 | 8.21 | 9.29 | 10.85 | 11.13 | 11.20 | 10.12 | 8.49 | 7.39 | 5.93 | 5.43 |
| 54 | 5.55 | 6.08 | 8.18 | 9.36 | 11.03 | 11.38 | 11.43 | 10.26 | 8.51 | 7.20 | 5.74 | 5.18 |
| 56 | 5.30 | 5.95 | 8.15 | 9.45 | 11.22 | 11.67 | 11.59 | 10.40 | 8.53 | 7.21 | 5.54 | 4.89 |
| 58 | 5.01 | 5.81 | 8.12 | 9.55 | 11.46 | 12.00 | 11.98 | 10.55 | 8.55 | 7.10 | 4.31 | 4.56 |
| 60 | 4.67 | 5.65 | 8.08 | 9.65 | 11.74 | 12.39 | 12.31 | 10.70 | 8.57 | 6.98 | 5.04 | 4.22 |
| | | | | | *SOUTH* | | | | | | | |
| 0 | 8.50 | 7.66 | 8.49 | 8.21 | 8.50 | 8.22 | 8.50 | 8.49 | 8.21 | 8.50 | 8.22 | 8.50 |
| 5 | 8.68 | 7.76 | 8.51 | 8.15 | 8.34 | 8.05 | 8.33 | 8.38 | 8.19 | 8.56 | 8.37 | 8.68 |
| 10 | 8.86 | 7.87 | 8.53 | 8.09 | 8.18 | 7.86 | 8.14 | 8.27 | 8.17 | 8.62 | 8.53 | 8.88 |
| 15 | 9.05 | 7.98 | 8.55 | 8.02 | 8.02 | 7.65 | 7.95 | 8.15 | 8.15 | 8.68 | 8.70 | 9.10 |
| 20 | 9.24 | 8.09 | 8.57 | 7.94 | 7.85 | 7.43 | 7.76 | 8.03 | 8.13 | 8.76 | 8.87 | 9.33 |
| 25 | 9.46 | 8.21 | 8.60 | 7.74 | 7.66 | 7.20 | 7.54 | 7.90 | 8.11 | 8.86 | 9.04 | 9.58 |
| 30 | 9.70 | 8.33 | 8.62 | 7.73 | 7.45 | 6.96 | 7.31 | 7.76 | 8.07 | 8.97 | 9.24 | 9.85 |
| 32 | 9.81 | 8.39 | 8.63 | 7.69 | 7.36 | 6.85 | 7.21 | 7.70 | 8.06 | 9.01 | 9.33 | 9.96 |
| 34 | 9.92 | 8.45 | 8.64 | 7.64 | 7.27 | 6.74 | 7.10 | 7.63 | 8.05 | 9.06 | 9.42 | 10.08 |
| 36 | 10.03 | 8.51 | 8.65 | 7.59 | 7.18 | 6.62 | 6.99 | 7.56 | 8.04 | 9.11 | 9351 | 10.21 |
| 38 | 10.15 | 8.57 | 8.66 | 7.54 | 7.08 | 6.50 | 6.87 | 7.49 | 8.03 | 9.16 | 9.61 | 10.34 |
| 40 | 10.27 | 8.63 | 8.67 | 7.49 | 6.97 | 6.37 | 6.76 | 7.41 | 8.02 | 9.21 | 9.71 | 10.49 |
| 42 | 10.40 | 8.70 | 8.68 | 7.44 | 6.85 | 6.23 | 6.64 | 7.33 | 8.01 | 9.26 | 9.8 | 10.64 |
| 44 | 10.54 | 8.78 | 8.69 | 7.38 | 6.73 | 6.08 | 6.51 | 7.25 | 7.99 | 9.31 | 9.94 | 10.80 |
| 46 | 10.69 | 8.86 | 8.90 | 7.32 | 6.61 | 5.92 | 6.37 | 7.16 | 7.96 | 9.37 | 10.07 | 10.97 |

## APPENDIX D
### PSYCHOMETRIC CONSTANT (Λ) FOR DIFFERENT ALTITUDES (Z)

$$\gamma = \frac{C_p P}{\epsilon \lambda} = 0.665 \times 10^{-3}$$

| Z (m) | Λ kPa/°C | z (m) | λ kPa/°C | z (m) | λ kPa/°C | z (m) | λ kPa/°C |
|---|---|---|---|---|---|---|---|
| 0 | 0.067 | 1000 | 0.060 | 2000 | 0.053 | 3000 | 0.047 |
| 100 | 0.067 | 1100 | 0.059 | 2100 | 0.052 | 3100 | 0.046 |
| 200 | 0.066 | 1200 | 0.058 | 2200 | 0.052 | 3200 | 0.046 |
| 300 | 0.065 | 1300 | 0.058 | 2300 | 0.051 | 3300 | 0.045 |
| 400 | 0.064 | 1400 | 0.057 | 2400 | 0.051 | 3400 | 0.045 |
| 500 | 0.064 | 1500 | 0.056 | 2500 | 0.050 | 3500 | 0.044 |
| 600 | 0.063 | 1600 | 0.056 | 2600 | 0.049 | 3600 | 0.043 |
| 700 | 0.062 | 1700 | 0.055 | 2700 | 0.049 | 3700 | 0.043 |
| 800 | 0.061 | 1800 | 0.054 | 2800 | 0.048 | 3800 | 0.042 |
| 900 | 0.061 | 1900 | 0.054 | 2900 | 0.047 | 3900 | 0.042 |
| 1000 | 0.060 | 2000 | 0.053 | 3000 | 0.047 | 4000 | 0.041 |

**Note:** Based on λ = 2.45 MJ kg$^{-1}$ at 20°C.

## APPENDIX E
### SATURATION VAPOR PRESSURE [E°(T)] FOR DIFFERENT TEMPERATURES (T)

$$e^0(T) = 0.6108 \exp\left[\frac{17.27T}{T + 273.3}\right]$$

| T °C | $e_s$ kPa | T °C | e°(T) kPa | T °C | e°(T) kPa | T °C | $e_s$ kPa |
|---|---|---|---|---|---|---|---|
| 1.0 | 0.657 | 13.0 | 1.498 | 25.0 | 3.168 | 37.0 | 6.275 |
| 1.5 | 0.681 | 13.5 | 1.547 | 25.5 | 3.263 | 37.5 | 6.448 |
| 2.0 | 0.706 | 14.0 | 1.599 | 26.0 | 3.361 | 38.0 | 6.625 |
| 2.5 | 0.731 | 14.5 | 1.651 | 26.5 | 3.462 | 38.5 | 6.806 |
| 3.0 | 0.758 | 15.0 | 1.705 | 27.0 | 3.565 | 39.0 | 6.991 |
| 3.5 | 0.785 | 15.5 | 1.761 | 27.5 | 3.671 | 39.5 | 7.181 |
| 4.0 | 0.813 | 16.0 | 1.818 | 28.0 | 3.780 | 40.0 | 7.376 |
| 4.5 | 0.842 | 16.5 | 1.877 | 28.5 | 3.891 | 40.5 | 7.574 |
| 5.0 | 0.872 | 17.0 | 1.938 | 29.0 | 4.006 | 41.0 | 7.778 |
| 5.5 | 0.903 | 17.5 | 2.000 | 29.5 | 4.123 | 41.5 | 7.986 |
| 6.0 | 0.935 | 18.0 | 2.064 | 30.0 | 4.243 | 42.0 | 8.199 |

$$e^0(T) = 0.6108 \exp\left[\frac{17.27T}{T+273.3}\right]$$

| T °C | $e_s$ kPa | T °C | $e^0(T)$ kPa | T °C | $e^0(T)$ kPa | T °C | $e_s$ kPa |
|---|---|---|---|---|---|---|---|
| 6.5 | 0.968 | 18.5 | 2.130 | 30.5 | 4.366 | 42.5 | 8.417 |
| 7.0 | 1.002 | 19.0 | 2.197 | 31.0 | 4.493 | 43.0 | 8.640 |
| 7.5 | 1.037 | 19.5 | 2.267 | 31.5 | 4.622 | 43.5 | 8.867 |
| 8.0 | 1.073 | 20.0 | 2.338 | 32.0 | 4.755 | 44.0 | 9.101 |
| 8.5 | 1.110 | 20.5 | 2.412 | 32.5 | 4.891 | 44.5 | 9.339 |
| 9.0 | 1.148 | 21.0 | 2.487 | 33.0 | 5.030 | 45.0 | 9.582 |
| 9.5 | 1.187 | 21.5 | 2.564 | 33.5 | 5.173 | 45.5 | 9.832 |
| 10.0 | 1.228 | 22.0 | 2.644 | 34.0 | 5.319 | 46.0 | 10.086 |
| 10.5 | 1.270 | 22.5 | 2.726 | 34.5 | 5.469 | 46.5 | 10.347 |
| 11.0 | 1.313 | 23.0 | 2.809 | 35.0 | 5.623 | 47.0 | 10.613 |
| 11.5 | 1.357 | 23.5 | 2.896 | 35.5 | 5.780 | 47.5 | 10.885 |
| 12.0 | 1.403 | 24.0 | 2.984 | 36.0 | 5.941 | 48.0 | 11.163 |
| 12.5 | 1.449 | 24.5 | 3.075 | 36.5 | 6.106 | 48.5 | 11.447 |

## APPENDIX F

## SLOPE OF VAPOR PRESSURE CURVE (Δ) FOR DIFFERENT TEMPERATURES (T)

$$\Delta = \frac{4098\left[0.6108 \exp\left(\frac{17.27T}{T+273.3}\right)\right]}{(T+237.3)^2}$$

| T °C | Δ kPa/°C | T °C | Δ kPa/°C | T °C | Δ kPa/°C | T °C | Δ kPa/°C |
|---|---|---|---|---|---|---|---|
| 1.0 | 0.047 | 13.0 | 0.098 | 25.0 | 0.189 | 37.0 | 0.342 |
| 1.5 | 0.049 | 13.5 | 0.101 | 25.5 | 0.194 | 37.5 | 0.350 |
| 2.0 | 0.050 | 14.0 | 0.104 | 26.0 | 0.199 | 38.0 | 0.358 |
| 2.5 | 0.052 | 14.5 | 0.107 | 26.5 | 0.204 | 38.5 | 0.367 |
| 3.0 | 0.054 | 15.0 | 0.110 | 27.0 | 0.209 | 39.0 | 0.375 |
| 3.5 | 0.055 | 15.5 | 0.113 | 27.5 | 0.215 | 39.5 | 0.384 |
| 4.0 | 0.057 | 16.0 | 0.116 | 28.0 | 0.220 | 40.0 | 0.393 |
| 4.5 | 0.059 | 16.5 | 0.119 | 28.5 | 0.226 | 40.5 | 0.402 |
| 5.0 | 0.061 | 17.0 | 0.123 | 29.0 | 0.231 | 41.0 | 0.412 |
| 5.5 | 0.063 | 17.5 | 0.126 | 29.5 | 0.237 | 41.5 | 0.421 |
| 6.0 | 0.065 | 18.0 | 0.130 | 30.0 | 0.243 | 42.0 | 0.431 |
| 6.5 | 0.067 | 18.5 | 0.133 | 30.5 | 0.249 | 42.5 | 0.441 |

$$\Delta = \dfrac{4098\left[0.6108\exp\left(\dfrac{17.27T}{T+273.3}\right)\right]}{\left(T+237.3\right)^{2}}$$

| T °C | Δ kPa/°C | T °C | Δ kPa/°C | T °C | Δ kPa/°C | T °C | Δ kPa/°C |
|---|---|---|---|---|---|---|---|
| 7.0 | 0.069 | 19.0 | 0.137 | 31.0 | 0.256 | 43.0 | 0.451 |
| 7.5 | 0.071 | 19.5 | 0.141 | 31.5 | 0.262 | 43.5 | 0.461 |
| 8.0 | 0.073 | 20.0 | 0.145 | 32.0 | 0.269 | 44.0 | 0.471 |
| 8.5 | 0.075 | 20.5 | 0.149 | 32.5 | 0.275 | 44.5 | 0.482 |
| 9.0 | 0.078 | 21.0 | 0.153 | 33.0 | 0.282 | 45.0 | 0.493 |
| 9.5 | 0.080 | 21.5 | 0.157 | 33.5 | 0.289 | 45.5 | 0.504 |
| 10.0 | 0.082 | 22.0 | 0.161 | 34.0 | 0.296 | 46.0 | 0.515 |
| 10.5 | 0.085 | 22.5 | 0.165 | 34.5 | 0.303 | 46.5 | 0.526 |
| 11.0 | 0.087 | 23.0 | 0.170 | 35.0 | 0.311 | 47.0 | 0.538 |
| 11.5 | 0.090 | 23.5 | 0.174 | 35.5 | 0.318 | 47.5 | 0.550 |
| 12.0 | 0.092 | 24.0 | 0.179 | 36.0 | 0.326 | 48.0 | 0.562 |
| 12.5 | 0.095 | 24.5 | 0.184 | 36.5 | 0.334 | 48.5 | 0.574 |

## APPENDIX G
## NUMBER OF THE DAY IN THE YEAR (JULIAN DAY)

| Day | Jan | Feb | March | April | May | June | July | Aug | Sept | Oct | Nov | Dec |
|---|---|---|---|---|---|---|---|---|---|---|---|---|
| 1 | 1 | 32 | 60 | 91 | 121 | 152 | 182 | 213 | 244 | 274 | 305 | 335 |
| 2 | 2 | 33 | 61 | 92 | 122 | 153 | 183 | 214 | 245 | 275 | 306 | 336 |
| 3 | 3 | 34 | 62 | 93 | 123 | 154 | 184 | 215 | 246 | 276 | 307 | 337 |
| 4 | 4 | 35 | 63 | 94 | 124 | 155 | 185 | 216 | 247 | 277 | 308 | 338 |
| 5 | 5 | 36 | 64 | 95 | 125 | 156 | 186 | 217 | 248 | 278 | 309 | 339 |
| 6 | 6 | 37 | 65 | 96 | 126 | 157 | 187 | 218 | 249 | 279 | 310 | 340 |
| 7 | 7 | 38 | 66 | 97 | 127 | 158 | 188 | 219 | 250 | 280 | 311 | 341 |
| 8 | 8 | 39 | 67 | 98 | 128 | 159 | 189 | 220 | 251 | 281 | 312 | 342 |
| 9 | 9 | 40 | 68 | 99 | 129 | 160 | 190 | 221 | 252 | 282 | 313 | 343 |
| 10 | 10 | 41 | 69 | 100 | 130 | 161 | 191 | 222 | 253 | 283 | 314 | 344 |
| 11 | 11 | 42 | 70 | 101 | 131 | 162 | 192 | 223 | 254 | 284 | 315 | 345 |
| 12 | 12 | 43 | 71 | 102 | 132 | 163 | 193 | 224 | 255 | 285 | 316 | 346 |
| 13 | 13 | 44 | 72 | 103 | 133 | 164 | 194 | 225 | 256 | 286 | 317 | 347 |
| 14 | 14 | 45 | 73 | 104 | 134 | 165 | 195 | 226 | 257 | 287 | 318 | 348 |
| 15 | 15 | 46 | 74 | 105 | 135 | 166 | 196 | 227 | 258 | 288 | 319 | 349 |
| 16 | 16 | 47 | 75 | 106 | 136 | 167 | 197 | 228 | 259 | 289 | 320 | 350 |
| 17 | 17 | 48 | 76 | 107 | 137 | 168 | 198 | 229 | 260 | 290 | 321 | 351 |

| Day | Jan | Feb | March | April | May | June | July | Aug | Sept | Oct | Nov | Dec |
|-----|-----|-----|-------|-------|-----|------|------|-----|------|-----|-----|-----|
| 18 | 18 | 49 | 77 | 108 | 138 | 169 | 199 | 230 | 261 | 291 | 322 | 352 |
| 19 | 19 | 50 | 78 | 109 | 139 | 170 | 200 | 231 | 262 | 292 | 323 | 353 |
| 20 | 20 | 51 | 79 | 110 | 140 | 171 | 201 | 232 | 263 | 293 | 324 | 354 |
| 21 | 21 | 52 | 80 | 111 | 141 | 172 | 202 | 233 | 264 | 294 | 325 | 355 |
| 22 | 22 | 53 | 81 | 112 | 142 | 173 | 203 | 234 | 265 | 295 | 326 | 356 |
| 23 | 23 | 54 | 82 | 113 | 143 | 174 | 204 | 235 | 266 | 296 | 327 | 357 |
| 24 | 24 | 55 | 83 | 114 | 144 | 175 | 205 | 236 | 267 | 297 | 328 | 358 |
| 25 | 25 | 56 | 84 | 115 | 145 | 176 | 206 | 237 | 268 | 298 | 329 | 359 |
| 26 | 26 | 57 | 85 | 116 | 146 | 177, | 207 | 238 | 269 | 299 | 330 | 360 |
| 27 | 27 | 58 | 86 | 117 | 147 | 178 | 208 | 239 | 270 | 300 | 331 | 361 |
| 28 | 28 | 59 | 87 | 118 | 148 | 179 | 209 | 240 | 271 | 301 | 332 | 362 |
| 29 | 29 | (60) | 88 | 119 | 149 | 180 | 210 | 241 | 272 | 302 | 333 | 363 |
| 30 | 30 | - | 89 | 120 | 150 | 181 | 211 | 242 | 273 | 303 | 334 | 364 |
| 31 | 31 | - | 90 | - | 151 | - | 212 | 243 | - | 304 | - | 365 |

## APPENDIX H

## $\Sigma^*(T_K)^4$ (STEFAN-BOLTZMANN LAW) AT DIFFERENT TEMPERATURES (T) WITH: $\Sigma^*(T_K)^4 = 4.903 \ Ч \ 10^{-9}$ MJ K$^{-4}$ M$^{-2}$ DAY$^{-1}$ AND $T_K$ = T[°C] + 273.16

| T (°C) | $\sigma^*(T_K)^4$ (MJ m$^{-2}$ d$^{-1}$) | T (°C) | $\sigma^*(T_K)^4$ (MJ m$^{-2}$ d$^{-1}$) | T (°C) | $\sigma^*(T_K)^4$ (MJ m$^{-2}$ d$^{-1}$) |
|--------|-----------------------------------------|--------|-----------------------------------------|--------|-----------------------------------------|
| 1.0 | 27.70 | 17.0 | 34.75 | 33.0 | 43.08 |
| 1.5 | 27.90 | 17.5 | 34.99 | 33.5 | 43.36 |
| 2.0 | 28.11 | 18.0 | 35.24 | 34.0 | 43.64 |
| 2.5 | 28.31 | 18.5 | 35.48 | 34.5 | 43.93 |
| 3.0 | 28.52 | 19.0 | 35.72 | 35.0 | 44.21 |
| 3.5 | 28.72 | 19.5 | 35.97 | 35.5 | 44.50 |
| 4.0 | 28.93 | 20.0 | 36.21 | 36.0 | 44.79 |
| 4.5 | 29.14 | 20.5 | 36.46 | 36.5 | 45.08 |
| 5.0 | 29.35 | 21.0 | 36.71 | 37.0 | 45.37 |
| 5.5 | 29.56 | 21.5 | 36.96 | 37.5 | 45.67 |
| 6.0 | 29.78 | 22.0 | 37.21 | 38.0 | 45.96 |
| 6.5 | 29.99 | 22.5 | 37.47 | 38.5 | 46.26 |
| 7.0 | 30.21 | 23.0 | 37.72 | 39.0 | 46.56 |
| 7.5 | 30.42 | 23.5 | 37.98 | 39.5 | 46.85 |
| 8.0 | 30.64 | 24.0 | 38.23 | 40.0 | 47.15 |
| 8.5 | 30.86 | 24.5 | 38.49 | 40.5 | 47.46 |
| 9.0 | 31.08 | 25.0 | 38.75 | 41.0 | 47.76 |
| 9.5 | 31.30 | 25.5 | 39.01 | 41.5 | 48.06 |

| T (°C) | $\sigma^*(T_K)^4$ (MJ m$^{-2}$ d$^{-1}$) | T (°C) | $\sigma^*(T_K)^4$ (MJ m$^{-2}$ d$^{-1}$) | T (°C) | $\sigma^*(T_K)^4$ (MJ m$^{-2}$ d$^{-1}$) |
|---|---|---|---|---|---|
| 10.0 | 31.52 | 26.0 | 39.27 | 42.0 | 48.37 |
| 10.5 | 31.74 | 26.5 | 39.53 | 42.5 | 48.68 |
| 11.0 | 31.97 | 27.0 | 39.80 | 43.0 | 48.99 |
| 11.5 | 32.19 | 27.5 | 40.06 | 43.5 | 49.30 |
| 12.0 | 32.42 | 28.0 | 40.33 | 44.0 | 49.61 |
| 12.5 | 32.65 | 28.5 | 40.60 | 44.5 | 49.92 |
| 13.0 | 32.88 | 29.0 | 40.87 | 45.0 | 50.24 |
| 13.5 | 33.11 | 29.5 | 41.14 | 45.5 | 50.56 |
| 14.0 | 33.34 | 30.0 | 41.41 | 46.0 | 50.87 |
| 14.5 | 33.57 | 30.5 | 41.69 | 46.5 | 51.19 |
| 15.0 | 33.81 | 31.0 | 41.96 | 47.0 | 51.51 |
| 15.5 | 34.04 | 31.5 | 42.24 | 47.5 | 51.84 |
| 16.0 | 34.28 | 32.0 | 42.52 | 48.0 | 52.16 |
| 16.5 | 34,52 | 32.5 | 42.80 | 48.5 | 52.49 |

## APPENDIX I

## THERMODYNAMIC PROPERTIES OF AIR AND WATER

### 1. Latent Heat of Vaporization ($\lambda$)

$\lambda = 2.501 - (2.361 \times 10^{-3})\, T$

where: $\lambda$, latent heat of vaporization [MJ kg$^{-1}$] T, air temperature [°C]

The value of the latent heat varies only slightly over normal temperature ranges. A single value may be taken (for ambient temperature = 20°C): $\lambda = 2.45$ MJ kg$^{-1}$.

### 2. Atmospheric Pressure (P)

$$P = P_0 \left( \frac{T_{K0} - \alpha(z - z_0)}{T_{K0}} \right)^{\frac{g}{\alpha_1 R}}$$

where:    P, atmospheric pressure at elevation z [kPa]

$P_0$, atmospheric pressure at sea level = 101.3 [kPa]

z, elevation [m]

$z_0$, elevation at reference level [m]

g, gravitational acceleration = 9.807 [m s$^{-2}$]

R, specific gas constant = 287 [J kg$^{-1}$ K$^{-1}$]

$\alpha_1$, constant lapse rate for moist air = 0.0065 [K m$^{-1}$]

$T_{K0}$, reference temperature [K] at elevation $z_0$ = 273.16 + T

T, means air temperature for the time period of calculation [°C]

When assuming $P_0$ = 101.3 [kPa] at $z_0$ = 0, and $T_{K0}$ = 293 [K] for T = 20 [°C], above equation reduces to:

$$P = 101.3\left(\frac{293 - 0.0065z}{293}\right)^{5.26}$$

## 3. Atmospheric Density (ρ)

$$\rho = \frac{1000P}{T_{Kv}R} = 3.486\frac{P}{T_{Kv}} \text{ and } T_{Kv} = T_K\left(1 - 0.378\frac{e_a}{P}\right)^{-1}$$

where: ρ, atmospheric density [kg m$^{-3}$]

R, specific gas constant = 287 [J kg$^{-1}$ K$^{-1}$]

$T_{Kv}$, virtual temperature [K]

$T_K$, absolute temperature [K]: $T_K$ = 273.16 + T [°C]

$e_a$, actual vapor pressure [kPa]

T, mean daily temperature for 24-hour calculation time steps.

For average conditions ($e_a$ in the range 1–5 kPa and P between 80 and 100 kPa), $T_{Kv}$ can be substituted by: $T_{Kv} \approx$ 1.01 (T + 273)

## 4. Saturation Vapor Pressure (e$_s$)

$$e^0(T) = 0.611\exp\left(\frac{17.27T}{T + 273.3}\right)$$

where: e°(T), saturation vapor pressure function [kPa]

T, air temperature [°C]

## 5. Slope Vapor Pressure Curve (Δ)

$$\Delta = \frac{4098e^0(T)}{(T + 273.3)^2} = \frac{2504\exp\left(\dfrac{17.27T}{T + 273.2}\right)}{(T + 273.3)^2}$$

where: Δ, slope vapor pressure curve [kPa C$^{-1}$]

T, air temperature [°C]

e°(T), saturation vapor pressure at temperature T [kPa]

In 24-hour calculations, Δ is calculated using mean daily air temperature. In hourly calculations T refers to the hourly mean, $T_{hr}$.

## 6. Psychrometric Constant (γ)

$$\gamma = \frac{C_pP}{\varepsilon\lambda}\times10^{-3} = 0.00163\frac{P}{\lambda}$$

where:   γ, psychrometric constant [kPa C$^{-1}$]

$c_{p,}$ specific heat of moist air = 1.013 [kJ kg$^{-1}$ °C$^{-1}$]

P, atmospheric pressure [kPa]: equations 2 or 4

ε, ratio molecular weight of water vapor/dry air = 0.622

λ, latent heat of vaporization [MJ kg$^{-1}$]

## 7. Dew Point Temperature ($T_{dew}$):

When it is not observed, $T_{dew}$ can be computed from $e_a$ by:

$$T_{dew} = \frac{116.91 + 237.3 In(e_a)}{16.78 - In(e_a)}$$

where: $T_{dew,}$ dew point temperature [°C]

$e_{a,}$ actual vapor pressure [kPa]

For the case of measurements with the Assmann psychrometer, $T_{dew}$ can be calculated from:

$$T_{dew} = (112 + 0.9 T_{wet}) \left( \frac{e_a}{e^0 (T_{wet})} \right)^{1/8} - 112 + 0.1 T_{wet}$$

## 8. Short Wave Radiation on a Clear-Sky Day ($R_{so}$)

The calculation of $R_{so}$ is required for computing net long wave radiation and for checking calibration of pyranometers and integrity of $R_{so}$ data. A good approximation for $R_{so}$ for daily and hourly periods is:

$R_{so} = (0.75 + 2 \times 10^{-5} z) R_a$

where: z, station elevation [m]

$R_{a,}$ extraterrestrial radiation [MJ m$^{-2}$ d$^{-1}$]

Equation is valid for station elevations less than 6000 m having low air turbidity. The equation was developed by linearzing Beer's radiation extinction law as a function of station elevation and assuming that the average angle of the sun above the horizon is about 50°.

For areas of high turbidity caused by pollution or airborne dust or for regions where the sun angle is significantly less than 50° so that the path length of radiation through the atmosphere is increased, an adoption of Beer's law can be employed where P is used to represent atmospheric mass:

$$R_{50} = R_a \left( \frac{-0.0018 P}{K_t \sin \phi} \right)$$

where: $K_t$ turbidity coefficient, $0 < K_t \leq 1.0$ where $K_t = 1.0$ for clean air and $K_t = 1.0$ for extremely turbid, dusty or polluted air.

  P, atmospheric pressure [kPa]

  Φ, angle of the sun above the horizon [rad]

  $R_{a,}$ extraterrestrial radiation [MJ m$^{-2}$ d$^{-1}$]

For hourly or shorter periods, $\Phi$ is calculated as:

$$\sin \Phi = \sin \varphi \sin \delta + \cos \varphi \cos \delta \cos \omega$$

where: $\varphi$, latitude [rad]

$\delta$, solar declination [rad]

$\omega$, solar time angle at midpoint of hourly or shorter period [rad]

For 24-hour periods, the mean daily sun angle, weighted according to $R_a$, can be approximated as:

$$\sin \phi_{24} = \sin\left[ 0.85 + 0.3\varphi\sin\left(\frac{2\pi}{365}J - 1.39\right) - 0.42\varphi^2 \right]$$

where: $\Phi_{24,}$ average $\Phi$ during the daylight period, weighted according to $R_a$ [rad]

$\varphi$, latitude [rad]

J, day in the year []

The $\Phi_{24}$ variable is used to represent the average sun angle during daylight hours and has been weighted to represent integrated 24-hour transmission effects on 24-hour $R_{so}$ by the atmosphere. $\Phi_{24}$ should be limited to $\geq 0$. In some situations, the estimation for $R_{so}$ can be improved by modifying to consider the effects of water vapor on short wave absorption, so that:

$$R_{so} = (K_B + K_D) R_a \text{ where:}$$

$$K_B = 0.98 \exp\left[ \frac{0.00146P}{K_t \sin \phi} - \left(\frac{W}{\sin \phi}\right)^{0.25} \right]$$

where: $K_{B,}$ the clearness index for direct beam radiation

$K_{D,}$ the corresponding index for diffuse beam radiation

$K_D = 0.35 - 0.33 K_B$ for $K_B \geq 0.15$

$K_D = 0.18 + 0.82 K_B$ for $K_B < 0.15$

$R_{a,}$ extraterrestrial radiation [MJ m$^{-2}$ d$^{-1}$]

$K_{t,}$ turbidity coefficient, $0 < K_t \leq 1.0$ where $K_t = 1.0$ for clean air and $K_t = 1.0$ for extremely turbid, dusty or polluted air.

P, atmospheric pressure [kPa]

$\Phi$, angle of the sun above the horizon [rad]

W, precipitable water in the atmosphere [mm] $= 0.14 e_a P + 2.1$

$e_{a,}$ actual vapor pressure [kPa]

P, atmospheric pressure [kPa]

# APPENDIX J
# PSYCHROMETRIC CHART AT SEA LEVEL

PSYCHROMETRIC CHART

Sea Level
Baro METRI PRESSURE 29.921 inches of

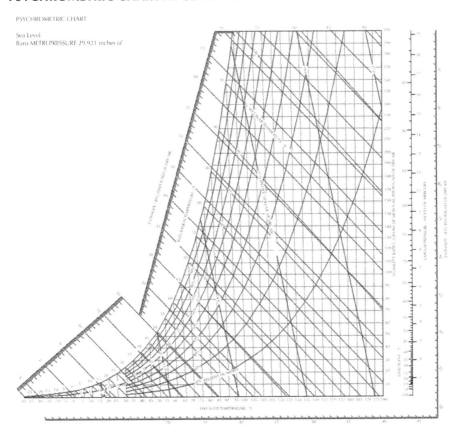

# Bibliography

## 1

**Literature cited**

1. Bidwell, R.G.S., 1979. *Plant physiology.* New York: McMillan. Pages 247–250.

2. Bonnet, J.A., 1968. *La Ciencia del Suelo.* Colegio de Ingenieros y Agrimensores de Puerto Rico. G.P.O. Apartado 3845, San Juan, P.R. 00936. Páginas 107–124.

3. Goyal, M.R., J.A. Santaella y L.E. Rivera, 1982. *El Tensiómetro: Su Uso, Intalación y Mantenimiento.* Servicio de Extensión Agrícola, Universidad de Puerto Rico, RUM. Páginas 1–17. IA 73–Serie 1.

4. *Glossary of Soil and Water Terms*, 1967. Special Publication SP-04-67. St. Joseph – Michigan: American Society of Agricultural and Biological Engineers.

5. *Glossary of Soil Science Terms*, 1979. Soil Science Society of America. 677 South Segoe Road. Madison-Wisconsin 53711, USA.

6. Hillel, D., 1982. *Introduction to soil physics.* New York: Academic Press.

7. Israelsen, Oscar y V.E. Hansen, 1964. *Principios y Aplicaciones del Riego.* Editorial Revelté, S. A. Barcelona.

8. Lugo López, M.A., 1953. *Moisture relationships of puerto rico soils.* Technical Paper No. 9. Agricultural Experiment Station, University of Puerto Rico, Río Piedras, Puerto Rico.

9. Muñoz, O. y V.R. Rodríguez, 1986. *Apreciación y Clasificación de Suelos en Puerto Rico.* Colegio de Ciencias Agrícolas, Universidad de Puerto Rico.

10. Noi, Y., 1967. *Relaciones Suelo-Agua.* Servicio de Extensión Agrícola, Israel, Departamento de Capacitación para el Extranjero, Israel.

11. Thorne, D.W. and M.D. Thorne, 1979. *Soil water and crop production.* Wesport, CT: AVI Publishing Company.

12. *Western Fertilizer Hand Book*, 1980. 6th ed., California Fertilizer Association. 2222 Watt Avenue, Sacramento, California 95825.

13. Wildman, W.E. and K.D. Gowans, 1978. *Soil physical environment and how it affects plant production.* Leaflet 2280. Division of Agricultural Sciences, University of California, Berkley, CA.

14. Withers, B. and S. Vipond, 1980. *Irrigation: Design and practice.* Ithaca, NY: Cornell University Press.

**Books/bulletins/journal and proceedings/reports**

Abegaonkar, M.P., R.N. Karekar and R.C. Ayer, 1999. A microwave micro-strip ring resonator as a moisture sensor for biomaterials: Application to wheat grains. *Measuring Science & Technology*, 10: 195–200.

Agehara, S. and D.D. Warncke, 2005. Soil moisture and temperature effects on nitrogen release from organic nitrogen sources. *Soil Science Society of America Journal*, 69: 1844–1855.

Amente, G., J.M. Baker and C.F. Reece, 2000. Estimation of soil solution electrical conductivity from bulk soil electrical conductivity in sandy soils. *Soil Science Society of America Journal*, 64: 1931–1939.

Baumhardt, R.L., R.J. Lascano and S.R. Evett, 2000. Soil material, temperature and salinity effects on calibration of multisensor capacitance probes. *Soil Science Society of America Journal*, 64: 1940–1946.

Bitteli, M., M. Flury and G.S. Campbell, 2003. A thermoelectric analyzer to measure the freezing and moisture characteristic of porous media. *Water Resources Research*, 39: 1041.

Bolvin, H., A. Chambarel and A. Chanzy, 2004. Three-dimensional numerical modeling of a capacitance probe: Application to measurement interpretation. *Soil Science Society of America Journal*, 68: 440–446.

Bryan, R.B., R.M. Hawke and D.L. Rockwell, 1999. Automated micro standpipe system for soil erosion research. *Soil Science Society of America Journal*, 63: 977–987.

Cantona, Y., A. Solé-Benet and F. Domingo, 2004. Temporal and spatial patterns of soil moisture in semiarid badlands of SE Spain. *Journal of Hydrology*, 285: 199–214.

Carcione, J.M. and G. Serian, 2000. An electromagnetic modeling tool for the detection of hydrocarbons in the subsoil. *Geophysica Prospecting*, 48: 231–256.

Chandler, D.G., M. Seyfried, M. Murdock and J.P. McNamara, 2004. Field calibration of water content reflectometers. *Soil Science Society of America Journal*, 68: 1501–1507.

Chertkov, V.Y., I. Ravina and V. Zadoenko, 2004. An approach for estimating the shrinkage geometry factor at moisture content. *Soil Science Society of America Journal*, 68: 1807–1817.

Chih-Pin, L., 2003. Frequency domain versus travel time analyses of TDR waveforms for soil moisture measurements. *Soil Science Society of America Journal*, 67: 720–729.

Constantini, E.A.C., F. Castelli, S. Raimondi and P. Lorenzoni, 2002. Asssessing soil moisture probe. *Soil Science Society of America Journal*, 66: 1889–1896.

Cornelis, W., J. Ronsyn, M. Van Meirvernne and R. Hartmann, 2001. Evaluation of pedotransfer functions for predicting the soil moisture retention curve. *Soil Science Society of America Journal*, 65: 638–648.

Das, B.S., J.M. Wraith and W.P. Inskeep, 1999. Nitrate concentrations in the root zone estimated using time domain reflectometry. *Soil Science Society of America Journal*, 63: 1561–1570.

Eigenberg, R.A., J.A. Nienaber, B.L. Woodbury and R.B. Ferguson, 2006. Soil conductivity as a measure of soil and crop status. *Soil Science Society of America Journal*, 70: 1600–1611.

Evett, S. and J.P. Laurent, 2000. Soil water components based on capacitance probes in a sandy soil. *Soil Science Society of America Journal*, 64: 311–318.

Evett, S., J.P. Laurent, P. Cepuder and C. Hignett, 2002. Neutron scattering, capacitance and TDR soil water content measurements on four continents. *17th World Congress of Soil Science (WCSS)*, August 14–21, Bangkok-Thailand, 1010–1021.

Friel, R. and D. Or, 1999. Frequency analysis of time-domain reflectometry (TDR) with application to dielectric spectroscopy of soil constituents. *Geophysics*, 64: 707–718.

Gish, T.J., W.P. Dulaney, K.J.S. Kung, C.S.T. Daughtry, J.A. Doolittle and P.T. Miller, 2002. Evaluating use of ground-penetrating radar for identifying subsurface flow pathways. *Soil Science Society of America Journal*, 66: 1620–1629.

Hook, W.R., T.P.A. Ferre and N.J. Livingston, 2004. The effects of salinity on the accuracy and uncertainty of water content measurement. *Soil Science Society of America Journal*, 68: 47–56.

Huisman, J.A., A.H. Weerts, T.J. Heimovaara and W. Bouten, 2002. Comparison of travel time analysis and inverse modeling for soil water content determination with time domain reflectometry. *Water Resources Research*, 38: 13–18.

Jones, S.B. and D. Or, 2004. Frequency domain analysis for extending time domain reflectometry water content measurement in highly saline soils. *Soil Science Society of America Journal*, 68: 1568–1577.

Lascano, Robert J., 2000. A general system to measure and calculate daily crop water use. *Agronomy Journal*, 92: 821–832.

Lebron, I., D.A. Robinson, S. Goldberg and S.M. Lesch, 2004. The dielectric permittivity of calcite and arid zone soils with carbonate minerals. *Soil Science Society of America Journal*, 68: 1549–1559.

Lee, J., R. Horton and D.B. Jaynes, 2000. A time domain reflectometry method to measure immobile water content and mass exchange coefficient. *Soil Science Society of America Journal*, 64: 1911–1917.

Lobell, D.B. and G.P. Asner, 2002. Moisture effects on soil reflectance. *Soil Science Society of America Journal*, 66: 722–727

Long, D.S., J.M. Wraith and G. Kegel, 2002. A heavy-duty time domain reflectometry soil moisture probe for use in intensive field sampling. *Soil Science Society of America Journal*, 66: 396–401.

Mamedov, A.I., C. Huang and G.J. Levy, 2006. Antecedent moisture Content and aging duration effects on seal formation and erosion in specific soils. *Soil Science Society of America Journal*, 70: 832–843.

Martínez-Fernández, J. and A. Ceballos, 2003. Temporal stability of oil moisture in a large-field experiment in spain. *Soil Science Society of America Journal*, 67: 1647–1656.

Matías, L.R., G.A. Bollero, R.G. Hoeft and D.G. Bullock, 2005. Spatial variability of the Illinois soil nitrogen test: Implications for soil sampling. *Agronomy Journal*, 97: 1485–1492.

Morgan, K.T., L.R. Parsons, T.A. Wheaton, D.J. Pitts and T.A. Obreza, 1999. Field calibration of a capacitance water content probe in fine sand soils. *Soil Science Society of America Journal*, 63: 987–989.

Nadler, A., A. Gamliel and I. Peret, 1999. Practical aspects of salinity effect on TDR-measured water content: A field study. *Soil Science Society of America Journal*, 63: 1070–1076.

Nadler, A., S.R. Green, I. Vogeler and B.E. Clothier, 2002. Horizontal and vertical TDR measurements of soil water content and electrical conductivity. *Soil Science Society of America Journal*, 66: 735–743.

Nissen, H., P. Moldrup and T. Olesen, 1999. Printed circuit board time domain reflectometry probe: Measurements of soil water content. *Soil Science Society of America Journal*, 64: 454–466.

Nissen, H., P. Moldrup, L. de Jonge and O. Jacobsen, 1999. Time domain reflectometry coil probe measurements of water content during fingered flow. *Soil Science Society of America Journal*, 63: 493–500.

Noborio, K., R. Horton and C.S. Tan, 1999. Time domain reflectometry probe for simultaneous measurement of soil matric potential and water content. *Soil Science Society of America Journal*, 63: 1500–1505.

O'Brien, J.J. and S.F. Oberbauser, 2001. An inexpensive portable meter for measuring soil moisture. *Soil Science Society of America Journal*, 65: 1081–1083.

Ochsner, T.E., R. Horton and T. Ren, 2001. Simultaneous water content, air-filled porosity and bulk density measurements with thermo-time domain reflectometry. *Soil Science Society of America Journal*, 65: 1618–1622.

Or, D. 2001. Who invented the tensiometer? *Soil Science Society of America Journal*, 65: 1–3.

Persson, M. and J. Wraith, 2002. Shaft mounted time domain reflectometry probe for water content and electrical conductivity measurements. *Vadose Zone Journal*, 1: 316–319.

Persson, M., B. Sivakumar, R. Berndtsson, O.H. Jacobsen and P. Schjonning, 2002. Predicting the dielectric constant-water content relationship using artificial neural networks. *Soil Science Society of America Journal*, 66: 1424–1429.

Quinones, H. and P. Ruelle, 2001. Operative calibration methodology of a TDR sensor for soil moisture monitoring under irrigated crops. *Subsurface Sensing Technologies and Applications*, 2: 31–45.

Ren, T., K. Noborio and R. Horton, 1999. Measuring soil water content, electrical conductivity and thermal properties with a thermo-time domain reflectometry probe. *Soil Science Society of America Journal*, 63: 450–457.

Robinson, D.A., 2004. Measurement of the solid dielectric permittivity of clay minerals and granular samples using a time domain reflectometry immersion method. *Vadose Zone Journal*, 3: 705–713.

Ruth, B., 1999. A capacitance sensor with planar sensitivity for monitoring soil water content. *Soil Science Society of America Journal*, 63: 48–54.

Seyfried, M.S. and M.D. Murdock, 2001. Response of a new soil water sensor to variable soil, water content and temperature. *Soil Science Society of America Journal*, 65: 28–34.

Seyfried, M.S. and M.D. Murdock, 2004. Measurement of soil water content with a 50-MHz soil dielectric sensor. *Soil Science Society of America Journal*, 68: 394–403.

Slima, M.B., R.Z. Morawski, A.W. Kraszewski and A. Barwicz, 1999. Calibration of a microwave system for measuring grain moisture content. *IEEE Transactions on Instrumentation and Measurement*, 48: 778–782.

Souza, C.F., D. Or and E.E. Matsura, 2004. A variable-volume TDR probe for measuring water content in large soil volumes. *Soil Science Society of America Journal*, 68: 25–31.

Starr, G., B. Lowery, E. Cooley and G. Hart, 1999. Soil water content determination using network analyzer reflectometry methods. *Soil Science Society of America Journal*, 63: 285–289.

Starr, G., B. Lowery, E. Cooley, 1999. Development of a resonant length technique for soil water content measurement. *Soil Science Society of America Journal*, 63: 278–285.

Suresh, S.J. and V.M. Naik, 2002. Theory of dielectric constant of aqueous solutions. *The Journal of Chemical Physics*, 116: 4212–4220.

Suwansawat, S. and C.H. Benson, 1999. Cell size for water content-dielectric constant calibrations for time domain reflectometry. *ASTM Geotechnical Testing Journal*, 22: 3–12.

Tabbagh, A., C. Camerlynck and P. Cosenza, 2000. Numerical modeling for investigating the physical meaning of the relationship between relative dielectric permittivity and water content of soils. *Water Resources Research*, 36: 2771–2776.

Tomer, M.D., B.E. Clothier, I. Vogeler and S. Green, 1999. A dielectric-water content relationship for sandy Volcanic Soils in New Zealand. *Soil Science Society of America Journal*, 63: 777–781.

Topp, G.C., S. Zegelin and I. White, 2000. Impacts of the real and imaginary components of relative permittivity on time domain reflectometry measurements in soils. *Soil Science Society of America Journal*, 64: 1244–1252.

Vaz, C.M.P. and J.W. Hopmans, 2001. Simultaneous measurement of soil penetration resistance and water content with a combined penetrometer-TDR moisture probe. *Soil Science Society of America Journal*, 65: 4–12.

Vaz, C.M.P., J.W. Hopmans, A. Macedo, L.H. Bassoi and D. Wildenschild, 2002. Soil water retention measurements using a combined tensiometer-coiled time domain reflectometry probe. *Soil Science Society of America Journal*, 66: 1752–1759.

Veldkamp, E. and J.J. O'Brien, 2000. Calibration of a frequency domain reflectometry sensor for humid tropical soils of volcanic origin. *Soil Science Society of America Journal*, 64: 1549–1553.

Walker, J.P. and P.R. Houser, 2002. Evaluation of the Ohm mapper instrument for soil moisture measurement. *Soil Science Society of America Journal*, 66: 728–734.

Wraith, J.M. and D. Or, 2001. Soil water characteristic determination from concurrent water content measurements in reference porous media. *Soil Science Society of America Journal*, 65: 1659–1666.

Xu, H., Z. Cai and H. Tsuruta, 2003. Soil moisture between rice-growing seasons affects methane emission, production and oxidation. *Soil Science Society of America Journal*, 67: 1147–1157.

Yu, C., A.W. Warrick and M.H. Conklin, 1999. Derived functions of time domain reflectometry for soil moisture measurement. Water Resources Research, 35: 1789–1796.

Zhang, N., G. Fan, K.H. Lee, G.J. Kluitenberg and T.M. Loughin, 2004. Simultaneous measurement of soil water content and salinity using a frequency-response method. *Soil Science Society of America Journal*, 68: 1515–1525.

**Web page links**

Bauder, T., I. Broner, J. Schneekloth and R. Waskom, 2006. Measurement of soil moisture. http://www.ext.colostate.edu/drought/soilmoist.html

Bittelli, M. Innovative methods for measuring soil water content. Department of Agro-Environmental Science and Technology, University of Bologna, Italy. www.distagenomics.unibo.it/wuemed/M_bittelli_innovative_methods.pdf

Croney, D. The electrical resistance method of measuring soil moisture. http://www.iop.org/EJ/abstract/0508-3443/2/4/301

Haman, D.Z. and T.H. Yeager, 2003. Controlling irrigation with tensiometers and time domain reflectometry (TDR). http://www.fngla.org/reports/25/finalreport.pdf

Hellman Editor. Irrigation scheduling of grapevines with evapotranspiration Data. http://winegrapes.tamu.edu/grow/scheduling.shtml

Herrera, Esteban. A practical way of measuring soil moisture. College of Agriculture and Home Economic, New Mexico State University. http://cahe.nmsu.edu/pubs/_h/h-637.html

ICT International. Soil moisture measurement Instrumentation. http://www.ictinternational.com.au/appnotes/ICT227.htm

*Leib, B.G., J.D. Jabro and G.R Matthews.* Field evaluation and performance comparison of soil moisture sensors. http://www.soilsci.com/pt/re/soilsci/abstract.

Measuring humidity using capacitive sensors. http://cr4.globalspec.com/thread/2318/Measuring-HUmidity-Using-Capacitive

Mecham, B.Q. Using soil moisture sensors to control landscape irrigation. http://www.ncwcd.org/ims/ims_info/usings1d.pdf

Metelerkamp, B., 2001. Sowacs (Soil water content sensors and measurement). www.sowacs.com.

Morris, M. Soil moisture monitoring: low-cost tools and methods. http://attra.ncat.org/attra-pub/soil_moisture.html

O'Brien, J.J. and S.F. Oberbauer. 2006 An inexpensive, portable meter for measuring soil moisture. Dept. of Biological Sciences, Florida International Univ., Miami-FL. http://soil.scijournals.org/cgi/content/full/65/4/1081

Prichard, T.L. Soil moisture management technology. University of California. http://www.cropinfo.net/AnnualReports/1997/instrumentation.wq.htm

Reed, Michael P. Finn. Measuring soil moisture in remotely sensed images. www.isprs.org/publications/related/ISRSE/html/papers/594.pdf

Schneekloth, J., T. Bauder, I. Broner and R. Waskom, 2002. Measurement of soil moisture. Colorado State University, Cooperative Extension, Drought and Fire Tip Sheets. www.ext.colostate.edu/drought/soilmoist.html

Shock, C. Instrumentation for soil moisture determination. University of California. http://ucce.ucdavis.edu.files.filelibrary/40/975.pdf

Soil moisture measurement instrumentation, 2006. http://www.ictinternational.com.au/appnotes/ICT227.htm

Sommers, L. and L. Wilding, 2006, 18th World Congress of Soil Science. http://www.iuss.org/bull108files/WCSS.htm

Valente, A., R. Morais, J. Boaventura Cunha, J.H. Correia and C. Couto, 2003. A multi-chip-module micro-system for soil moisture measures. date.hu/efita2003/centre/pdf/1002.pdf#search=%22methods%20%20measure%20soil%20moisture%22

Van der Gulik, T. Irrigation scheduling techniques. Ministry of Agriculture and Lands, Canada. www.agf.gov.bc.ca/resmgmt/publist/500series/577100-1.pdf#search=%22methods%20%20measure%20soil%20moisture%22

Werner, H. Measuring soil moisture for irrigation water management. South Dakota State University. http://agbiopubs.sdstate.edu/articles/FS876.pdf#search=%22methods%20%20measure%20soil%20moisture%22

Ziemer, R.R and N.A. Macgillivray. Measuring moisture near soil surface: minor differences due to neutron source type. http://www.fs.fed.us/psw/rsl/projects/water/Ziemer67.PDF#search=%22methods%20%20measure%20soil%20moisture%22

# 2

## Literature cited

1. Allen, R.G., 1986. A Penman for all seasons. *Journal Irrigation and Drainage Division of ASCE*, 112(4): 348–368.

2. Allen, R.G. and W.O. Pruitt, 1986. Rational use of the FAO Blaney - Criddle formula. *Journal Irrigation and Drainage Division of ASCE*, 112(2): 139–155.

3. *Climatic Summary of the United States*. U.S. Dept. of Commerce, U.S. Printing Office Washington, D.C. Report 86–45, Page. 40.

4. Doorenbos, J. and W.O. Pruitt, 1977. *Crop water requirements*. FAO irrigation and drainage Division of ASCE. Paper 24, Food and agriculture organization of the United Nations, Rome. Pages 1–156.

5. Doorenbos, J. and W.O. Pruitt, 1979. *Yield response to water*. FAO Irrigation and Drainage. Paper 33, Food and Agriculture Organization of the United Nations, Rome.

6. Goyal, M.R., 1988. Potential evapotranspiration for the South Coast of Puerto Rico with the Hargreaves – Samani technique. *Journal of Agriculture of the University of Puerto Rico*, 72(1): 41–50

7. Hargreaves, G. H., 1974. Estimation of potential and crop evapotranspiration. *Transactions of ASAE*, 17: 701–704.

8. Hargreaves, G.H. and S.A. Samani, 1985. Reference crop evapotranspiration from temperature. *Applied Engineering in Agriculture*. ASAE, 1(2): 96–99.

9. Hargreaves, G.H. and Z.A. Samani, 1986. *World water for agriculture precipitation management*. International Irrigation Center, Utah State University, Logan, UT, USA. Pages. 1–617.

10. Jensen, M. E., 1980. Design and Operation of Farm Irrigation Systems. *ASAE Monograph #3*, American Society of Agricultural Engineers, Chapter 6: 189–225.

11. Jones, J.W., L.H. Allen, S.F. Shih, J.S. Rogers, L.C. Hammond, A.G. Smajstrala and J.D. Martsolf, 1984. *Estimated and measured evapotranspiration for florida climatic, crops and soils.* Agricultural Experiment Station, Institute of Food and Agricultural Sciences, University of Florida, Gainesville. Bulletin No. 840, Pages 1–65.

12. Linacre, E.T., 1977. A simple formula for estimating evapotranspiration rates in various climates, using temperature data along. *Agricultural Meteorology*, 18: 409–424.

13. Michael, A.M., 1978. *Irrigation theory and practice.* New Delhi: Vikas Publishing, Chapter 7: 448–584.

14. Penman, H.L., 1949. The dependence of transpiring on climate and soil conditions. *Journal of Soil Sciences*, 1: 74–89.

15. Rosenberg, N.J., B.L. Blad and S.B. Verma, 1983. *Microclimate: The biological environment.* A Wiley- Interscience. Chapter 7: 209–287.

16. Salih, A.M. and U. Sendil, 1984. Evapotranspiration under extremely arid climates. *Journal of Irrigation and Drainage Division of ASCE*, 110 (IR3): 289–303.

17. Shih, S.F., 1984. Data requirement for evapotranspiration estimation. *Journal of Irrigation and Drainage Division of ASCE*, 110(IR3): 263–274.

18. Wright, J.L., 1982. New evapotranspiration crop coefficients. *Journal of Irrigation and Drainage Division of ASCE*, 108 (IR1): 57–74.

**2.2 Books/bulletins/journals and proceedings/reports**

**General references**

Ajayi, A.E. and A.A. Olufayo, 2004. Evaluation of two temperature stress indices to estimate grain sorghum yield and evapotranspiration. *Soil Science Society of America Journal*, 68: 1282–1287.

Allen, L.H., Jr., Deyun Pan, K.J. Boote, N.B. Pickering and J.W. Jones, 2003. Carbon dioxide and temperature effects on evapotranspiration and water use efficiency of soybean. *Agronomy Journal*, 95: 1071–81.

Allen, L.H., Jr., D. Pan, K.J. Boote, N.B. Pickering and J.W. Jones, 2003. Carbon dioxide and temperature effects on evapotranspiration and water use efficiency of soybean. *Soil Science Society of America Journal*, 67: 1071–1081.

Allen, R.G., 2005. *crop evapotranspiration: Guidelines for computing crop water requirements.* Food & Agriculture Organization of the United States. Publication 23765. Pages 1–328.

Allen, R.G., M. Smith, A. Perrier and L.S. Pereira, 1994. An update for the definition of reference evapotranspiration. *ICID Bulletin*, 43(2): 1–34.

Allen, Richard G., I.A. Walter, R.L. Elliot, T.A. Howell, D. Itenfisu, M.E. Jensen and R.L. Snyder, eds, 2005. *The ASCE Standardized Reference Evapotranspiration Equation.* The American Society of Civil Engineers, Reston: Virginia Publication 23136. Pages 1–216.

Baird, K.J., 2006. Development of a new methodology for estimating groundwater evapotranspiration. ProQuest / UMI. Pages 1–194.

Beyazgul, M., Y. Kayan and F. Engelsman, 2000. Estimation methods for crop water requirements in the Gediz Basin of Western Turkey. *Journal of Hydrology*, (Amsterdam), 229: 19–26.

Bonan, G.B., 2002. *Ecological climatology: Concepts and applications.* Cambridge University Press. Pages 141–146.

Boote, K.J., J.W. Jones, W.D. Batchelor, E.D. Nafziger and O. Myers, 2003. Genetic coefficients in the CROPGRO–Soybean model: Links to field performance and genomics, *Agronomy Journal*, 95: 32–51.

Bremer, D.J., 2003. Evaluation of microlysimeters used in turfgrass evapotranspiration studies using the dual-probe heat-pulse technique. *Soil Science Society of America Journal*, 67: 1625–1632.

Caballero, R., A. Bustos and R. Román, 2001. Soil salinity under traditional and improved irrigation schedules in Central Spain. *Soil Science Society of America Journal*, 65: 1210–1218.

Cason, K., 2002. Elevated drip irrigation system works. *Countryside & Small Stock Journal*, 87(3): 1–20.

Chamran, F., P.E. Gessler and O.A. Chadwick, 2002. Spatially explicit treatment of soil-water dynamics along a semiarid catena. *Soil Science Society of America Journal*, 66: 1571–1583.

Dasberg, S., 2000. *Drip irrigation*. New York: Springer-Verlag Telos. Pages 85–97.

Donatelli, M., C.O. Stöckle, R.L. Nelson and G. Bellocchi, 2003. ET-CSDLL: A dynamic link library for the computation of reference and crop evapotranspiration. *Soil Science Society of America Journal*, 67: 1334–1336.

Doorenbos, J. and W.O. Pruitt, 1975. *Guidelines for predicting crop water requirements*. Irrigation and Drainage Paper 24, Food and Agriculture Organization of the United Nations, Rome. Page 179.

Droogers, P., 2000. Estimating actual **evapotranspiration** using a detailed agro-hydrological model. *Journal of Hydrology*, (Amsterdam), 229: 50–58.

Fan, T., B.A. Stewart, W.A. Payne, W. Yong, J. Luo and Y. Gao, 2005. Long-term fertilizer and water availability effects on cereal yield and soil chemical properties in Northwest China. *Soil Science Society of America Journal*, 69: 842–855.

Fares, A. and A.K. Alva, 2000. Evaluating the capacitance probes for optimal irrigation of citrus through soil moisture monitoring in an entisol profile. *Irrigation Science*, 19: 57–64.

Frank, A.B., 2003. Evapotranspiration from northern semiarid grasslands. *Soil Science Society of America Journal*, 67: 1504–1509.

Harmsen, E.W., 2003. Fifty years of crop evapotranspiration studies in Puerto Rico. *Journal of Soil and Water Conservation*, 58(4): 214–223.

Hunt, L.A., M.P. Reynolds, K.D. Sayre, S. Rajaram, J.W. White and W. Yan, 2003. Crop modeling and the identification of stable coefficients that may reflect significant groups of genes, *Agronomy Journal*, 95: 20–31.

Ibáñez, M. and F. Castellví, 2000. Simplifying daily evapotranspiration estimates over short full-canopy crops. *Soil Science Society of America Journal*, 64: 628–632.

Jenkinson, B.J., D.P. Franzmeier and W.C. Lynn, 2002. Soil hydrology on an end moraine and a dissected till plain in West-Central Indiana. *Soil Science Society of America Journal*, 66: 1367–1376.

Jensen, M.E., R.D. Burman and R.G. Allen. 1990. *Evapotranspiration and irrigation water requirements*. ASCE Manuals and Reports on Engineering Practices No. 70, American Society of Civil Engineers, New York. p. 360.

Jones, F.E., 1992. *Evaporation of water*. CRC Press. Chapter 9: 123–140.

Kadlec, R.H., 2006. *Water temperature and evapotranspiration in surface flow wetlands in hot arid climate*. Elsevier. Pages 1–12.

Kite, G.W. and P. Droogers, 2000. Comparing evapotranspiration estimates from satellites, hydrological models and field data. Journal of Hydrology, (Amsterdam), 229: 3–18.

Lazarovitch, N., A. Ben-Gal and U. Shani, 2006. An automated rotating lysimeter system for greenhouse evapotranspiration studies. *Soil Science Society of America Journal*, 70: 801–804.

Levitt, D.G., M.J. Hartmann, K.C. Kisiel, C.W. Criswell, P. Dwain-Farley and C. Christensen, 2005. Comparison of the water balance of an asphalt cover and an evapotranspiration cover at technical area 49 at the Los Alamos National Laboratory. *Soil Science Society of America Journal*, 69: 789–797.

Mahmood, N., N. Shah, M. Ross and J. Vomacka, 2005. Evapotranspiration of two vegetation covers in a shallow water table environment. *Soil Science Society of America Journal*, 69: 492–499.

Mahmood, R. and K.G. Hubbard, 2003. Simulating sensitivity of soil moisture and evapotranspiration under heterogeneous soils and land uses. Journal of Hydrology, (Amsterdam), 280: 72–90.

Mansell, M.G., 2003. *Rural and urban hydrology*. London: Thomas Telford. Pages 52–179.

Martin, D.L. and J.R. Gilley, 1993. Irrigation Water Requirements. In *The SCS national engineering handbook*, Chapter 2: 284. Washington, DC: Soil Conservation Service.

Mavi, H.S., 2004. *Agrometeorology: Principles and applications of climate studies in agriculture*. New York: Haworth Press. pp. 76–83.

Mazor, I. and E. Mazor, 2003. *Chemical and isotopic groundwater hydrology*. Marcel Dekker. Page 20–25.

Mo, X., S. Liu, Z. Lin and W. Zhao, 2004. Simulating temporal and spatial variation of **evapotranspiration** over the Lushi basin. Journal of Hydrology, (Amsterdam), 285: 125–142.

Monteith, J.L. and M.H. Unsworth, 1990. *Principles of environmental physics*, 2nd ed., London: Edward Arnold.

Monteith, J.L., 1965. *Evaporation and environment*. 19th Symposia of the Society for Experimental Biology, Cambridge: Cambridge University Press, Pages 205–234.

Morgan, K.T., L.R. Parsona, T.A. Wheaton, D.J. Pitts and T.A. Oberza, 1999. Field calibration of a capacitance water content probe in fine sand soils. *Soil Science Society of America Journal*, 63: 987–989.

Morgan, K.T., T.A. Obreza, J.M.S. Scholberg, L.R. Parsons and T.A. Wheaton, 2006. Citrus water uptake dynamics on a sandy Florida entisol. *Soil Science Society of America Journal*, 70: 90–97.

Nachabe M., N. Shah, M. Ross and J. Vomacka, 2005. Evapotranspiration of two vegetation covers in a shallow water table environment. *Soil Science Society of America Journal*, 69(1): 492–499.

Naden, P.S., E.M. Blyth, P. Broadhurst, C.D. Watts and I.R. Wright, 2000. Modeling the spatial variation in soil moisture at the landscape scale: An application to five areas of ecological interest in the UK. *Hydrological Processes*, 14:785–809.

Nyhan, J.W., 2005. A seven-year water balance study of an evapotranspiration landfill cover varying in slope for semiarid regions. *Soil Science Society of America Journal*, 69: 466–480.

Payne, W.A., 1999. Shallow tillage with a traditional West African hoe to conserve oil water. *Soil Science Society of America Journal*, 63: 972–976.

Penman, H.L., 1948. Natural evaporation from open water, bare soil and grass. *Proceedigs of the Royal Society A*, 193: 120–146.

Penman, H.L., 1963. *Vegetation and hydrology*. Tech. Comm. No. 53, Commonwealth Bureau of Soils, Harpenden, England. Pages 125.

Pereira, L.S. and R.G. Allen, 1998. Crop water requirements. In *Handbook of agricultural engineering*, Chapter 1.5.1. CIGR and ASABE.

Pereira, L.S., A. Perrier, R.G. Allen and I. Alves, 1996. Evapotranspiration: Review of concepts and future trends. *Journal of Irrigation and Drainage Engineering*, ASCE 25.

Perrier, A., 1978. Importance des définitions de l'évapotranspiration dans le domaine pratique de la mesure, de l'estimation of de la notion de coefficients culturaux. *XV' Journal of Hydraulics, Société Hydrotechnique de France*, Question IV, Rapport 1:1–7 (in French).

Perrier, A., 1982. Land surface processes: vegetation. In *Land surface processes in atmospheric general circulation models*, ed. P.S. Eagleson, 395–448. Cambridge, MA: Cambridge University Press,.

Perrier, A., 1985. Updated evapotranspiration and crop water requirement definitions. In: Perrier, A. and Riou, C. (eds) *Crop water requirements* (ICID Int. Conf., Paris, Sept. 1984). INRA, Paris. Pages 885–887.

Rijtema, P. E., 1965. *Analysis of actual evapotranspiration*. Agricultural Research Report No. 69, Centre for Agricultural Publishing and Documentation, Wageningen.

Robock, A., K.Y. Vinnokov, G. Srinivasan, J.K. Entin, S. Hollinger, N.A. Spernskaya, S. Liu and A. Namkhai, 2000. The global soil moisture data bank. *Bulletin of the American Meteorological Society*, 81:1281–1299.

Ryszkowski, L., 2002. *Landscape ecology in agroecosystems management*. CRC Press. Pages 76–81.

Sau, F., K.J. Boote, W. McNair-Bostick, J.W. Jones and M. Inés-Mínguez, 2004. Testing and improving evapotranspiration and soil water balance of the DSSAT crop models. Agronomy Journal, 96: 1243–1257.

Shahin, M.A., 2002. *Hydrology and water resources of Africa*. Springer. Pages 157–212.

Sherrod, L.A., G.A. Peterson, D.G. Westfall and L.R. Ahuja, 2005. Soil organic carbon pools after 12 years in no-till dry land agroecosystems. *Soil Science Society of America Journal*, 69: 1600–1608.

Shock, C.C., 2001. *Drip irrigation: An introduction*. Oregon State University, Extension Service. Pages 25–32.

Simmers, I., 2003. *Understanding water in a dry environment: Hydrological processes in arid and semi-arid zones*. Taylor Francis: UK. Pages 92–99.

Slatyer, R.O. and I.C. McIlroy, 1961. Evaporation and the principle of its measurement. In

*Practical meteorology*, CSIRO (Australia) and UNESCO, Paris.

Smith, M., R.G. Allen, J.L. Monteith, A. Perrier, L. Pereira and A. Segeren, 1992. *Report of the expert consultation on procedures for revision of FAO guidelines for prediction of crop water requirements*. UN-FAO, Rome, Italy. Pages 54.

Sommer, R., H. Fölster, K. Vielhauer, E.J. Maklouf-Carvalho and P.L.G. Vlek, 2003. Deep soil water dynamics and depletion by secondary vegetation in the Eastern Amazon. *Soil Science Society of America Journal*, 67: 1672–1686.

Suleiman, A. and R. Crago, 2004. Hourly and daytime evapotranspiration from grassland using radiometric surface temperatures. *Soil Science Society of America Journal*, 68: 384–390.

Sumner, D. M., 2001. *Evapotranspiration from a cypress and pine forest subjected to natural fires, Volusia County Florida, 1998–99*. Water Resource. Investigations Rep. 01-4245. USGS, Reston, VA.

Thokal, R.T., D.M. Mahale and A. Powar, 2004. *Drip irrigation systems: Clogging and prevention*. Rajasthan (India): Pointer Publisher. Pages 32–40.

Thorsten, W., F. Stewart, H. Gupta, E. Bogh, L. Bastidas, C. Nobre and O. Galvao, 2005. *Regional hydrological impacts of climatic change: Impact assessment and decision making*. IAHS Press. Pages 35.

Tolk, J.A., T.A. Howell and S.R. Evett, 2006. Nighttime evapotranspiration from alfalfa and cotton in a semiarid climate. *Agronomy Journal*, 98: 730–736.

Vázquez, Z.R.F. and J. Feyen, 2001. *Effect of potential evapotranspiration estimates on the performance of the mike she code applied to a medium sized catchment*. Annual Meeting of American Society of Agricultural and Biological Engineers, 012037.

Ward, A.D and S. Trimble, 2004. *Environmental hydrology*. Lewis Publishers. Chapter 4: 83–115.

Yang, J., B. Li and S. Liu, 2000. A large weighing lysimeter for evapotranspiration and soil water-groundwater exchange studies. *Hydrological Processes*, 14:1887–1897.

Zhang, Y., Q. Yu, C. Liu, J. Jiang and X. Zhang, 2004. Estimation of winter wheat evapotranspiration under water stress with two semiempirical approaches. *Soil Science Society of America Journal*, 68: 159–168.

Zhu, Y., R.H. Fox and J.D. Toth, 2002. Leachate collection efficiency of zero-tension pan and passive capillary fiberglass wick lysimeters. *Soil Science Society of America Journal*, 66: 37–43.

**Evapotranspiration equations**

Allen, R.G. and W.O. Pruitt, 1986. Rational use of the FAO Blaney-Criddle formula. *Journal of Irrigation and Drainage Engineering, ASCE*, 112(IR2): 139–155.

Allen, R.G. and W.O. Pruitt, 1991. FAO-24 reference evapotranspiration factors. *Journal of Irrigation and Drainage Engineering, ASCE*, 117(5): 758–773.

Allen, R.G., 1986. A penman for all seasons. *Journal of Irrigation and Drainage Engineering*, 112(4): 348–368.

Allen, R.G., 1992. *Evaluation of a temperature difference method for computing grass reference evapotranspiration*. Report submitted to UN-FAO Water Resources Development and Management Service, Land and Water Development Division, Rome. Pages 50.

Allen, R.G., 1995. *Evaluation of procedures for estimating grass reference evapotranspiration using air temperature data only*. Report prepared for FAO, Water Resources Development and Management Service, FAO, Rome.

Allen, R.G., W.O. Pruitt, J.A. Businger, L.J. Fritschen, M.E. Jensen and F.H. Quinn, 1996. Evaporation and Transpiration. In *ASCE handbook of hydrology*. Chapter 4: 125–252, New York, NY.

Batchelor, C. H., 1984. The accuracy of evapotranspiration functions estimated with the FAO modified Penman equation. *Irrigation Science*, 5(4): 223–234.

Blaney, H.F. and W.D. Criddle, 1950. *Determining water requirements in irrigated areas from climatological and irrigation data*. USDA Soil Conservation Service SCS-TP96. Pages 44.

Brutsaert, W.H., 1982. *Evaporation into the atmosphere*. Dordrecht, Holland: R. Deidel Publishing Company.

Burman, R. and L.O. Pochop, 1994. *Evaporation, evapotranspiration and climatic data*. Amsterdam: Elsevier Science B.V.

Businger, J.A., 1956. Some remarks on Penman's equations for the evapotranspiration. *Netherlands Journal of Agricultural Science*, 4: 77.

Castrignanò, A., A. de Caro and E. Tarantino, 1985. Verifica sulla validità di alcuni metodi empirici di stima dell'evapotraspirazione potenziale nel Metapontino. (Verification of validity of several empirical methods of estimating potential evapotranspiration in southern Italy). L'Irrigazione XXXII (4): 23–28 (in Italian).

Chiew, F.H.S., N.N. Kamadalasa, H.M. Malano and T.A. McMahon, 1995. Penman-Monteith: FAO-24 reference crop evapotranspiration and class-A pan data in Australia. *Agricultural Water Management*, 28: 9–21.

Choisnel, E., O. de Villele and F. Lacroze, 1992. Une approche uniformisée du calcul de l'évapotranspiration potentielle pour l'ensemble des pays de la Communauté Européenne, Com. Commun. Européennes, EUR 14223 FR, Luxembourg, Pages 176.

Christiansen, J.E., 1968. Pan evaporation and evapotranspiration from climatic data. *Journal of Irrigation and Drainage Engineering, ASCE*, 94: 243–265.

Cuenca, R.H. and M.T. Nicholson, 1982. Application of the Penman equation wind function. *Journal of Irrigation and Drainage Engineering, ASCE*, 108(1): 13–23.

Doorenbos, J. and A.H. Kassam, 1979. *Yield response to water*. FAO Irrigation and Drainage. Paper No. 33, FAO, Rome, Italy. Pages 193.

Doorenbos, J. and W.O. Pruitt, 1975. *Guidelines for predicting crop water requirements*. Irrigation and Drainage, Paper 24, Food and Agriculture Organization of the United Nations, Rome, Pages 179.

Feddes, R.A., 1987. *Crop factors in relation to Makkink reference crop evapotranspiration*. Technical Bulletin Institute for Land and Water Management Research. No. 67, pp. 33–45.

Frére, M. and G.F. Popov, 1979. *Agrometeorological crop monitoring and forecasting*. FAO Plant Production and Protection. Paper No. 17. FAO, Rome, Italy. Pages 38–43.

Frevert, D.K., R.W. Hill and B.C. Braaten, 1983. Estimation of FAO evapotranspiration coefficients. *Journal of Irrigation and Drainage Engineering*, ASCE, 109(2): 265–270.

George, W., W.O. Pruitt and A. Dong, 1985. Evapotranspiration modeling. In *California irrigation Management Information System*. Final Report, by R. Snyder, D.W. Henderson, W.O. Pruitt and A. Dong, Calif. Dept. Water Resour. Contract. No. B53812. Land, Air and Water Resources Pap. 10013-A, Univ. Calif., Davis. Pages III-36–III-59.

Gosse, G., A. Perrier and B. Itier, 1977. Etude de l'évapotranspiration réelle d'une culture de blé dans le bassin parisien. *Annual Agronomy*, 28(5): 521–541. (in French).

Gunston, H. and C.H. Batchelor, 1983. A comparison of the Priestley–Taylor and Penman methods for estimating reference crop evapotranspiration in tropical countries. *Agricultural Water Management*, 6:65–77.

Hargreaves, G.H., 1983. Discussion of application of Penman wind function by R.H. Cuenca and M.J.J. Nicholson. *Journal of Irrigation and Drainage Engineering*, ASCE, 109(2): 277–278.

Hargreaves, G.L., G.H. Hargreaves and J.P. Riley, 1985. Agricultural benefits for Senegal river basin. *Journal of Irrigation and Drainage Engineering*, ASCE, 111: 113–124.

Hashemi, F. and M.T. Habibian, 1979. Limitations of temperature based methods in estimating crop evapotranspiration in arid-zone agricultural development project. *Agricultural Meteorology*, 20: 237–247.

Hatfield, J.L. and M. Fuchs, 1990. Evapotranspiration models. In *Management of farm irrigation systems*, eds. G.J. Hoffman, T.A. Howell, and K.H. Solomon, 33–59. St. Joseph, MI: ASAE.

Howell, T.A., A.D. Schneider, and M.E. Jensen, 1991. History of lysimeter design and use for evapotranspiration measurements. In *Lysimeters for evapotranspiration and environmental measurements*, eds. R.G. Allen, T.A. Howell, W.O. Pruitt, L.A. Walter and M.E. Jensen, 1–9. New York, NY: ASCE.

Itier, B. 1996. Measurement and estimation of evapotranspiration. In *Sustainability of irrigated agriculture*, eds. L.S. Pereira, R.A. Feddes, J.R. Gilley, B. Leseffre, 171–191. Dordrecht: Kluwer.

Itier, B. and A. Perrier, 1976. Présentation d'une étude analytique de l'advection: I. Advection liée aux variations horizontales de concentration et de température. *Annual Agronomy*, 27(2): 111–140.

Itier, B., Y. Brunet, K.J. Mcaneney and J.P. Lagouarde, 1994. Downwind evolution of scalar fluxes and surface resistance under conditions of local adection. Part I: A reappraisal of boundary conditions. Agricultural and Forest Meteorology, 71: 211–255.

Jensen, M.E. 1974. *Consumptive use of water and irrigation water requirements*. Rep. Tech. Com. on Irrig. Water Requirements. Irrigation and Drainage Engineering Division, ASCE, Pages 227.

Jensen, M.E. and H.R. Haise, 1963. Estimating evapotranspiration from solar radiation. *Journal of Irrigation and Drainage Engineering Division*, ASCE, 89:15–41.

Jensen, M.E., R.D. Burman and R.G. Allen, 1990. *Evapotranspiration and irrigation water requirements*. ASCE Manuals and Reports on Engineering Practices No. 70, New York, NY: ASCE, Page 360.

Katerji, N. and A. Perrier, 1983. Modélization de l'évapotranspiration réelle ETR d'une parcelle de luzerne: rôle d'un coefficient cultural. *Agronomie*, 3(6): 513–521 (in French).

Makkink, G.F., 1957. Testing the Penman formula by means of lysimeters. *Journal of the Institution of Water Engineers*, 11(3): 277–288.

McNaughton, K.G. and P.G. Jarvis, 1984. Using fee Penman-Monteith equation predictively. *Agricultural Water Management*, 8: 263–278.

Monteith, J.L. 1985. Evaporation from land surfaces: progress in analysis and prediction since 1948. In *Advances in Evapotranspiration: Proceedings of the ASAE Conference on Evapotranspiration*, Decenmber 16–17, Chicago, Ill. St. Joseph, MI: ASAE. Pages 4–12.

Monteith, J.L., 1973. *Principles of environmental physics*. London: Edward Arnold.

Monteith, J.L., 1981. Evaporation and surface temperature. *Quarterly Journal of the Royal Meteorological Society*, 107: 1–27.

Pelton, W.L., K.M. King and C.B. Tanner, 1960. An evaluation of the Thornthwaite and mean temperature methods for determining potential evapotranspiration. Agronomy Journal, 52: 387–395.

Penman, H.L., 1948. Natural evaporation from open water, bare soil and grass. *Proceedings of the Royal Society A*, 193: 120–146.

Penman, H.L., 1963. *Vegetation and hydrology*. Tech. Comm. No. 53, Commonwealth Bureau of Soils, Harpenden, England. Page 125.

Pereira, L.S. and M. Smith, 1989. Proposed procedures for revision of guidelines for predicting crop water requirements. *Land and water use division*, FAO Rome. Pages 36.

Phene, C.J., D.A. Dark and G.E. Cardon, 1996. Real time calculation of crop evapotranspiration using an automated pan evaporation system. In *Evaporation and irrigation scheduling*, eds. C.R. Camp, E.J. Sadler and R.E. Yoder, 189–194. ASCE

Piper, B.S., 1989. Sensitivity of Penman estimates of evaporation to errors in input data. *Agricultural Water Management*, 15:279–300.

Priestley, C.H.B. and R.J. Taylor, 1972. On the assessment of surface heat flux and evaporation using large scale parameters. *Monthly Weather Review*, 100: 81–92.

Pruitt, W.O., 1996. Empirical method of estimating evapotranspiration using primarily evaporation pans. *Proceedings of the Conference on Evapotranspiration and its Role in Water Resources Management*. December, Chicago. New York: ASAE, Pages 57–61.

Rosenberg, N.J., B.L. Blad and S.B. Verma, 1983. *Microclimate: The biological environment*, 2nd ed. New York: John Wiley.

Seemann, J., Y.I. Chirkov, J. Lomas and B. Primault, 1979. *Agrometeorology*. Berlin, Heidelberg: Springer Verlag.

Seguin, B., Y. Brunet and A. Perrier, 1982. Estimation of evaporation: A review of existing methods and recent developments. *European Geologic Society Symposium on Evaporation*. Leeds, U.K. Pages 21.

Sharma, M.L. 1985. Estimating evapotranspiration. In *Advances in irrigation*, ed. D. Hillel, 3: 213–281. New York: Academic Press.

Stewart, J.B., 1983. A discussion of the relationships between the principal forms of the combination equation for estimating evapotranspiration. *Agricultural Meteorology*, 30: 111–127.

Stewart, J.B., 1983. A discussion of the relationships between the principal forms of the combination equation for estimating evaportranspirations. *Agricultural Meteorology*, 30: 111–127.

Tanner, C.B. and M. Fuchs, 1968. Evaporation from unsaturated surfaces: A generalized combination equation. *Journal of Geophysical Research*, 73(4): 1299–1304.

Tanner, C.B. and W.L. Pelton, 1960. Potential evapotranspiration estimates by the approximate energy balance of Penman. *Journal of Geophysical Research*, 65(10): 3391–3413.

Thompson, N., L.A. Barrie and M. Ayles, 1981. *The meteorological office rainfall and evaporation calculation system: MORECS*. Hydrological Memorandum 45, Hydrometeorological Services, London, Page 66.

Thornthwaite, C.W., 1948. An approach toward a rational classification of climate. *Geography Review*, pp. 38, 55.

Turc, L., 1961. Evaluation des besoins en eau d'irrigation, évapotranspiration potentielle, formule climatique simplifiée et mise a jour. (in French). *Annual Agronomy*, 12:13–49.

Watts, P.J. and N.H. Hancock, 1985. Evaporation and potential evaporation: A practical approach for agricultural engineers. *Mechanical Engineering Translation*, 10(4): 231–240.

Wright, J.L., 1982. New evapotranspiration crop coefficients. *Journal of Irrigation and Drainage Engineering*, ASCE, 108 (2): 57–74.

Wright, J.L., 1988. Daily and seasonal evapotranspiration and yield of irrigated alfalfa in southern Idaho. *Agronomy Journal*, 80: 662–669.

**Evapotranspiration and weather measurements**

Allen, R.G., 1996. Assessing integrity of weather data for use in reference evapotranspiration estimation. *Journal of Irrigation and Drainage Engineering Division*, ASCE, 122(2): 97–106.

Allen, R.G., W.O. Pruitt and M.E. Jensen, 1991. Environmental requirements for lysimeters. In *Proceedings of the ASCE International Symposium on Lysimetry*, eds. R.G. Allen, T.A. Howell, W.O. Pruitt, L.A. Walter and M.E. Jensen, Lysimeters for Evapotranspiration and Environmental Measurements, Honolulu, HA. New York, NY: ASCE. Pages 170–181.

Bastiaanssen, W.G.M., 1995. *Regionalization of surface flux densities and moisture indicators in composite terrain*. Doctoral thesis, Wageningen Agricultural University, Wageningen, Page 273.

Beard, J.R., 1985. An assessment of water use by turfgrass. In *Turfgrass Water Conservation*, eds. V.A. Gibeault and S.T. Cockerham. Publ. 21405, University of California Division of Agricultural and Natural Resources, Berkley, CA. Pages. 45–60.

Biran, I., B. Bravdo, I. Bushkin-Harav and E. Rawitz, 1981. Water consumption and growth rate of 11 turfgrasses as affected by mowing height, irrigation frequency and soil moisture. *Agronomy Journal*, 73:85–90.

Blad, B.L. and N.J. Rosenberg, 1974. Lysimetric calibration of the Bowen-ratio energy balance method for evapotranspiration estimation in the Central Great plains. *Journal Applied Meteorology*, 13(2): 227–236.

Brutsaert, W.H., 1982. *Evaporation into the atmosphere*. Dordrecht, Holland: R. Deidel Publishing Company.

Businger, J.A. and A.M. Yaglom, 1971. Introduction to Obukhov's paper on turbulence in an atmosphere with a non-uniform temperature. *Boundary-Layer Meteorology*, 2: 3–6.

Businger, J.A., 1988. A note on the Businger-Dyer profiles. *Boundary-Layer Meteorology*, 42: 145–151.

Campbell, G.S. 1977. *An introduction to environmental biophysics*. N.Y. Springer Verlag, p. 159.

Carrijo, O.A. and R.H. Cuenca, 1992. Precision of evapotranspiration estimates using neutron probe. *Journal of Irrigation and Drainage Engineering*, ASCE, 118(6): 943–953.

Dolman, A.J. and J.B. Stewart, 1987. Modelling forest transpiration from climatological data. In *Forest hydrology and watershed management*, eds. R.H. Swanson, P.Y. Bernier and P.D. Woodard, 167: 319–327. IAHS Publ.

Fritschen, L.J. and C.L. Fritschen, 1991. *Design and evaluation of net radiometers*. Paper presented at the 7th Symposium on Meteorology Observations and Instrumentation, New Orleans, LA., USA. Page 5.

Gash, J.H.C., W.J. Shuttleworth, C.R. Lloyd, J.C. André, J.P. Goutorbe and J. Gelpe, 1989.

Micrometeorological measurements in Les Landes forest during HAPEX-MOBILHY. *Agricultural and Forest Meteorology*, 46:131–147.

Grant, D.R., 1975. Comparison of evaporation from barley with Penman estimates. *Agricultural Meteorology*, 15: 49–60.

Itier, B. 1996. Measurement and estimation of evapotranspiration. In *Sustainability of irrigated agriculture*, eds. L.S. Pereira, R.A. Feddes, J. R.Gilley and B. Leseffr, 171–191. Dordrecht: Kluwer Academic Publishers.

Itier, B. and A. Perrier, 1976. Présentation d'une étude analytique de l'advection: I. Advection liée aux variations horizontales de concentration et de température. *Annual Agronomy*, 27(2): 111–140.

Itier, B., Y. Brunet, K.J. Mcaneney and J.P. Lagouarde, 1994. Downwind evolution of scalar fluxes and surface resistance under conditions of local adection. Part I: A reappraisal of boundary conditions. *Agricultural and Forest Meteorology*, 71: 211–255.

Kizer, M.A., R.L. Elliott and J.F. Stone, 1990. Hourly ET model calibration with eddy flux and energy balance data. *Journal of Irrigation and Drainage Engineering*, ASCE, 116(2): 172–181.

Marsh, A.W., R.A. Strohman, S. Spaulding, V. Younger and V. Gibeault, 1980. Turf grass irrigation research at the University of California: warm and cool season grasses tested for water needs. *Irrigation Journal*, Pages 20–21, 32–33.

Meyer, W.S. and L. Mateos, 1990. Effects of soil type on soybean crop water use in weighing lysimeters. II: Effect of lysimeter canopy height discontinuity on evaporation. *Irrigation Science*, 11:233–237.

Neale, C.M.U., E.G. Kruse and R.E. Yoder, 1991. Field experience with hydraulic weighing lysimeters. In *Lysimeters for evapotranspiration and environmental measurements*, eds. R.G. Allen et al., 160–169. New York, NY: ASCE.

Pearce, A.J., J.H.C. Gash and J.B. Stewart, 1980. Rainfall interception in a forest stand estimated from grassland meteorological data. *Journal of Hydrology*, 46: 147–163.

Perrier, A. and A. Tuzet, 1991. Land surface processes: Description, theoretical approaches, and physical laws underlying their measurements. In *Land surface evaporation: measurement and parameterization*, eds. T.J. Schmugge and J.C. André, 145–155. Berlin: Springer-Verlag.

Perrier, A., B. Itier, J.M. Bertolini and N. Katerji, 1976. A new device for continuous recording of the energy balance of natural surfaces. *Agricultural Meteorology*, 16(1): 71–85.

Perrier, A., N. Katerji, G. Gosse and B. Itier, 1980. In situ study of evapotranspiration rates for a wheat crop. *Agricultural Meteorology*, 21: 295–311.

Perrier, A., P. Archer, and B. de Pablos, 1974. Etude de l'évapotranspiration réelle et maximele de diverses cultures. I: Dispositif et mesure. *Annual Agronomy*, 25(3): 229–243.

Pruitt, W.O. and F.J. Lourence, 1985. Experiences in lysimetry for ET and surface drag measurements. *Advances in Evapotranspiration*. St. Joseph, MI: ASAE, Pages 51–69.

Pruitt, W.O., 1991. Development of crop coefficients using lysimeters. In *Proceedigs of the ASCE International Symposium on Lysimetry*, eds. R.G. Allen, et al., *Lysimeters for Evapotranspiration and Environmental Measurements*, Honolulu, HA. New York, NY: ASCE, Pages 182–190.

Pruitt, W.O., D.L. Morgan and F.J. Lourence, 1973. Momentum and mass transfers in the surface boundary layer. *Quarterly Journal of the Royal Meteorological Society*, 99: 370–386.

Rosenberg, N.J., B.L. Blad and S.B. Verma, 1983. *Microclimate: The biological environment*, 2nd ed., New York: John Wiley.

Schulze, K.,1995. *Report of expert meeting for the preparation of an intercomparison of instruments and procedures for measurement and estimation of evaporation and evapotranspiration.* World Meteorological Organization, Commission for Instruments and Methods of Observation. Geneva, Switzerland. Pages 30.

Seemann, J., Y.I. Chirkov, J. Lomas and B. Primault, 1979. *Agrometeorology*. Berlin, Heidelberg: Springer Verlag.

Shuttleworth, W.J. and J.S. Wallace, 1985. Evaporation from sparse crops: An energy combination theory. *Quarterly Journal of the Royal Meteorological Society*, 111: 839–853.

Shuttleworth, W.J., 1993. Evaporation. In *Handbook of hydrology*, 4.1–4.53 New York: McGraw Hill.

Slatyer, R.O. and I.C. McIlroy, 1961. Evaporation and the principle of its measurement. In *Practical meteorology*. Paris: CSIRO (Australia) and UNESCO.

Stewart, J.B. and L.W. Gay, 1989. Preliminary modeling of transpiration from the FIFE in Kansas. *Agricultural and Forest Meteorology*, 48: 305–315.

Stringer, W.C., D.D. Wolf and R.E. Baser, 1981. Summer regrowth of tall fescue: Stubble characteristics and microenvironment. *Agronomy Journal*, 73: 96–100.

Tarantino, E., 1991. Grass reference measurements in Italy In *Lysimeters for evapotranspiration and environmental measurements*, eds. R.G. Allen, T.A. Howell, W.O. Pruitt, L.A. Walter and M.E. Jensen, 200–209. New York, NY: ASCE.

Thom, A.S., J.L. Thony and M. Vauclin, 1981. On the proper employment of evaporation pans and atmometers in estimating potential transpiration. *Quarterly Journal of the Royal Meteorological Society*, 107: 711–736.

Walter, L.A., E. Siemer, L.R. Dirks, J.P. Quinian and R.D. Burman, 1991. Lysimeters vs. buffer areas: evapotranspiration and agronomic comparisons. In *Lysimeters for evapotranspiration and environmental measurements*, eds. R.G. Allen, T.A. Howell, W.O. Pruitt, L.A. Walter and M.E. Jensen, 10–18. New York, NY: ASCE.

Wehner, D.J. and T.L. Watschke, 1981. Heat tolerance of Kentucky bluegrasses, perennial ryegrasses and annual bluegrass. *Agronomy Journal*, 73:79–84.

WMO, 1983. *Guide to meteorological instruments and observing practices*, 5th ed., WMO n 8. Geneva: WMO Publication.

Wright, J.L., 1991. Using lysimeters to develop evapotranspiration crop coefficients. In *Lysimeters for evapotranspiration and environmental measurements*, eds. R.G. Allen, T.A. Howell, W.O. Pruitt, L.A. Walter and M.E. Jensen, 191–199. New York, NY: ASCE.

**Parameters in evapotranspiration equations**

Allen, R.G., M.E. Jensen, J.L. Wright and R.D. Burman, 1989. Operational estimates of reference evapotranspiration. *Agronomy Journal*, 81: 650–662.

Allen, R.G., M. Smith, A. Perrier and L.S. Pereira, 1994. An update for the definition of reference evapotranspiration. *ICID Bulletin*, 43(2): 1–34.

Allen, R.G. 1995. *Evaluation of procedures for estimating mean monthly solar radiation from air temperature*. Report prepared for FAO, Water Resources Development and Management Service, FAO, Rome.

Bosen, J.F., 1958. An approximation formula to compute relative humidity from dry bulb and dew point temperatures. *Monthly Weather Review*, 86(12): 486.

Brunt, D. 1952. *Physical and dynamical meteorology*, 2nd ed., Cambridge: Cambridge University Press. Pages 428.

Brutsaert, W.H., 1982. *Evaporation into the atmosphere*. Dordrecht, Holland: R. Deidel Publishing Company.

Burman, R.D., M.E. Jensen and R.G. Allen, 1987. Thermodynamic factors in evapotranspiration. *Proceedings of the Irrigation and Drainage Engineering Specialty Conference*, Portland, Ore., ASCE, Pages 28–30.

Burman, R. and L.O. Pochop, 1994. *Evaporation, evapotranspiration and climatic data*. Amsterdam: Elsevier Science B.V.

Businger, J.A., 1988. A note on the Businger-Dyer profiles. *Boundary-Layer Meteorology*, 42: 145–151.

Choudhury, B.J., S.B. Idso and R.J. Reginato, 1987. Analysis of an empirical model for soil heat flux under a growing wheat crop for estimating evaporation by an infrared temperature based energy balance equation. *Agricultural and Forest Meteorology*, 39: 283–297.

Clothier, B.E., K.L. Clawson, P.J. Pinter, M.S. Moran, R.J. Reginato and R.D. Jackson, 1986. Estimates of soil heat flux from net radiation during the growth of alfalfa. *Agricultural and Forest Meteorology*, 37: 319–329.

Duffie, J.A. and W.A. Beckman, 1991. *Solar engineering of thermal processes*, 2nd ed., New York, NY: John Wiley and Sons. Pages 994.

Harrison, L.P., 1963. Fundamentals concepts and definitions relating to humidity. In *Humidity and moisture*, ed. A. Wexler, Vol. 3. NY.: Reinhold Publishing Co.

Dyer, A.J., 1974. A review of flux-profile relationships. *Boundary Layer Meterology*, 7: 363–372.

Dyer, A.J. and B.B. Hicks, 1970. Flux-gradient relationships in the constant flux layer. *Quarterly Journal of the Royal Meteorological Society*, 96: 715–721.

Frevert, D.K., R.W. Hill and B.C. Braaten, 1983. Estimation of FAO evapotranspiration coefficients. *Journal of Irrigation and Drainage Engineering*, ASCE, 109(2): 265–270.

Garratt, J.R., 1992. *The atmospheric boundary layer*. Cambridge: Cambridge University Press, Pages 316.

Garratt, J.R. and B.B. Hicks, 1973. Momentum, heat and water vapour transfer to and from natural and artificial surfaces. *Quarterly Journal of the Royal Meteorological Society*, 99: 680–687.

George, W., W.O. Pruitt and A. Dong, 1985. Evapotranspiration modeling. In *California irrigation management information system,* Final Report by R. Snyder, D.W. Henderson, W.O. Pruitt and A. Dong, California. Department Water Resources. Contract. No. B53812. Land, Air and Water Resources Pap. 10013-A, University California, Davis, III-36–III-59.

Gosse, G., A. Perrier and B. Itier, 1977. Etude de l'évapotranspiration réelle d'une culture de blé dans le bassin parisien. *Annual Agronomy*, 28(5): 521–541. (in French).

Harrison, L.P., 1963. Fundamentals concepts and definitions relating to humidity. In *Humidity and moisture*, ed. A. Wexler, Vol. 3. NY: Reinhold Publishing Co.

Hashemi, F. and M.T. Habibian, 1979. Limitations of temperature based methods in estimating crop evapotranspiration in arid-zone agricultural development project. Agricultural Meteorology, 20: 237–247.

Hatfield, J.L. and M. Fuchs, 1990. Evapotranspiration models. In *Management of farm irrigation systems*, eds. G.J. Hoffman, T.A. Howell and K.H. Solomon, 33–59. St. Joseph, MI: ASAE.

Hottel, H.C., 1976. A simple model for estimating the transmittance of direct solar radiation through clear atmospheres. *Solar energy*, 18: 129.

Idso, S.B. and R.B. Jackson, 1969. Thermal radiation from the atmosphere. *Journal of Geophysical Research*, 74: 5397–5403.

Jensen, J.R., 1988. Effect of asymmetric, daily air temperature and humidity waves on calculation of reference evapotranspiration. *Proceedings fo the European Economic Community Workshop on Management of Water Resources in Cash Crops and in Alternative Production Systems.* Brussels, Belgium. Pages 12.

List, R. ., 1984. *Smithsonian meteorological tables*, 6th rev. ed., Washington, DC: Smithsonian Institution, Pages 539.

Liu, B.Y.H. and R.C. Jorden, 1960. The interrelationship and characteristic distribution of direct diffuse and total solar radiation. *Solar Energy*, 4(3): 1–19.

Matias, P.G.M., 1992. *SWATCHP: A model for a continuous simulation of hydrologic processes in a system vegetation - soil - aquifer - river*. PhD dissertation, Technical University of Lisbon.

Monteith, J.L. and M.H. Unsworth, 1990. *Principles of environmental physics*, 2nd ed., London: Edward Arnold.

Monteith, J.L., 1965. Evaporation and Environment. *19th Symposia of the Society for Experimental Biology*, Cambridge: Cambridge University Press, 19: 205–234.

Murray, F.W., 1967. On the computation of saturation vapor pressure. *Journal of Applied Meteorology*, 6: 203–204.

Penman, H.L., 1963. *Vegetation and hydrology*. Tech. Comm. No. 53, Commonwealth Bureau of Soils, Harpenden, England. Pages 125.

Pereira, L.S. and M. Smith, 1989. *Proposed procedures for revision of guidelines for predicting crop water requirements*. Land and Water Use Division, FAO Rome. Pages 36.

Pruitt, W.O. and J. Doorenbos, 1977. *Background and development of methods to predict reference crop evapotranspiration*. Appendix II in FAO-ID-24. Pages 108–119.

Shaw, R.H. and A.R. Pereira, 1982. Aerodynamic roughness of a plant canopy: A numerical experiment. *Agricultural Meteorology*, 26: 51–65.

Stewart, J.B., 1988. Modelling surface conductance of pine forest. *Agricultural and Forest Meteorology*, 43: 19–35.

Shuttleworth, W.J., 1993. Evaporation. In *Handbook of hydrology*, ed. D.R. Maidment, 4.1–4.53. New York: McGraw Hill.

Shuttleworth, W.J. and J.S. Wallace, 1985. Evaporation from sparse crops: An energy combination theory. *Quarterly Journal of the Royal Meteorological Society*, 111: 839–853.

Smith, M., R.G. Allen, J.L. Monteith, A. Perrier, L. Pereira and A. Segeren, 1992. *Report of the expert consultation on procedures for revision of FAO guidelines for prediction of crop water requirements*. UN-FAO, Rome, Italy, Pages 54.

Stewart, J.B., 1989. On the use of the Penman-Monteith equation for determining areal evapotranspiration. In *Estimation of Areal Evapotranspiration*. IAHS, 177: 3–12.

Stewart, J.B. and S.B. Verma, 1992. Comparison of surface fluxes and conductances at two contrasting sites within the FIFE area. *Journal of Geophysical Research*, 97(17): 18623–18628.

Szeicz, G. and I.F. Long, 1969. Surface resistance of crop canopies. *Water Resources Research*, 5: 622 –633.

Thom, A.S., 1971. Momentum absorption by vegetation. *Quarterly Journal of the Royal Meteorological Society*, 97: 414–428.

Thom, A.S., 1972. Momentum, mass and heat exchange of vegetation. *Quarterly Journal of the Royal Meteorological Society*, 98: 124–134.

Thom, A.S. and H.R. Oliver, 1977. On Penman's equation for estimating regional evaporation. *Quarterly Journal of the Royal Meteorological Society*, 103: 345–357.

Thom, A.S., J.B. Stewart, H.R. Oliver and J.H.C. Gash, 1975. Comparison of aerodynamic and energy budget estimates of fluxes over a pine forest. *Quarterly Journal of the Royal Meteorological Society*, 101: 93–105.

Van Bavel, C.H., L.J. Fritschen and W.E. Reeves, 1963. Transpiration of sudangrass as an externally controlled process. *Science*, 141: 269–270.

Verma, S.B., 1989. Aerodynamic resistances to transfers of heat, mass and momentum. In *Estimation of areal evapotranspiration*, eds. T.A. Black, D.L. Spittlehouse, M.D. Novak and D.T. Price, 13–20. IAHS Pub. No. 177.

Webb, E.K., 1970. Profile relationships: the log-linear range and extension to strong stability. *Quarterly Journal of the Royal Meteorological Society*, 96: 67–90.

Wallace, J.S., J.M. Roberts and M.V.K. Sivakuma, 1990. The estimation of transpiration from sparse dryland millet using stomatal conductance and vegetation area indices. *Agricultural and Forest Meteorology*, 51: 35–49.

**Analysis of weather and evapotranspiration data**

Allen, R.G. and C.E. Brockway, 1983. Estimating consumptive use on a statewide basis. *Irrigation and Drainage Specialty Conference at Jackson*, WY. New York, NY: ASCE, Pages 79–89.

Allen, R.G. and Wright, J.L. 1977. Translating wind measurements from weather stations to agricultural crops. *Journal of Hydrologic Engineering*, ASCE, 2(1): 26–35.

Allen, R.G., 1996. Assessing integrity of weather data for use in reference evapotranspiration estimation. *Journal of Irrigation and Drainage Engineering Division*, ASCE, 122(2): 97–106.

Allen, R.G., 1997. Self-calibrating method for estimating solar radiation from air temperature. *Journal of Hydrologic Engineering*, ASCE, 2(2): 56–67.

Allen, R.G., C.E. Brockway and J.L. Wright, 1983. Weather station siting and consumptive use estimates. *Journal of Water Resources Engineering and Management Division*, ASCE, 109(2): 134–146.

Burman, R.D., J.L. Wright and M.E. Jensen, 1975. Changes in climate and estimated evaporation across a large irrigated area in Idaho. *Transactions of ASAE*, 18(6): 1089–1091, 1093.

Pereira, L.S., 1998. *Water and soil management for sustainable agriculture in the Huang-Huai-Hai rivers plain (North China)*. Final Report of EC Research Contract CT93-250, Instituto Superior de Agronomia, Lisbon.

Pruitt, W.O. and B.D. Swann, 1986. Evapotranspiration studies. In: N.S.W.: *Daily vs. hourly meteorological data*. Irrigation '86, Darling downs Institute of Advanced Education, Toowoomba, Queensland, Australia, Pages 29.

Pruitt, W.O. and J. Doorenbos, 1977. *Background and development of methods to predict reference crop evapotranspiration* (ETo). Appendix 11, In: FAO-ID-24, Pages 108–119.

Rosenberg, N.J., B.L. Blad and S.B. Verma, 1983. *Microclimate: The biological environment*, 2nd ed., New York: John Wiley.

Snyder, R.L. and W.O. Pruitt, 1992. Evapotranspiration data management in California. In

*Proceedings of the Irrigation and Drainage sessions of ASCE Water Forum '92*, ed. T. Engman, New York, NY: ASCE, Pages 128–133.

**Crop evapotranspiration**

Alves, I.L., 1995. *Modeling crop evapotranspiration: Canopy and aerodynamic resistances.* PhD Dissertation, ISA, University of Tecnical Lisboa.

Bevan, K., 1979. A sensitivity analysis of the Penman-Monteith actual evapotranspiration estimates. Journal of Hydrology, 44: 169–190.

Howell, T.A., S.R. Evett, J.A. Tolk, A.D. Schneider and J.L. Steiner, 1996. Evapotranspiration of corn: Southern High Plains. In *Evapotranspiration and irrigation scheduling*, eds. C.R. Camp, E.J. Sadler, and R.E. Yoder, 158–166. New York: ASAE.

Jensen, M.E., 1968. Water consumption by agricultural plants. In *Water deficits and plant growth*, ed. T. Kozlowski, 1–22. New York: Academic Press.

Jensen, M.E., 1974. *Consumptive use of water and irrigation water requirements.* A Report prepared by the Technical Committee on Irrigation Water Requirements. Irrigation and Drainage Division, ASCE, Pages 227.

Lee, R., 1980. *Forest hydrology.* New York: Columbia University Press.

Perrier, A., N. Katerji, G. Gosse and B. Itier, 1980. In situ study of evapotranspiration rates for a wheat crop. *Agricultural Meterology*, 21:295–311. (in French).

Rijtema, P.E., 1965. *Analysis of actual evapotranspiration.* Agricultural Research Report No. 69, Centre for Agricultural Publishing and Documentation, Wageningen.

Shuttleworth, W.J., 1993. Evaporation. In *Handbook of hydrology*, ed. D.R. Maidment, 4.1–4.53. New York: McGraw Hill.

**Crop coefficients**

Abtew, W. and J. Obeysekera, 1995. Lysimeter study of evapotranspiration of cattails and comparison of three estimation methods. *Transactions of ASAE*, 38(1): 121–129.

Allen, R.G., J. Prueger and R.W. Hill, 1992. Evapotranspiration from isolated stands of hydrophytes: Cattail and Bulrush. *Transactions of ASAE*, 35(4): 1191–1198.

Allen, R.G., M. Smith, L.S. Pereira and W.O. Pruitt, 1997. Proposed revision to the FAO procedure for estimating crop water requirements. In *Proceedings 2nd International Symposium on Irrigation of Horticultural Crops*, ed. K.S. Chartzoulakes. ISHS. Acta Horticulturae, I: 17–33.

Allen, R.G., R.W. Hill and S. Vemulapali, 1994. *Evapotranspiration parameters for variably-sized wetlands.* Paper presented at the 1994 Summer Meeting of ASAE No. 942132, Pages 24.

Burman, R.D., P.R. Nixon, J.L. Wright and W.O. Pruitt, 1980. Water requirements. In *Design and operation of farm irrigation systems*, ed. M.E. Jensen, 189–232. St. Joseph, MI: American Society of Agricultural Engineers.

Doorenbos, J. and A.H. Kassam, 1979. *Yield response to water.* FAO Irrigation and Drainage Paper No. 33, FAO, Rome, Italy. Pages 193.

Doorenbos, J. and W.O. Pruitt, 1977. *Crop water requirements.* Irrigation and Drainage Paper No. 24, (rev.) FAO, Rome, Italy. Pages 144.

Elliott, R.L., S.L. Harp, G.D. Grosz and M.A. Kizer, 1988. Crop coefficients for peanut evapotranspiration. *Agricultural Water Management*, 15: 155–164.

Feddes, R.A., 1987. *Crop factors in relation to Makkink reference crop evapotranspiration.* Technical Bulletin Institute for Land and Water Management Research. No. 67, Pages 33–45.

Fereres, E., 1981. *Drip irrigation management.* Cooperative Extension, University of California, Berkeley, CA, Leaflet No. 21259.

Grattan, S.R., W. Bowers, A. Dong, R.L. Snyder, J.J. Carroll and W. George, 1998. New crop coefficients estimate water use of vegetables, row crops. *California Agriculture*, 52(1): 16–21.

Howell, T.A., D.A. Bucks, D.A. Goldhamer and J.M. Lima, 1986. Management Principles: 4.1 Irrigation Scheduling. In *Trickle irrigation for crop production: Design, operation and management*, eds. F.S. Nakayama and D.A. Bucks, Elsevier.

Howell, T.A., J.L. Steiner, A.D. Schneider and S.R. Evett, 1995. Evapotranspiration of irrigated winter wheat: Southern high plains. *Transactions of ASAE*, 38(3): 745–759.

Jensen, M.E., 1974. *Consumptive use of water and irrigation water requirements*. A Report prepared by the Technical Committee on Irrigation Water Requirements. Irrigation and Drainage Division, ASCE, Pages 227.

Kolar, J.J. and R.A. Kohl, 1976. *Irrigating alfalfa for seed production*. University of Idaho Agricultural Experiment Station Current Information Series 357, Moscow, Idaho. Pages 3.

Liu, Y., J.L. Teixeira, H.J. Zhang and L.S. Pereira, 1998. Model validation and crop coefficients for irrigation scheduling in the North China plain. *Agricultural Water Management*, 36: 233–246.

Neale, C.M.U., 1987. *Development of reflectance based crop coefficients for corn*. Unpublished PhD Dissertation, Agricultural Engineering, Colorado State University, Fort Collins, USA.

Neale, C.M.U., W.C. Bausch and D.F. Heerman, 1989. Development of reflectance-based crop coefficients for corn. *Transactions of ASAE*, 32(6): 1891–1899.

Pastor, M. and F. Orgaz, 1994. Riego deficitario del olivar: Los programas de recorte de riego en olivar. *Agricultura*, 746: 768–776 (in Spanish).

Pereira, L.S., A. Perrier, R.G. Allen and I. Alves, 1996. Evapotranspiration: Review of concepts and future trends. In *Evapotranspiration and irrigation scheduling*, eds. C.R. Camp, E.J. Sadler, R.E. Yoder. 109–115. New York: ASAE.

Pereira, L.S., J.L. Teixeira, L.A. Pereira, M.I. Ferreira and R.M. Fernando, 1987. Simulation models of crop response to irrigation management: research approaches and needs. In *Simulation models for cropping systems in relation to water management*, ed. J. Feyen, 19–36. Luxembourg: Commission of the European Communities, EUR 10869.

Pruitt, W.O., 1986. Traditional methods. *Evapotranspiration research priorities for the next decade*. ASAE Paper No. 86-2629. Pages 23.

Pruitt, W.O., E. Fereres, P.E. Martin, H. Singh, D.W. Henderson, R.M. Hagan, E. Tarantino and B. Chandio, 1984. Microclimate, evapotranspiration, and water-use efficiency for drip- and furrow-irrigated tomatoes. *Proceedings of the 12th Congress, International Commission on Irrigation and Drainage*, Ft. Collins, CO., Pages 367–394.

Rogers, J.S., L.H. Allen and D.J. Calvert, 1983. Evapotranspiration for humid regions: Developing citrus grove, grass cover. *Transactions of ASAE*, 26(6): 1778–1783, 1792.

Snyder, R.L., B.J. Lanini, D.A. Shaw and W.O. Pruitt, 1989. *Using reference evapotranspiration and crop coefficients to estimate crop evapotranspiration for agronomic crops, grasses, and vegetable crops*. Cooperative Extension, University of California, Berkeley, CA, Leaflet No. 21427, Pages 12.

Snyder, R.L., B.J. Lanini, D.A. Shaw and W.O. Pruitt, 1989. *Using reference evapotranspiration and crop coefficients to estimate crop evapotranspiration for trees and vines*. Cooperative Extension, University of California, Berkeley, CA, Leaflet No. 21428, Pages 8.

Wright, J.L. and M.E. Jensen, 1972. Peak water requirements of crops in southern Idaho. *Journal of Irrigation and Drainage Division*, ASCE, 96 (IR1): 193–201.

Wright, J.L., 1981. Crop coefficients for estimates of daily crop evapotranspiration. *Irrigation Scheduling for Water and Energy Conservation in the 80s*, December 1981, ASAE.

Wright, J.L., 1982. New evapotranspiration crop coefficients. *Journal of Irrigation and Drainage Division*. ASCE, 108: 57–74.

Wright, J.L., 1990. *Evapotranspiration data for dry, edible beans at Kimberly, Idaho*. Unpublished data, USDA-ARS, Kimberly, Idaho.

**Length of crop growth stages**

Allen, R.G. and F.N. Gichuki, 1989. Effects of Projected $CO_2$: Induced climatic changes on irrigation water requirements in the Great Plains States (Texas, Oklahoma, Kansas, and Nebraska). In *The Potential Effects of Global Climate Change on the United States: Appendix C - Agriculture. Vol. 1. EPA-230-05-89-053*, eds. J.B. Smith and D.A. Tirpak, 6: 1–42. Washington, DC: U.S. Environmental Protection Agency, Office of Policy, Planning and Evaluation.

Dingkuhn, M., 1994. Climatic determinants of irrigated rice performance in the Sahel. III. Characterizing environments by simulating crop phenology. *Agricultural Systems*, 48: 435–456.

Doorenbos, J. and W.O. Pruitt, 1977. *Crop water requirements*. Irrigation and Drainage Paper No. 24, (rev.) FAO, Rome, Italy. Pages 144.

Everson, D.O., M. Faubion and D.E. Amos, 1978. *Freezing temperatures and growing seasons in Idaho*. University of Idaho Agricultural Experiment, Pages 18.

Kruse E.G. and H.R. Haise, 1974. *Water use by native grasses in high altitude Colorado meadows*. USDA Agricultural Research Service, Western Region report ARS-W-6-1974. Pages 60.

O'Halloran, T.F., 1997. *Reported crop acreages by month for the imperial irrigation district*. Imperial Irrigation District, Imperial, CA, USA.

Ritchie, J.T. and B.S. Johnson, 1990. Soil and plant factors affecting evaporation. In *Irrigation of agricultural crops*, eds. B.A. Stewart and D.R. Nielsen, Chapter 13:363–390, Agronomy Series 30. American Society of Agronomy.

Ritchie, J.T. and D.S. NeSmith, 1991. Temperature and crop development. In *Modeling plant and soil systems*, eds. R.J. Hanks and J.T. Ritchie, Chapter 2: 5–29, Agronomy Series No. 31, Madison, Wise: American Society of Agronomy.

Ritchie, J.T., 1991. Wheat phasic development. In *Modeling plant and soil systems*, eds. R.J. Hanks and J.T. Ritchie, Chapter 3: 31–54, Agronomy Series No. 31. Madison, Wise: American Society Agronomy.

Snyder, R.L., B.J. Lanini, D.A. Shaw and W.O. Pruitt, 1989. *Using reference evapotranspiration and crop coefficients to estimate crop evapotranspiration for agronomic crops, grasses, and vegetable crops*. Cooperative Extension, University of California, Berkeley, CA, Leaflet No. 21427, Pages 12.

Snyder, R.L., B.J. Lanini, D.A. Shaw and W.O. Pruitt, 1989. *Using reference evapotranspiration and crop coefficients to estimate crop evapotranspiration for trees and vines*. Cooperative Extension, University of California, Berkeley, CA, Leaflet No. 21428, Pages 8.

Wright, J.L., 1982. New evapotranspiration crop coefficients. *Journal of Irrigation and Drainage Division*, ASCE, 108 (IR2): 57–74.

**Non-growing season evapotranspiration**

Flerchinger, G.N. and F.B. Pierson, 1991. Modeling plant canopy effects on variability of soil temperature and water. *Agricultural and Forest Meteorology*, 56: 227–246.

Liu, Y., J.L. Teixeira, H.J. Zhang and L.S. Pereira, 1998. Model validation and crop coefficients for irrigation scheduling in the North China. *Agricultural Water Management*, 36: 233–246.

Running, S.W. and J.C. Coughlan, 1988. A general model of forest ecosystem processes for regional applications: I. Hydrologic balance, canopy gas exchange and primary production processes. *Ecological Modeling*, 42: 125–154.

Saxton, K.E., H.P. Johnson and R.H. Shaw, 1974. Modeling evapotranspiration and soil moisture. *Transactions of ASAE*, 17(4): 673–677.

Sinclair, T.R., 1984. Leaf area development in field-grown soybeans. *Agronomy Journal*, 76:141–146.

**Salinity impacts on evapotranspiration**

Ayers, R.S. and D.W. Westcot, 1985. *Water quality for agriculture*. Irrigation and Drainage Paper 29, Rev. 1. Food and Agriculture Organization of the United Nations, Rome. Pages 17.

Doorenbos, J. and A.H. Kassam, 1979. *Yield response to water*. Irrigation and Drainage Paper 33. Food and Agriculture Organization of the United Nations, Rome. Pages 193.

Hanks, R.J., 1984. Prediction of crop yield and water consumption under saline conditions. In *Soil salinity under irrigation: Processes and management*, eds. Shainberg, I. and J. Shalhevet, Section 8.2: 272–283. Berlin, Springer-Verlag.

Hoffman, G.J., J.A. Jobes and W.J. Alves, 1983. Response to tall fescue to irrigation water salinity, leaching fraction, and irrigation frequency. *Agricultural Water Management*, 7: 439–456.

Letey, J. and A. Dinar 1986. Simulated crop-water production functions for several crops when irrigated with saline waters. *Hilgardia*, 54(1): 1–32.

Letey, J., A. Dinar, and K.C. Knapp, 1985. Crop-water production function model for saline irrigation waters. *Soil Science Society of America Journal*, 49: 1005–1009.

Maas, E.V., 1990. Crop salt tolerance. In *Agricultural salinity assessment and management*, ed. K.K. Tanji, 262–304, ASCE Manuals and Reports on Engineering Practice No. 71. New York: ASCE.

Mieri, A., 1984. Plant response to salinity: experimental methodology and application to the field. In *Soil salinity under irrigation: Processes and management*, eds. I. Shainberg and J. Shalhevet, Section 8.3: 284–297. Berlin: Springer-Verlag.

Oster, J.D., 1994. Irrigation with poor quality water: Review article. *Agricultural water management*, 25: 271–297.

Oster, J.D., I. Shainberg and I.P. Abrol, 1996. Reclamation of salt-affected soil. In *Soil erosion, conservation, and rehabilitation*, ed. M. Agassi, Chapter 14: 315–335. New York: Marcel Dekker, Inc.

Rhodes, J.D., A. Kandiah and A.M. Mashali, 1992. *The use of saline waters for crop production*. Irrigation and Drainage Paper 48. Food and Agriculture Organization of the United Nations, Rome, Pages 133.

Shalhevet, J., 1984. Management of irrigation with brackish water. In *Soil salinity under irrigation: Processes and management*, eds. Shainberg, I. and J. Shalhevet, Section 8.4: 298–318. Berlin: Springer-Verlag.

Shalhevet, J., 1994. Using water of marginal quality for crop production: Major issues - review article. *Agricultural water management*, 25: 233–269.

Stewart, J.I., R.M. Hagan and W.O. Pruitt, 1976. Salinity effects on corn yield, evapotranspiration, leaching fraction, and irrigation efficiency. In ed. H.E. Dregne. *Managing Saline Water for Irrigation: Proceedings of the International Conference on Managing Saline Water for Irrigation: Planning for the Future*, Lubbock, TX, Pages 316–331.

Tanji, K.K., 1990. *Agricultural salinity assessment and management*. ASCE Manuals and Reports on Engineering Practice No. 71 ASCE, New York, Pages 113–137.

**Factors affecting evapotranspiration**

Allen, I.H., P. Jones and J.W. Jones, 1985. Rising atmospheric $CO_2$ and evapotranspiration. *Proceedings of the National Conference on Advances in Evapotranspiration*, December 16–17, Chicago, IL. St. Joseph, MI: ASAE.

Burman, R.D., J.L. Wright and M.E. Jensen, 1975. Changes in climate and estimated evaporation across a large irrigated area in Idaho. *Transactions of ASAE*, 18(6): 1089–1091, 1093.

Doorenbos, J. and A.H. Kassam, 1979. *Yield response to water*. FAO Irrigation and Drainage Paper No. 33, FAO, Rome, Italy. Pages 193.

Loommis, R.S. and W.A. Williams, 1969. Productivity and the morphology of crop stand patterns with leaves. In *Physiological aspects of crop yield*, ed. J.D. Eastin, 27–47. ASA, CSSA and SSSA, Madison, WI.

Rhoades, J D., A. Kandiah and A.M. Mashali, 1992. *The use of saline waters for crop production*. Irrigation and Drainage Paper. FAO, Rome.

Shuttleworth, W.J. and J.S. Wallace, 1985. Evaporation from sparse crops - an energy combination theory. *Quarterly Journal of the Royal Meteorological Society*, 111: 839–853

Wallace, J.S., J.M. Roberts and M.V.K. Sivakuma, 1990. The estimation of transpiration from sparse dryland millet using stomatal condutance and vegetation area indices. *Agricultural and Forest Meteorology*, 51: 35–49.

**Soil water balance and irrigation scheduling**

Bastiaanssen, W.G.M., 1995. *Regionalization of surface flux densities and moisture indicators in composite terrain*. Doctoral thesis, Wageningen Agricultural University, Wageningen, Pages 273.

Belmans, C., J.G. Wesseling and R.A. Feddes, 1983. Simulation model of the water balance of a cropped soil: SWATRE I. Hydrology, 63: 271–286.

Camp, C.R., E.J. Sadler and R.E. Yoder, 1996. Evapotranspiration and irrigation scheduling. *Proceedings of the International Conference on Evapotranspiration and Irrigation Scheduling*. St. Joseph, MI: American Society of Agricultural Engineers, ISBN 0-929355-82-2. Pages 1166.

Doorenbos, J. and A.H. Kassam, 1979. *Yield response to water*. FAO Irrigation and Drainage Paper No. 33, FAO, Rome, Italy. Pages 193.

Jordan, W.R. and J.T. Ritchie, 1971. Influence of soil water stress on evaporation, root absortion and internal water status of cotton. *Plant Physiology*, 48: 783–788.

Kabat, P., B.J. van den Broek and R.A. Feddes, 1992. SWACROP: A water management and crop production simulation model. *ICID Bulletins*, 41(2): 61–84.

Merriam, J.L., 1966. A management control concept for determining the economical depth and frequency of irrigation. *Transactions of the American Society of Agricultural and Biological Engineers*, 9: 492–498.

Pereira, L.S., B.J. van den Broek, P. Kabat and R.G. Allen, 1995. *Crop-water simulation models in practice*. Wageningen: Wageningen Academic Publishers, Pages 339.

Pereira, L.S., M.A. Ait Kadi Perrier and P. Kabat, 1992. Crop water models. *Special issue of the ICID Bulletin*, Pages 200.

Raes, D., H. Lemmens, P. Van Aelst, M. Vanden Bulcke and M. Smith, 1988. *IRSIS: Irrigation scheduling information system*. Reference Manual No. 3. Institute for Land and Water Management, Belgium: Katholieke Universiteit Leuven, (1&2): 119 & 71.

Smith, M., 1992. *CROPWAT: A computer program for irrigation planning and management*. FAO Irrigation and Drainage Paper 46, FAO, Rome.

Teixeira, J.L., M.P. Farrajota and L.S. Pereira, 1995. PROREG simulation software to design demand in irrigation projects. In *Crop-water simulation models in practice*, eds. L.S. Pereira, B.J. van den Broek, P. Kabat, and R. G. Allen, 273–285. Wageningen: Wageningen Academic Publishers.

Tuzet, A., A. Perrier and C. Masaad, 1992. Crop water budget estimation of irrigation requirement. *ICID Bulletins*, 41(2): 1–17.

**Web page links**

Al-Kaisi, M., 2000. Crop water use or evapotranspiration. In ed. John VanDyk. http://www.ent.iastate.edu/Ipm/Icm/2000/5-29-2000/wateruse.html

Allen, R.G., L.S. Pereira, D. Raes and M. Smith, 1998. *Crop Evapotranspiration - Guidelines for Computing Crop Water Requirements*. http://www.fao.org/docrep/X0490E/X0490E00.htm

Austin Lawn Sprinklers Association, 1999. *Technical information - Using evapotranspiration data*. ALSA. http://www.alsaustin.org/tech-evapo.htm>

Brown, P.W., 1999. *AZMET evapotranspiration estimates: A tool for improving water management of turfgrass*. AZMET. http://ag.arizona.edu/azmet/et1.htm

Chemin, Y., 2004. Evapotranspiration of crops by remote sensing using the energy balance based algorithms. Pages 1–10. http://www.iwmi.cgiar.org/Assessment/files_new/publications/Workshop

DeGaetano A.T., K.L. Eggleston and W.W. Knapp, 2001. Daily evapotranspiration and soil moisture estimates for the North Eastern United States. Pages 1–11. http://www.nrcc.cornell.edu/reports/RR_94-1.html

Dingman, S.L., 1994, *Physical hydrology*. http://www.uregina.ca/~sauchyn/geog327/

Farnsworth, R.K., E.S. Thompson and E.L. Peck, 1996. *Evaporation, evapotranspiration and water use*. Evaporation Atlas for the Continental 48 United States, National Weather Service, NOAA, Washington, DC. http://www.engineering.usu.edu/uwrl/atlas/ch3/index.html

Geophysics Review 1995. *Evapotranspiration*. U.S. National Report to IUGG, Vol. 33. American Geophysical Union. http://www.agu.org/revgeophys/engman00/node6.html

Hanson, R.L., 1991. *Evapotranspiration and droughts*. National Water Summary 1988–89: Hydrologic Events and Floods and Droughts, U.S. Geological Survey Water-Supply Paper 2375. Pages 99–104. http://geochange.er.usgs.gov/sw/changes/natural/et/

Igbadun, H.E., H.F. Mahoo, A.K. Tarimo and B.A. Salim, 2006. Performance of two temperature–based reference evapotranspiration models. In ed. Baanda A. Salim. http://cigr-ejournal.tamu.edu/submissions/volume.htm

Irmak, S. and D.Z. Haman, 2003. Evapotranspiration: Potential or Reference. University of Florida. http://edis.ifas.ufl.edu/AE256

Irmak, S., 2001. Evapotranspiration research. In ed. Derrel Martin. Pages 1–4. http://bse.unl.edu/Research/evapo.htm

Kaisi, M.A., 2000. *Crop water use or evapotranspiration*. Department of Entomology, Iowa State University. http://www.ent.iastate.edu/Ipm/Icm/2000/5-29-2000/wateruse.html

Lu, J., G. Sun, S. G. McNulty and D. Amatya, 2005. A comparison of six potential evapotranspiration methods for regional use in the Southeastern United States. In ed. Jianbiao Lu. Pages 621–633. http://www.srs.fs.usda.gov/pubs

Mao, L.M., M.J. Bergman and C.C. Tai, 2002. Evapotranspiration measurements and estimation of three wetland environments in the upper St. Johns River, Florida. Pages 1–9. http://findarticles.com/p/articles/mi_qa4038/is_200210/ai_n9090867

Pidwirny, M., 2006. Introduction to the Hydrosphere. Chapter 8: *Actual and potential evapotranspiration.* Universiy of British Columbia, Okanagan. http://www.physicalgeography.net/fundamentals/8j.html

Sumner, D.M., 2006. Evapotranspiration from successional vegetation in a deforested area of the lake Wales Ridge, Florida. In ed. D. M. Sumner. Pages 1–46. http://fl.water.usgs.gov/PDF_files/wri96_4244_sumner.pdf

White, M. A. and G. E. Host, 2001. An assessment of potential evapotranspiration for the northern lake states. Pages 1–3. http://www.nrri.umn.edu/gla/pet.htm

Zimmermann, N.E., 2000. Calculation of potential evapotranspiration. In ed. Niklaus E. Zimmermann. Chapter 3: 27–33. http://www.wsl.ch/staff/niklaus.zimmermann/programs/aml3_3.html

# 3

## Literature cited

1. Anónimo, 1981. Questions and answers about tensiometers. Division of Agricultural Sciences, University of California.
2. Bottcher, A.B. and L.W. Miller, 1982. Automatic tensiometer scanner for rapid measurements. *Transactions of ASAE*, 25(5): 1338–1342
3. Fischbach, P.E., T.L. Thompson and L.E. Stetson, 1978. Electric controls for automatic surface irrigation systems with reuse system. *Transactions of ASAE.*
4. Golberg, D., B. Gornat and D. Rimon, 1976. *Drip irrigation: Principles, design and agricultural practices.* Israel: Drip Irrigation Scientific Publications.
5. Irrometer moisture indicator – reference book. Riverside, CA: Irrometer Co.
6. Leslie Long, F., 1982. A new solid – state device for reading tensiometers. *Soil Science*, 133(3): 131–132.
7. Marthaler, H.P., W. Vogelsanger, F. Richard and P.J. Wirenga, 1983. A pressure transducer for field tensiometers. *Soil Science Society of America Journal*, 47: 634–637.
8. Operating instructions–soil moisture tester. Delmhorst Instrument Co.
9. Puckett, W.E. and J.H. Dane, 1981. Testing tensiometers by a vacuum method. Soil Science, 132: 444–445.
10. Ross, D.S., R.C. Funt, C.W. Reynolds, D.C. Coston, H.H. Fries and N.J. Smith. *Trickle irrigation—an introduction.* The Northeast Regional Agricultural Engineering Service, Cornell University, Ithaca, N.Y.
11. Smajstrla, A.G., D.S. Dalton and F.X. Duran. *Tensiometers for soil moisture measurement and irrigation scheduling.* Circular 487 Florida Cooperative Extension Service, Gainesville – FL.

### Books/bulletins/journal and proceedings/reports

Abbate, E., J.L. Dardanelli, M.G. Cantarero and M. Maturano, 2004. Climatic and water availability effects on water-use efficiency in wheat P. Suer. *Crop Science Journal*, 44: 474–483.

Abramson, L.W., S. Thomas, S. Sharma and G.M. Boyce, 2001. *Slope stability and stabilization methods.* John Wiley. Pages 138–157, 248, 263.

Alam, M. and D.H. Rogers, 1997. *Tensiometer use in scheduling iirrigation.* Kansas State University Agricultural Experiment Station and Coop. Extension Service. Manhattan, Kansas. Pages 321–332.

Albright, W.H., G.W. Gee, G.V. Wilson and M.J. Fayer, 2002. *Alternative cover assessment project Phase I Report.* DRI-41183. Desert Research Institute. Reno, NV. Pages 533–556.

Alon, B.G., N. Lazorovitch and U. Shani, 2004. Subsurface drip irrigation in gravel-filled cavities. *Vadose Zone Journal*, 55:2345–2654.

Assouline S., M. Möller, S. Cohen, M.A. Ben-Hur Grava, K. Narkis and A. Silber, 2006. Soil-plant system response to pulsed drip irrigation and salinity: Bell pepper case study. *Soil Science Society of America Journal*, 70: 1924–1927.

Ayodele E. A., A. Ayorinde and A. J. Olufayo, 2004. Evaluation of two temperature stress indices to estimate grain sorghum yield and evapotranspiration. Agronomy Journal, 96: 1282–1287.

Brutsaert, Wilfried, 2005. *Hydrology: An introduction.* Cambridge: Cambridge University Press. Pages 253–275, 305, 362, 452–463.

Cardenas-Laihacar B., M.D. Dukes and G.L. Miller, 2005. *Sensor-based control of irrigation in Bermuda grass.* Gainsville, FL: Agricultural and Biological Engineering Deptartment, University of Florida. Pages 23–55.

Caruso, M. and A. Tarantino, 2004. *A shearbox for testing unsaturated soils at medium to high degrees of saturation.* Géotechnique Technical Note, 54, No. 4, Pages 281–284.

David, C. and K. Nielsen, 2004. Forage yield and quality under varying water availability. *Agronomy Journal,* 96: 204–213.

Meek, D.W. and J. W. Singer, 2004. Estimation of duration indices for repeated tensiometer reading. Agronomy Journal, 96: 1787–1790.

Goldman, Amy P., 2004. *The compleat squash: A passionate grower's guide to pumpkins, squashes, and gourds.* Artisan. Page 18. October.

Hemond, Harold F., L. Fechner and J. Elizabeth, 1999. *Chemical fate and transport in the environment.* Academic Press. Page 224.

Hensley, D. and J. Deputy, 1999. *Using tensiometers for measuring soil water and scheduling irrigation.* College of Tropical Agriculture and Human Resources, University of Hawaii at Manoa. Pages 644–667.

Hoffmann, C., A. Tarantino and L. Mongiovì, 2006. Thermal Effects on Response of High Suction Tensiometer. *Proceedings 4th International Conference on Unsaturated Soils,* April 2–5, Phoenix, USA, Pages 423–456.

Hoitink, D.J., K.W. Burk, J.V. Ramsdell and W.J. Shaw, 2003. *Hanford Site Climatological Data Summary 2002 with Historical Data.* PNNL-14242. Pacific Northwest National Laboratory, Richland, WA. Pages 645–665.

Hubbell, J.M. and J. Buck Sisson, 2004. Comments on tensiometer modification for diminishing errors due to the fluctuating inner water column. *Soil Science Society of America Journal,* 68(2): 709

Hutchinson, P.A., 2006. Routine measurement of the soil water potential gradient near satura-

tion using a pair of tube tensiometers. *Australian Journal of Soil Research,* 39(5): 1147–1156.

Lanier, E., L. David, J. Jordan, S.Barnes, J. Matthews, G.L. Grabow, W.J. Griffin Jr., J.E. Bailey, P.D. Johnson, J.F. Spears and R. Wells, 2004. Disease management in overhead sprinkler and subsurface drip irrigation systems for peanut. Agronomy Journal, 96:756–798.

Lindeburg, M.R., 2003. *Environmental engineering reference manual for the PE Exam.* California, CA: Professional Publications, Page 302

Maarten, P., R. Raangs and H. Verjeij, 2000. *Response of the osmotic tensiometer to varying temperature.* Laboratory for Inorganic Materials Science Department of Chemical Technology and MESA Research Institue, University of Twente. Pages 231–234.

Maman N., D.J. Lyon, S.C. Mason, T.D. Galusha and R. Higgins, 2003. Pearl millet and grain sorghum yield response to water supply in Nebraska. Agronomy Journal, 95: 1618–1624.

Masumoto, Mas David, 1999. *Harvest son: Planting roots in American soil.* W.W. Norton & Company. Pages 175.

McElroy, D.L. and J.M. Hubbell, 2004. Evaluation of the conceptual flow model for a deep vadose zone system using advanced tensiometers. *Vadose Zone Journal,* 55: 170–182.

Meek, D.W. and J.W. Singer, 2004. Estimation of duration indices for repeated tensiometer readings. Agronomy Journal, 96: 1787–1790.

Neitzel, D.A., 2003. *Hanford site national environmental policy act (NEPA) characterization.* PNL-6415, Rev. 15. Pacific Northwest National Laboratory, Richland, WA. Pages 44–56.

Or, Dani, 2001. Who invented the tensiometer? *Soil Science Society of America Journal,* 65(1): 1.

Osmond, D.L. and D.H. Hardy, 2004. Characterization of turf practices in five North Carolina communities. *Journal of Environmental Quality,* 33: 565–575.

Panagiotis, V. and M. Sakellariou-Makrantonaki, 2005. Intermittent water application through surface and subsurface drip irrigation. *ASAE Annual Meeting.* Pages 22–78.

Payne, W.A., C. Chen and D.A. Ball, 2004. Agronomic potential of narrow-leafed and white lupins in the Inland Pacific Northwest. *Agronomy Journal*, 96: 1501–1508.

Pielou, E.C., 2000. *Fresh water.* University of Chicago Press. Pages 67–68

Pikul, J.L. Jr., J.K. Aase and V.L. Cochran, 2004. Water use and biomass production of oat–pea hay and lentil in a semiarid climate. *Agronomy Journal*, 96: 298–304.

Polomski, Bob, 2001. *Month-by-month gardening in the Carolinas.* Franklin, TN: Cool Springs Press. Pages 241.

Preece, J.E. and P.E. Read, 2005. *The biology of horticulture: An introductory textbook.* John Wiley & Sons. Pages 221–222.

Reich, L., 2001. *Weedless gardening.* Workman Publishing Company. Page 48.

Sadras, V.O. and D.K. Roget, 2004. Production and environmental aspects of cropping intensification in a semiarid environment of southeastern Australia. *Agronomy Journal*, 96: 236–243.

Siemens, J. and M. Kaupenjohann, 2004. Comparison of three methods for field measurement of solute leaching in a sandy soil. *Soil Science Society of America Journal*, 68: 1191–1196.

Sisson, J.B., G.W. Gee, J.M. Hubbell, W.L. Bratton, J.C. Ritter, A.L. Ward and T.G. Caldwell, 2002. *Advances in Tensiometry for Long-Term Monitoring of Soil Water Pressures.* Geo Sciences Research Department, Idaho National Engineering and Environmental Laboratory, Bechter, Idaho. Pages 34–45.

Tarantino, A. and L. Mongiovì, 2003. *Calibration of tensiometer for direct measurement of matrix suction.* Géotechnique Technical Note, 53(1):137–141.

Tarantino, A. and L. Mongiovì, 2001. *Experimental methodology and cavitation mechanisms in tensiometer measurements.* Geotechnical and Geological Engineering, 19(3): 189–210.

Tarantino, A., 2003. Direct Measurement of Soil Water Tension. *Proceedings 3rd International Conference on Unsaturated Soils* (JFT Jucá, TMP de Campos, FAM Marinho, editors.), Recife, Brasil. Pages 44–65.

Tarantino, A., 2004. Discussion on: *Tensiometer saturation and the reliable measurement of soil*

*suction,* by W.A. Take and M. D. Bolton'. Géotechnique, 54(3):229–232.

Thomas, J.T. and A. Husein, 2003. Application of soil fumigants through micro-irrigation systems. *ASAE Annual Meeting.* Pages 45–68.

Thomson, S.J. and B. Blake Ross, 1996. *Using soil moisture sensors for making irrigation management decisions in Virginia.* Extension Agricultural Engineers, Virginia Tech. Pages 67–78.

Tiner, R.W., 1999. *Wetland indicators: A guide to wetland identification, delineation, classification, and mapping.* CRC Press. Pages 136, 237.

Toker, N.K., 2002. *Improvements and reliability of MIT tensiometers and studies on soil moisture characteristic curves.* MSc Dissertation, Massachusetts Institute of Technology, Boston, US. Pages 544–566.

Wang, C.H., 1999. *The capacitive drop tensiometer: A novel multi analysis technique for measuring the properties of liquids.* College of Precision Instrument and Optoelectronic Engineering, Tianjin University. Pages 345–349.

Ward, A.D. and S.W. Trimble, 2003. *Environmental hydrology.* CRC Press. Pages 77–78.

Welsh, P., 1999. *Pat Welsh's Southern California gardening: A month-by-month guide completely revised and updated.* Chronicle Books. Pages 182, 132.

White, R.E., 2003. *Soils for fine wines.* USA: Oxford University Press. Pages 145–163.

**Web page links**

Bond, W.J. and P.A. Hutchinson, 2006. Principles, implementation, installation and operation of the tube tensiometer drainage meter. http://www.clw.csiro.au/publications/science/2006/sr18-06.pdf

Drip irrigation: history. http://en.wikipedia.org/wiki/Drip_irrigation    http://en.wikipedia.org/wiki/Talk:Drip_irrigation

Drip Irrigation Guide: Subsurface. http://www.netafimusa.net/downloads/LND/LSUBGD_Drip_Irrigation_Guide.pdf

Graham, Andrew, 2004. New surface tensiometer – AquaPi. http://www.edie.net/news/news_story.asp?id=8149

Hla, A.K. and F. Thomas Scherer, 2003. Introduction to micro-irrigation. North Dakota State

University, NDSU Extension Service. Pages 2–24. http://www.ext.nodak.edu/extpubs/ageng/irrigate/ae1243w.htm

Holtz, W., 2002. Tensiometer (KSV Sigma 701). http://microlab.berkeley.edu/labmanual/chap8/8.53.html

Hutchinson, Paul, 2002. Tube tensiometer installation and operation guide. http://www.clw.csiro.au/publications/technical2002/tr12-02.pdf

Irrigation - micro, sprinkler and subsurface, 2004. Department of Primary Industries. Agriculture and Food Home. Pages 11–15. http://www.dpi.vic.gov.au/dpi/nrenfa.nsf/LinkView

Kietzman, S. 2006. Conjecture Corporation. http://www.wisegeek.com/what-is-drip-irrigation.htm

King, B., 2006. Irrigation Agrologist, Sask Water, Outlook, Saskatchewan: Trickle Irrigation. Pages 20–22. http://www.quantumlynx.com/water/back/vol2no1/v21st4.htm

Knox, J.W. and Weatherhead, E.K. 2003. Trickle irrigation in England & Wales. http://www.silsoe.cranfi…ation.htm

Leith, Heiner, 1998. Tensiometer-based irrigation for nursery and greenhouse plant production. http://lieth.ucdavis.edu/Extension/CAN-TensiomSum98/index.htm

Ling, Peter. A review of soil moisture sensors. http://floriculture.osu.edu/archive/apr05/Sensors.html

Pinnoc, Derek R., 2005. Operating instructions for an electronic tensiometer irrometer models R-RSuU. http://www.usu.edu/cpl/PDF/TensiometerOperatingInstructions2.pdf

Rogers, Danny H., Loyd R. Stone and Richard D. Black, 1997. Tensiometer use in scheduling irrigation. http://www.oznet.ksu.edu/library/ageng2/l796.pdf

Shannon, Kietzman, 2006. Conjecture Corporation. Pages 2–22. http://www.wisegeek.com/what-is-drip-irrigation.htm

Smajstrla, Allen G. and Dalton S. Harrison, http://edis.ifas.ufl.edu/AE146 - FOOTNOTE_2 2002. Tensiometers for soil moisture measurement and irrigation scheduling. April. http://edis.ifas.ufl.edu/AE146

Southern Nevada Water Authority. 2006. Pages 15–21. http://www.snwa.com/html/land_irrig_drip.html

Stryker, J., 2005. Drip irrigation design guidelines. Jess Stryker's Irrigation Tutorials. Pages 12–34. http://www.irrigationtutorials.com/dripguide.htm

Taber, Henry G., 2006. Tensiometers tips for vegetables. June. http://www.public.iastate.edu/~taber/Extension/tensiometer%20tips/tens_tips.html

Törmälä, Sauli, 2006. Sigma703D Tensiometer. http://www.ksvltd.com/content/index/sigma703D

Von Unold, Georg, 2006. Change of plus minus sign for Tensiometer signals! http://www-www.ums-muc.de/datenmanagment/news_en/singlenews_english.html?L=1tx_ttnews%5Btt_news%5D=42&tx_ttnews%5BbackPid%5D=46&Hash=ofe8ecef62

Wilson C. and M. Bauer, 2006. Drip irrigation for home gardens. Pages 34–41. http://www.ext.colostate.edu/Pubs/garden/04702.html

# 4

## Literature cited

1. Bebr, R., 1954. *Agricultural hydraulics.* S.A. Barcelona, Madrid: Salvat Publishing, Chapter 8: Pages 212–220.

2. Jensen, M.E., 1980. *Design and operation of farm irrigation systems.* St. Joseph MI: Monograph #3 by American Society of Agricultural Engineers, Pages 829.

3. Israelsonn, W. and V.E. Hansen, 1965. *Principles and applications of the irrigation.*:S.A. Barcelona-Madrid: Reverte Editorial.

4. On-Farm Irrigation Committee of Irrigation and Drainage Division, 1987. *Selection of irrigation for methods agriculture.* New York, NY: American Society of Civil Engineers, Pages 95.

5. Reed, A.D., 1980. *Irrigation costs.* Leaflet 2875 October by Division of Agricultural Science, Berkley-CA: University of California, Pages 1–10.

6. Resh, H.M., 1985. *Hydroponics food production*, 3rd ed., Santa Barbarian, CA: Word Bridge Press.

7. Magazine "Agriculture of the Americas". 1983 and 1984 July and February of 1988.

8. Solomon, K. H., 1988. *Selection of the irrigation system*. Publ. Núm. 880702 – Institute of Agricultural Technology of California. Pages 1–11.

9. Turner, J.H. and C.L. Anderson, 1980. *Planning an irrigation system*. Athens-GA: American for Association Vocational Instructional Materials, Pages 120.

**Books/bulletins/journal and proceedings/reports**

Abrisqueta, J.M, A. Ruiz and J.A Franco, 2001. Water balance of apricot trees (Prunus armeniaca L. cv. B·lida) under drip irrigation. *Agricultural Water Management Journal*, 50(3): 211–227.

Aragües, R., E. Playán, R. Ortiz and A. Royo, 1999. A new drip-injection irrigation system for crop salt tolerance evaluation. *Soil Science Society of America Journal.*, 63(1): 1397–1403.

Assouline, S., 2002. The effects of microdrip and conventional drip irrigation on water distribution and uptake. *Soil Science Society of America Journal*, 66: 1630–1636.

Assouline, S., M. Möller, S. Cohen, M. Ben-Hur, A. Grava, K. Narkis and A. Silver, 2006. Soil-plant system response to pulsed drip irrigation and salinity: bell pepper case study. *Soil Science Society of America Journal*, 70(1): 1556–1568.

Aw, D., G. Diemer, 2005. *Making a large irrigation scheme work: A case study from Mali*. World Bank Publications, Infrastructure and Water Management. Pages 7–9.

Ayars, J.E., R.A. Schoneman, F. Dale, B. Meso and P. Shouse, 2001. Managing subsurface drip irrigation in the presence of shallow ground water. *Agricultural Water Management Journal*, 47(3): 243–264.

Battam, M., D. Boughton, P. Hulme and B. Sutton, 2001. Drip irrigated cotton: Observing wetting patterns. *Irrigation Journal*, 51(4): 13–16.

Ben Gal, A., N. Lazorovitch and U. Shani, 2004. Subsurface drip irrigation in gravel-filled cavities. *Vadose Zone Journal*, (3): 1407–1413.

Benjamin, J.G., L.K. Porter, H.R. Ahuja and G. Butters, 1998. Nitrogen movements with furrow irrigation method and fertilizer band placement. *Soil Science Society of America Journal*, 62(1): 1103–1108.

Bonachela, S., F. Orgaz, F.J. Villalobos and E. Fereres, 2001. Soil evaporation from drip-irrigated olive orchards. *Irrigation Science*, 20(2): 65–71.

Breazeale, D., J. Neyfeld, G. Myer and J. Davidson, 2000. Breakeven analysis of alfalfa seed production using subsurface drip irrigation. *Journal of Applied Irrigation Science*, 35(1): 91–99.

Bronson, K.F., A.B. Onken, J.W. Keeling, J.D. Booker and H.A. Torbert, 2001. Nitrogen response in cotton as affected by tillage system and irrigation level. *Soil Science Society of America Journal*, 65(1): 1153–1163.

Burt, C.M., 2000. *Drip and micro irrigation for trees, vines, and row crops*. San Luis Obispo, California: Irrigation Training and Research Center.

Burt, C.M. and S.W. Styles, 2000. Updating drip irrigation knowledge. *Irrigation Journal*, 50(2): 8–10.

Cason, K., 2003. Elevated drip irrigation system works. *Countryside & Small Stock Journal*, 87(5): 20.

Cassel Shamasarkar, F., S. Sharmasarkar, L.J. Held, S.D. Miller, G.F. Vance and R. Zhang, 2001. Agroeconomic analyses of drip irrigation for sugar beet production. *Agronomy Journal*, 93: 517–523.

Cook, F.J., P. Fitch and P.J. Thorburn, 2006. Modeling trickle irrigation: Comparison of analytical and numerical models for estimation of wetting front position with time. CSIRO. Pages 1–8.

Crescimanno, G. and P. Garofalo, 2006. Management of irrigation with saline water in cracking clay soils. *Soil Science Society of America Journal*, 70(1): 1774–1787.

El-Bably, A.Z., 2002. Effect of irrigation and nutrition of copper and molybdenum on Egyptian clover (*Trifolium alexandrnium 1.*). *Agronomy Journal*, 94(1): 1066–1070.

El-Hafedh, A.V.O.M., H. Daghari and M. Maalej, 2001. Analysis of several discharge rate-spacing-duration combinations in drip irrigation system. *Agricultural Water Management Journal*, 52(1): 33–52.

Feng, G.L., A. Meiri and J. Letey, 2003. Evaluation of a model for irrigation management under saline conditions: I. Effects on plant growth. *Soil Science Society of America Journal*, 67(1): 71–76.

Franco, J.A., J.M. Abrisqueta, A. Hernansaez and F. Morena, 2000. Water balance in a young almond orchard under drip irrigation with water of low quality. *Agricultural Water Management Journal*, 43(1): 75–98.

Handley, D., 2000. Drip irrigation: The system of choice for California grape growers. *Irrigation Journal*, 50(2): 22–24.

Howell, T.A., A.D. Schneider and D.A. Dusek, 2002. Effects on furrow diking on corn response to limited and full sprinkler irrigation. *Soil Science Society of America Journal*, 66(1): 222–227.

Huang, Z.B., S. Assouline, J. Zilberman and M. Ben-Hur, 2000. Tillage and saline irrigation effects on water and salt distribution in a sloping field. *Soil Science Society of America Journal*, 64(1): 2096–2102.

Jordan, J.E., R.H. White, D.M. Victor, T.C. Hale, J.C. Thomas and M.C. Engelke, 2003. Effect of irrigation frequency on turf quality, shoot density and root length of five bentgrass cultivars. *Crop Science*, 43(1): 282–287.

Kabutha, C, H. Blank and B. Van Koppen, 2000. *Drip irrigation kits for small-holder farmers in Kenya: Experience and a way forward*. Paper presented at the 2000 Micro Irrigation Conference, Cape Town, South Africa. Pages 1–8.

Kang, Y. and H. Jun, 2002. Effect of sprinkler irrigation on field micro-climate. *ASAE Annual Meeting*. Paes 1–15.

Lang, S., 2002. *Hillside landscaping*. It is all the Planning. Sunset Books Inc. Pages 79–90.

Lanier, J.E., L. David, J. Jordan, S. Barnes, J. Matthews, G.L. Grabow, W.J. Griffin, Jr., J.E. Bailey, P. D. Johnson, J.F. Spears and R. Wells, 2004. Disease management in overhead sprinkler and subsurface drip irrigation systems for peanut. *Agronomy Journal*, 96: 1058–1065.

Laosheng, W.U., 2000. Drip irrigation using low-quality water. *Irrigation Journal*, 50(3): 18–20.

Manoliadis, O.G. 2001. Analysis of irrigation systems using sustainability-related criteria. *Journal of Environmental Quality*, 30: 1150–1153.

Mehmet, O. and Biçak, 2002, *Modern and traditional irrigation technologies in the eastern mediterranean*. IDRC. Drip Irrigation in Northern Cyprus, Pages 73–83.

Mmolowa, K. and D. Or, 2003. Experimental and numerical evaluation of analytical volume balance model for soil water dynamics under drip irrigation. *Soil Science Society of America Journal*, 67: 1657–1671.

Moines, D., 2006. *All About Sprinklers & Drip Systems*. Meredith Books, Iowa. Pages 1–127.

Ngigi, S.N., J.N. Thome, D.W. Waweru and H.G. Blank, 2001. *Low-cost irrigation for poverty reduction: An evaluation of low-head drip irrigation technologies in Kenya*. In IWMI. Annual report 2000–2001. Colombo, Sri Lanka: IWMI. Pages 23–29.

Pair, Claude H., 1983. *Irrigation*. 5th Edition. Irrigation Association.

Plaut, Z. and M. Ben-Hur, 2005. Irrigation management of peanut with a moving sprinkler system: Runoff, yield and water use efficiency. *Agronomy Journal*, 97(1): 1202–1209.

Postel, S. and P. Polak, 2005. *Drip irrigation for small farmers: A new initiative to alleviate hunger and poverty*. New Delhi, India: International Development Enterprises. Pages 1–21.

Priyanjith, K.J.K.T., D.S.P. Kuruppuarachchi and H.A. Gunathilaka, 2002. Field Evaluation of Drip Irrigation System for Small Scale Banana (Musa) Orchards. Makandura, Sri Lanka: Wayamba University of Sri Lanka. *Faculty of Agriculture and Plantation Management*. Pages 185–193.

Royo, A., R. Aragües, E. Playán and R. Ortiz, 2000. Salinity-grain yield response functions of barley cultivars assessed with a drip-injection irrigation system. *Soil Science Society of America Journal*, 64(1): 359–365.

Sharmasarkar, F.C., S. Sharmasarkar, L.J. Held, S.D. Miller, G.F. Vance and R. Zhang, 2001. Agroeconomic analyses of drip irrigation for sugarbeet production. *Agronomy Journal*, 93(1): 517–523.

Skaggs, R.K., 2001. Predicting drip irrigation use and adoption in a desert region. *Agricultural Water Management Journal*, 51(2): 125–142.

Thompson, T.L., T.A. Doerge and R.E. Godin, 2002. Subsurface drip irrigation and fertigation of broccoli: I. yield, quality and nitrogen uptake. *Soil Science Society of America Journal*, 66: 186–192.

Tinglu, F., B.A. Stewart, W.A. Payne, Y. Wang, S. Song, J. Luo and C.A. Robinson, 2005. Supplemental irrigation and water-yield relationships for plasticure crops in the loess plateau of China. *Agronomy Journal*, 97(1): 177–188.

Tiwari, K.N., A. Singh and P.K. Mal, 2002. Effect of drip irrigation on yield of cabbage (*Brassica oleracea* L. var. capitata) under mulch and non-mulch conditions. *Agricultural Water Management*, 58(1): 19–28.

Trooien, T.P., F.R. Lamm, L.R. Stone, M.D. Alam, H. Rogers and G.A. Clark, 2000. Subsurface drip irrigation using livestock wastewater: Drip line flow rates. *Applied Engineering in Agriculture*, 16(5): 505–508.

Valiantzas, John D. 2002. Continuous outflow variation along irrigation laterals: Effect of the number of outlets. *Journal Irrigation and Drainage Engineering*, 128(1): 34–42.

Yuan, B.Z., Y. Kang and S. Nishiyama, 2001. Drip irrigation scheduling for tomatoes in unheated greenhouses. *Irrigation Science*, 20(3): 149–154.

Zayani K., A. Aloini and F. Lebdi, 2001. Design of drip line in irrigation systems using the energy drop ratio approach. *Transactions of ASAE*, 44(5): 1127–1133.

**Web page links**

Broner, I., 2005. Center-pivot irrigation systems. http://www.ext.colostate.edu/PUBS/crops/04704.html

Chiappini, R., 2001. Energy Efficient Crop Irrigation, U.S Department of Energy. Pages 1–2. www.eere.energy.gov/inventions/pdfs/nwpreag.pdf

Davis, R. and R. Hirji, 2003. Water Conservation: Irrigation. http://wwwwds.worldbank.org/servlet/WDSContentServer/WDSP/IB/2003/08/02/000094946_03071804010122/Rendered/PDF/multi0page.pdf.

Guidelines For Landscape Drip Irrigation system, 2004. http://isw03.cityofmesa.org/utilities/conservation/pdf/dripguide-v701-0503.pdf#search=%22monthly%20average%20drip%20irrigation%22.

Hanson, Fipps and Martin, 2000. Drip irrigation of row crops: What is the state of the art? 4th Decennial Irrigation Symposium. ASAE. Pages 1–13. http://www.oznet.kstate.edu/sdi/Abstracts/Drip Irrigation of Row Crops.htm

Hill, W.R. and B. Kitchen, 2002. Sprinklers, crop water use, and irrigation time. Utah State University Extension Service. Pages 1–10. www.extension.usu.edu/files/engrpubs/biewm27.pdf

Kopp, K. and J. Hoover, 2004. Irrigation system maintenance. Utah State University Extension. Pages 1–5. www.extension.usu.edu/files/factsheets/irrigation.pdf

Loudon, T., 1999. Sprinkler irrigation systems handbook. Michigan State University extension. http://www.egr.msu.edu?age/aenewsletter/ae_7_99/sprinkler_irrigation_systems_han.htm

Monthly Irrigation Amount, 2004. http://www.nespal.org/SIRP/Research/2005.03.Fact_Sheet_02SW.pdf.

Narain, S., 2003. Drip irrigation the long and short of it. Center for Science and Environment. Page 1. www.cseindia.org/dtesupplement/water20031115/drip_irrigation.pdf

Resh, 2006. Drip irrigation or micro-irrigation. Homegrown Hydroponics Inc. http://www.hydroponics.com/gardens/grownintechniques.html

Roberts, M., 2005. Effects of wastewater application on trees. Kansas State University. Pages 1–8. www.asabe.org/membership/students/Roberts.doc

Scherer, T., 2005. Selecting a sprinkler irrigation system. htpp.//www.ext.nodak.edu/extpubs/ageng/irrigate/ae91w.htm

Selker, John, 2004. Pesticides in Southern Willamette Valley groundwater. Oregon State University. Pages 1–4. http://extension.oregonstate.edu/catalog/pdf/ec/ec1565.pdf

Stryker, Jess, 2005. Drip irrigation system design guidelines. Jess Stryker's Free Landscape Irrigation Tutorials for Homeowners and Landscape Professionals, ed. Jess Stryker. http://www.irrigationtutorials.com/dripguide.htm

Weber, C.E., M.D. Butler, C.K. Campbell, B.A. Holliday and J. Klanzer, 2004. Management guide for drip irrigation on seed carrots in central Oregon. Oregon State University Extension Service. Pages 1–7. http://extension.oregonstate.edu/catalog/pdf/em/em8880-e.pdf

Wright, J., 2002. Comprehensive guide to sprinkler irrigation systems. http://www.bae.umn.edu/extens/ennotes/enmay02/mwps30.htm

Zazueta, F.S., A.G. Smajstrla and G.A. Clark, 2006. Irrigation system controllers. http.//edis.ifas.ufl.edu/AE077

# 5

## Literature cited

1. Goldberg, D., B. Gornat and D. Rimon, 1976. *Drip irrigation: Principles, design and agricultural practices.* Kfar Shmaryahu, Israel: Drip Irrigation Scientific Publications.

2. Goyal, M.R., J.A. Santaella y L.E. Rivera, 1982. El tensiómetro: su uso, instalación y mantenimiento. Colegio de Ciencias Agrícolas, Servicio de Extensión Agrícola, Recinto Universitario de Mayagüez, Río Piedras-E.U.A, A.

3. Israelson, D.W. y V.E. Hansen, 1965. *Principios y Aplicaciones del Riego.* Editorial Reverte. S. A., Barcelona-España.

4. Jensen, M.E., 1980. *Design and operation of farm irrigation systems.* St. Joseph, MI: ASAE Monograph No. 3, American Society of Agricultural Engineers.

5. Ross, D.S., R.C. Funt, C.W. Reynolds, D.S. Coston, H.H. Fries and N.J. Smith, 1978. *Trickle irrigation and introduction.* The Northeast Regional Agricultural Engineering Service (NRAES), NRAE-4, Cornell University, Ithaca, NY-USA.

6. Ross, D.S., R.A. Parsons, W.R. De Tar, H.H. Fries, D. D. Davis, C.W. Reynolds, H.E. Carpenter and E.D. Markwardt, 1980. *Trickle irrigation in the Eastern United States.* Cooperative Extension Service NRAES-4 Cornell University, Ithaca, NY-USA.

## Books/bullentins/journals and proceedings/reports

Agassi, M., J. Tarchitzky, R. Keren, Y. Chen, D. Goldstein and E. Fizik, 2003. Effects of prolonged irrigation with treated municipal effluent on runoff rate. *Journal of Environmental Quality*, 32(3): 1053–1057.

Aragües, R., E. Playan, R. Ortiz and A. Royo, 1999. A new drip-injection irrigation system for crop salt tolerances evaluation. *Soil Science Society of America Journal*, 63(5): 1397–1403.

Assouline, S., 2002. The effect of micro drip and conventional drip irrigation on water distribution and uptake. *Soil Science Society of America Journal*, 66(5): 1630–1636.

Assouline, S., M. Möller, S. Cohen, M. Ben-Hur, A. Grava, K. Narkis and A. Silber, 2006. Soil plant system response to pulsed drip irrigation and salinity: bell pepper case study. *Soil Science Society of America Journal*, 70(5): 1556–1568.

Assouline, S., S. Cohen, D. Meerbach, T. Harodi and M. Rosner, 2002. Microdrip irrigation of field crops: Effect on yield, water uptake and drainage in sweet corn. *Soil Science Society of America Journal*, 66(1): 228–235.

Ayars, J.E., C.J. Phene, R.B. Hutmacher, K.R. Davis, R.A. Schoneman, S.S. Vail and R.M. Mead, 1999. Subsurface drip irrigation of row crops: A review of 15 years of research at the Water Management Research Laboratory. *Agricultural Water Management*, 42(2): 1–27.

Ayars, J.E., R.A. Shoneman, F. Dale, B. Meso and P. Shouse, 2001. Managing subsurface drip irrigation in the presence of shallow ground water. *Agricultural Water Management*, 47(3): 243–264.

Blass, S., 1973. *Water in strife and action* (Hebrew). Israel: Massada limited

Burt, C.M., 2005. Leaching of accumulated soil salinity under drip irrigation. *Transactions of the ASAE*, 48(6): 2115–2121.

Burt, C., 1999. *Drip and micro irrigation for trees, vines, and row crops.* California, CA: Irrigation Training and Research Center, Pages 7–8

Burton, D.J., F.H. Harned, B.J. Lesikar, J.F. Prochaska and R.J. Suchecki, 2001. On-site wastewater treatment. *Proceedings of the Ninth*

*National Symposium on Individual and Small Community Sewage Systems*, St. Joseph-MI. ASAE 701P0009, Fort Worth, TX, USA. Pages 606–616.

Camp, C.R., 1998. Subsurface drip irrigation: A review. *Transactions of ASAE*, 41(5):1353–1367.

Camp, C.R., F.R. Lamm, R.G. Evans and C.J. Phene. 2000. Subsurface drip irrigation – past, present and future. In *Proceedings of the 4th Decennial National Irrigation Symposium*, eds. R.G. Evans, B.L. Benham, and T.P. Trooien, November 14–16, Phoenix AZ. St. Joseph MI: ASAE, Pages 363–372.

Cassel, S.F., S. Sharmasarkar, L.J. Held, S.D. Miller, G.F. Vance and R. Zhang, 2001. Agroeconomic analyses of drip irrigation for sugar beet production. *Agronomy Journal*, 93(3): 517–523.

Colombo, A., 2006. Plant water accessibility function: A design and management tool for trickle irrigation. *Agricultural Water Management*, 82 (1/2): 45–62.

De Kreij, Van der Burg, A.M.M. and W.T. Runia, 2003. Drip irrigation emitter clogging in Dutch greenhouses as affected by methane and organic acids. *Agricultural Water Management*, 60: 73–85

Drotleff, L., 2006. The Father of drip. *American Vegetable Grower*, 54 (5): 18.

Dukes, M.D. and J.M. Scholberg, 2005. Soil moisture controlled subsurface drip irrigation on sandy soils. *Applied Engineering in Agriculture*, 21(1): 89–101.

Enciso, J., 2004. *Installing a subsurface drip irrigation system for row crops.* Texas Cooperative Extension, Texas A&M University System, College Station, TX-USA.

Hanson, B., 2006. Keeping the lines open. *American Vegetable Grower*, 54(5): 20.

Hills, D.J. and M.J. Brenes, 2001. Microirrigation of wastewater effluent using drip tape. *Applied Engineering in Agriculture*, 17(3): 303–308.

Holzapfel, E.A., C. Lopez and J.P. Joublan, 2001. Efecto del agua y fertirrigacion en el desarrollo y producción de naranjos cv. *Thompson Navel. Agricultura Técnica*, 61(1): 51–60.

Lamm, F.R., J.E. Ayers and F.S. Nakayama, 2006. *Microirrigation for crop production: Design, operation and management.* Boston, MA-USA: Elsevier.

Lanier, J.E., D.L. Jordan, J. Stephen Barnes, J. Mathews, Gary L. Grabow, William J. Griffin Jr., Jack E. Bailey, P. Dewayne Johnson, Janet F. Spears and Randy Wells, 2004. Disease management in overhead sprinkler and subsurface drip irrigation systems for peanut. *Agronomy Journal*, 96(4): 1058–1065.

Li, G.Y., 2006. Influence of geometrical parameters of labyrinth flow path of drip emitters on hydraulic and anti-clogging performance. *Transactions of the ASAE*, 49(3): 637–643.

Mamedov, A.I., I. Shainberg and G.J. Levy, 2000. Irrigation with Influent water: effects of rain fall energy on soil infiltration. *Soil Science Society of American Journal*, 64(2): 732–737.

Mane, M.S., 2006. *Principles of drip irrigation.* New Delhi, India: Asiatic Books. Pages 1–168

Mehmet, O. and H.A Biçak, 2002. Modern and traditional irrigation technologies in the Eastern Mediterranean. IDRC, Drip Irrigation in Northern Cyprus, pages 73–83.

Mmolawa, K. and D. Or, 2003. Experimental and numerical evaluation of analytical volume balance model for soil water dynamics under drip irrigation. *Soil Science Society of American Journal*, 67(6): 1657–1671.

Mofoke, A.L.E., 2006. Yield of tomato grown under continuous-flow drip irrigation in Bauchi state of Nigeria. *Agricultural Water Management*, 84(1/2): 166–172.

Nujamudeen, M.S. and P.B. Dharmasena, 2002. Performance of chillie under drip irrigation with mulch. *Annual of the Sri Lanka Department of Agriculture*, 4: 89–94.

Phene, C.J., R. Yue, I-Pai Wu, J.E. Ayars, R.A. Schoneman and B. Meso. 1992. *Distribution uniformity of subsurface drip irrigation systems.* ASAE Paper No. 92-2569, St. Joseph, MI: ASAE. Pages 1–14.

Phillip, D., C.P. Brown, V.G. Allen and D.B. Webster, 2006. Influence of irrigation on mineral concentration in three bluestem species. *Crop Science Journal*, 46(5): 2033–2040.

Postel, S. and P. Polak, 2005. *Drip irrigation for small farmers: A new initiative to alleviate hun-*

*ger and poverty*. New Delhi-India: International Development Enterprises.

Schneider, A.D. and T.A. Howell, 1999. LEPA and spray irrigation for grain crops. *Journal of Irrigation and Drainage Engineering*, 125(4): 167–172.

Schulz, M.A., 2000. *Subsurface drip irrigation for broad acre crops*. Department of Natural Resources and Environment, State of Victoria, Australia.

Sijali, I.V., 2001. *Drip irrigation: Options for smallholder farmers in eastern and Southern Africa*. Regional Land Management Unit, Nairobi-Kenya.

Singh, D.K., T.B. S. Rajput, H.S. Sikarwar, R.W. Shoo and T. Ahmad, 2006. Simulation of soil wetting pattern with subsurface drip irrigation from line source. *Agricultural Water Management*, 83(1–2): 130–134.

Singh, H.P., 2000. Microirrigation. *International Conference on Micro and Sprinkler Irrigation Systems*. New Delhi-India: Central Board of Irrigation and Power.

Sivanappan, R.K., 1994. Prospects of micro-irrigation in India. *Irrigation and Drainage Engineering*, 8: 49–58.

*Sprinklers & Drip Systems*. Sunset Publishing Corporation, Edition 2006. Pages 1–98.

Styles, S.W. and C.M. Burt, 1999. *Drip and micro irrigation for trees, vines and row crops*. California: Irrigation Training and Research Center. Pages 1–292

Thompson, T.L., T.A. Doerge and R.E. Godin, 2000. Nitrogen and water interactions in subsurface drip-Irrigated cauliflower: II. Agronomic, economic and environmental outcomes. *Soil Science Society of American Journal*, 64: 412–418.

Thompson, T.L., T.A. Doerge and R.E. Godin, 2002. Subsurface drip irrigation and fertigation of broccoli: II. Agronomic, economic, and environmental outcomes. *Soil Science Society of America Journal*, 66: 178–185.

Wang, D., M.C. Shannon, C.M. Grieve, P.J. Shouse and D.L. Suarez, 2002. Ion partitioning among soil and plant component under drip, furrow, and sprinkler irrigation regimes: Field and modeling assessments. *Journal of Environmental Quality*, 31(50): 1684–1693.

Zegbe, J.A., M.H. Behboudian and B.E. Clothier, 2004. Emitter and filter tests for waste water reuse by drip irrigation. *Agricultural Water Management*, 68(2): 135–149.

**Web page links**

ATI-NET, CATI, 2000. The purpose of this forum is to provide an online niche that focuses on the technical and management aspects of micro-irrigation systems. http://www.microirrigationforum.com

Bar, I., 1999. Guidelines for drip irrigation and fertigation of pines and hardwoods. http://www.woodycrops.org/paducah/bar.html

Cuykendall, C.H., G.B. White, B.E. Shaffer, A.N. Lakso and R.M. Dunst, 1999. Economics of drip irrigation for juice grape vineyards in New York State. Pages.4–10. www.aem.cornell.edu/research/researchpdf/rb9901.pdf

Drip irrigation parts, systems and supplies. Planning and Installation Guide, 2005. http://www.dripdepot.com/planning_installation_guide_1.html?id=wfjajmWE

Drip Rite Irrigation Products, 2006. New Press website. http://www.dripirr.com/services/system_guide/specifications/trees.htm

Geoflow: subsurface drip systems. The ultimate subsurface irrigation system. http://www.geoflow.com/agriculture.html

Gil-Marin, J.A., 2001. Forma y dimensiones del bulbo húmedo con fines de diseño de riego por goteo en dos suelos típicos de sabana. Revista Científica UDO Agrícola, Universidad de Oriente Press, 1(1): 42–47. http://www.bioline.org.br/request?cg01007

Gil-Marin, J.A., L. Khan and R. Hernandez, 2002. Evaluación del comportamiento hidráulico de varios emisores importados para riego por goteo. Revista Científica UDO Agrícola, Universidad de Oriente Press, 2(1): 64–72. http://www.bioline.org.br/request?cg02008

Haman, D. Z. and F. T. Izuno, 2003. Principles of micro irrigation. University of Florida, Institute of Food and Agricultural Science, Document AE70. http://edis.ifas.ufl.edu/WI007

Handson, B.R., D.M. May and L.J. Schwankl, 2003. Effect of irrigation frequency on subsurface drip-irrigated vegetables. *Extension Newsletter* 13(2): 1–2. http://aggie-horticulture.tamu.

edu/extension/newsletters/vpmnews/feb03/art-5feb.html

Harris, Graham, 2006. Sub-surface drip irrigation (SDI) - Advantages and limitations. May 25.

Delivery, Toowoomba – Australia. http://www2.dpi.qld.gov.au/fieldcrops/17650.html

Holden, Janelle, 2001. Drip-irrigation projects nourishing area fruit trees. http://www.cortez-journal.com/archives/1news1661.htm

Hook, J., 2005. Irrigation Water & Power Development in the East National Environmentally Sound Production Agriculture Laboratory. Pages 1–37. www.sandia.gov/energy-water/EastPresentations/IrrigWaterPower.pdf

ICT International, 2006. Soil Science, Plant Science and Meteorology Equipment. http://www.ictinternational.com.au/faqjetfill.htm

Jaeger, W.K., 2004. Energy Pricing and Irrigated Agriculture in the Upper Klamath Basin. pages 1–11. http://extension.oregonstate.edu/catalog/pdf/em/em8846-e.pdf

Jain Irrigation System Ltd., 2006. Drip Irrigation Systems. http://www.jains.com/irrigation/drip%20irrigation%20system.htm

Lamm, F.R., 2002. Advantages and disadvantages of subsurface drip irrigation. Kansas State University, Northwest Research-Extension Center. http://www.oznet.ksu.edu/sdi/Reports/2002/ADofSDI.pdf

Laosheng, Wu, 2000. Drip irrigation using low-quality water. Pages 1–5. http://esce.ucr.edu/soilwater/summer2000/soilwatersummer_2000.htm

Northern Garden Supply. How- to tutorials about drip irrigation. http://www.dripirrigation.ca/HowTo.asp

Qassim, A. 2003. Subsurface drip irrigation: Situation analysis. ISIA, Tatura - Australia. http://www.wca-infonet.org/cds_upload/1058151636725_SUBSURFACE_IRRIGATION.pdf

Raine, S.R, J.P. Foley and C.R. Henkel, 2000. Drip irrigation in the Australian cotton industry: a scoping study. National Centre for Engineering in Agriculture, Publication 179757/1, USQ, Toowoomba-Australia. Pages 7–64. http://www.ncea.org.au/Irrigation/downloads/DripIrrigation.pdf

Roberts, M. 2005. Effects of wastewater application on trees. Kansas State University. Pages 1–8. www.asabe.org/membership/students/Roberts.doc

Salcedo F., R. Barrios, M. Garcia and T. Valdez, 2005. Distribución de agua en un sistema de microaspersión sobre un ultisol cultivado con Lima Tahití en el estado Monagas-Venezuela. Revista Científica UDO Agrícola, Universidad de Oriente Press, 5(1): 64–72.

http://www.bioline.org.br/request?cg05012

Sanders, D.C., 2001. Components. North Carolina State University, Horticulture Information Leaflets, HIL-33-A.

http://www.ces.ncsu.edu/depts/hort/hil/hil-33-a.html

Sanders, D.C., 2001. Drip or trickle irrigation systems: An outline of components. Extension Horticultural Specialist. North Carolina State University, Department of Horticultural Science, College of Agriculture & Life Sciences. http://www.ces.ncsu.edu/depts/hort/hil/hil-33-a.html

Seaman, G., 2006. Drip irrigation. http://www.eartheasy.com/grow_drip-irrigation.htm

Sellés G., R. Ferreyra, G. Contreras, R. Ahumada, J. Valenzuela and R. Bravo, 2003. Manejo de riego por goteo en uva de mesa cv. Thpmpson seedless cultivada en suelos de textura fina. Agricultura Técnica, Instituto de Investigaciones Agropecuarias, INIA, 63(2): 180–192. http://www.bioline.org.br/abstract?id=at03019&lang=en

Shock, C.C., 2006. Drip irrigation: An introduction. Oregon State University, Cooperative Extension Services, Bulletin EM 8782. http://www.cropinfo.net/drip.htm

Shock, C.C., R.J. Flock, E.B.G. Feibert, A.B. Pereira and M. O'Neill, 2005. Drip irrigation guide for growers of hybrid Poplar. Oregon State University Extension Service. EM 8902, pages 1–6. http://extension.oregonstate.edu/catalog/pdf/em/em8902.pdf

Shock, C.C., R.J. Flock, E.B.G. Feibert, C.A. Shock, A.B. Pereira and L.B. Jensen, 2005. Irrigation monitoring using soil water tension. Oregon State University Extension Service. EM 8900. Pages 1–6. http://extension.oregonstate.edu/catalog/pdf/em/em8900.pdf

Shock, C.C., R.J. Flock, E.P Eldredge, A.P. Pereira and L.B. Jensen, 2006. Drip irrigation guide for potatoes in the treasure valley. Oregon State University Extension Service. EM 8912. Pages 1–8. http://extension.oregonstate.edu/catalog/pdf/ec/ec1368.pdf

Smarts Values Online, 2006. Free drip and micro watering handbook online. http://dripwatering.com/os-load.asp?load=os-catalog&catalogID=162&trail=7,162

Stryker J. 2006. Drip irrigation design guidelines. http://www.irrigationtutorials.com/dripguide.htm

Tekinel, Osman and Riza Kamber, 2006. Traditional irrigation technologies. Chapter II: Trickle irrigation experiments in Turkey. http://www.idrc.ca/en/ev-42826-201-1-DO_TOPIC.html

The Urban Farmer Store, 2006. Introduction to drip irrigation. http://www.urbanfarmerstore.com/drip/dripintro.html#applications

**Subsurface drip irrigation (SDI)**

Adamsen, F.J., 1989. Irrigation method and water quality effect on peanut yield and grade. *Agronomy Journal*, 81(4): 589–593.

Adamsen, F.J., 1992. Irrigation method and water quality effects on corn yield in the Mid-Atlantic Coastal Plain. *Agronomy Journal*, 84(5): 837–843.

Anonymous, 1998. Farm and Ranch Irrigation Survey. In: Vol. 3. 1997 Census of Agriculture. United States Department of Agriculture, National Agriciculture Statistics Service, Washington, DC.

Anonymous, 2000. 1999 Annual irrigation survey. *Irrigation Journal*, 50(1):16–31.

ASAE Standards, 1996. *S526.1. Soil and water terminology*, 43rd ed., St. Joseph, Mich.: ASAE.

ASAE Standards, 1999a. *S526.1: Soil and water terminology*, 46th ed., St. Joseph, MI: ASAE.

ASAE Standards, 1999b. *EP-458: Field evaluation of microirrigation systems*. St. Joseph, MI:ASAE.

Ayars, J.E., C.J. Phene, R.B. Hutmacher, K.R. Davis, R.A. Schoneman, S.S. Vail and R.M. Mead, 1999. Subsurface drip irrigation of row crops: a review of 15 years of research at the Water Management Research Laboratory. *Agricultural Water Management*, 42:1–27.

Ayars, J.E., C.J. Phene, R.A. Schoneman, B. Meso, F. Dale and J. Penland, 1995. Impact of bed location on the operation of subsurface drip irrigation systems. *In Proceedings of the Fifth International Microirrigation Congress*, ed. F.R. Lamm, pp. 141–146. St. Joseph, MI: ASAE.

Ayars, J.E., C.J. Phene, R.A. Schoneman, B. Meso, F. Dale and J. Penland, 1995. Impact of bed location on the operation of subsurface drip irrigation systems. *In Proceedings 5th International Microirrigation Congress*, ed. F.R. Lamm, 141–146. St. Joseph, MI: ASAE.

Barth, H.K., 1995. Resource conservation and preservation through a new subsurface irrigation system. *In Proceedings 5th International Microirrigation Congress*, ed. F.R. Lamm, 168–174. St. Joseph, MI: ASAE.

Bar-Yosef, B., C.J. Phene and R.B. Hutmacher, 1991. Plants response to subsurface trickle fertigation. BARD Project No. 1-1116-86 Final Report. Bet Dagan, Israel: BARD. Bar-Yosef, B., B. Sagiv, and T. Markovitch. 1989. Sweet corn response to surface and subsurface trickle phosphorus fertigation. *Agronomy Journal* 81(3): 443–447.

Batchelor, C.H., C.J. Lovell, M. Murata and S.P. McGrath, 1994. Improving water use effectiveness by subsurface irrigation. *Aspects of Applied Biology*, 38: 269–278.

Ben-Asher, J. and C.J. Phene, 1993. Analysis of surface and subsurface drip irrigation using a numerical model. In *Subsurface Drip Irrigation-Theory, Practices and Application*, 185–202. CATI Pub. No. 92 1001. Fresno, CA: California State University.

Bogle, C.R., T.K. Hartz and C. Nunez, 1989. Comparison of subsurface trickle and furrow irrigation on plastic-mulched and bare soil for tomato production. Journal of Amercian Society for Horticultural Sciences, 114(1): 40–43.

Bosch, D.J., N.L. Powell and F.S. Wright, 1992. An economic comparison of subsurface microirrigation with center pivot sprinkler irrigation. *Journal of Production Agriculture*, 5(4): 431–437.

Bracy, R.P., R.J. Edling and E.B. Moser, 1995. Drip-irrigation management and fertilization of bell pepper in a humid area. *In Proceedings 5th International Microirrigation Congress*, ed. F.R. Lamm, 181–186. St. Joseph, MI: ASAE.

Braud, H.J., 1970. Subsurface irrigation in the Southeast. *In Proceedings of the National Irrigation Symposium*, E1-E9. St. Joseph, MI: ASAE.

Brown, K.W., J.C. Thomas, S. Friedman and A. Meiri, 1996. Wetting patterns associated with directed subsurface irrigation. *In Proceedings International Conference on Evapotranspiration and Irrigation Scheduling*, eds. C.R. Camp, E.J. Sadler, and R.E. Yoder, 806–811. St. Joseph, MI: ASAE.

Bucks, D.A., 1995. Historical developments in microirrigation. *In Proceedings of 5th International Microirrigation Congress*, ed. F.R. Lamm, 1–5. St. Joseph, MI: ASAE.

Bucks, D.A., and S. Davis, 1986. Historical Development. *In Trickle Irrigation for Crop Production*, eds. F.S. Nakayama, and D.A. Bucks, 1–26. New York, NY: Elsevier.

Bucks, D.A., L.J. Erie, O.F French, F.S. Nakayama and W.D. Pew, 1981. Subsurface trickle irrigation management with multiple cropping. *Transactions of the ASAE*, 24(6): 1482–1489.

Bucks, D.A., F.S. Nakayama and A.W. Warrick, 1982. Principles, practices, and potentialities of trickle (drip) irrigation. *In Advances in Irrigation*, ed. D. Hillel, 219–299. New York, NY: Academic Press. Pages 302

Bui, W., 1990. Performance of "Turbo Model" drip irrigation tubes. *In Proceedigs 3rd National Irrigation Symposum*, 198–203. St. Joseph, MI: ASAE.

Bui, W., and R.V. Osgood. 1990. Subsurface irrigation trial for alfalfa in Hawaii. In Proceedigs 3rd National Irrigation Symposium, 658–660. St. Joseph, MI: ASAE.

Burt, C.E., 1995. Is buried drip the future with permanent crops? *Irrigation Business Technology* 3(1): 20–22.

Burt, C.M., S. Styles, R.E. Walker and J. Parrish, 1999. *Irrigation evaluation software*. Irrigation Training and Research Center, Cal. Poly. San Luis Obispo, CA.

Caldwell, D.S., W.E. Spurgeon and H.L. Manges, 1994. Frequency of irrigation for subsurface drip-irrigated corn. *Transactions of the ASAE*, 37(4): 1099–1103.

Camp, C.R., 1998. Subsurface drip irrigation: A review. *Transactions of ASAE*, 41(5):1353–1367.

Camp, C.R., P.J. Bauer and P.G. Hunt, 1997a. Subsurface drip irrigation lateral spacing and management for cotton in the southeastern Coastal Plain. *Transactions of the ASAE*, 40(4): 993–999.

Camp, C.R., J.T. Garrett, E.J. Sadler and W.J. Busscher, 1993. Microirrigation management for doublecropped vegetables in a humid area. *Transactions of the ASAE*, 36(6): 1639–1644.

Camp, C.R., E.J. Sadler and W.J. Busscher, 1989. Subsurface and alternate-middle micro irrigation for the southeastern Coastal Plain. *Transactions of the ASAE*, 32(2): 451–456.

Camp, C.R., E.J. Sadler and W.J. Busscher, 1997b. A comparison of uniformity measures for drip irrigation systems. *Transactions of the ASAE*, 40(4): 1013–1020.

Cavanaugh, P., 1992. Raisin grower buries drip. *Grape Grower*, 24 (4): 6–8.

Chase, R.G., 1985a. Phosphorus application through a subsurface trickle system. *In Proceedings 3rd International Drip/Trickle Irrigation Congress*, 1: 393–400. St. Joseph, MI: ASAE.

Chase, R.G., 1985b. Subsurface trickle irrigation in a continuous cropping system. *In Proceedings of the 3rd International Drip/Trickle Irrigation Congress*, 2: 909–914. St. Joseph, MI: ASAE.

Clark, G.A., C.D. Stanley and D.N. Maynard, 1993. Surface vs. subsurface drip irrigation of tomatoes on a sandy soil. *Proceedings of the Florida State Horticultural Society*, 106: 210–212.

Clark, G.A., C.D. Stanley, D.N. Maynard, G.J. Hochmuth, E.A. Hanlon and D.Z. Haman, 1991. Water and fertilizer management of microirrigated fresh market tomatoes. *Transactions of the ASAE*, 34(2): 429–435.

Coelho, F.E. and D. Or, 1996. A parametric model for two-dimensional water uptake intensity by corn roots under drip irrigation. *Soil Science Society of America Journal* 60(4): 1039–1049.

Darusman, A.H. Khan, L.R. Stone and E.R. Lamm, 1997a. Water flux below the root zone vs. dripline spacing in drip-irrigated corn. *Soil Science Society of America Journal* 61(6): 1755–1760.

Darusman, A.H. Khan, L.R. Stone, and F.R. Lamm. 1997b. Water flux below the root zone vs. irrigation amount in drip-irrigated corn. *Agronomy Journal*, 89(3): 375–379.

Davis, K.R., C.J. Phene, R.L. McCormick, R.B. Hutmacher and D.W. Meek, 1985. Trickle frequency and installation depth effects on tomatoes. *In Proceedings 3rd International Drip/Trickle Irrigation Congress*, 2: 896–902. St. Joseph, Mich.: ASAE.

Davis, S, 1967. Subsurface irrigation-How soon a reality? *Agricultural Engineering*, 48(11): 654–655.

Davis, S., and S.D. Nelson. 1970a. Subsurface irrigation today and tomorrow in California. In *Proceedings of the National Irrigation Symposium*, Hl-H8. St. Joseph, MI: ASAE.

Davis, S. and S.D. Nelson, 1970b. Subsurface irrigation easily automated. *Journal of the Irrigation and Drainage Division*, ASCE 96(IR1): 47–51.

Davis, S. and W.J. Pugh, 1974. Drip irrigation: Surface and subsurface compared with sprinkler and furrow. In *Proceedings of the 2nd International Drip Irrigation Congress*, 109–114. Riverside, CA: University of California.

DeTar, W.R., G.T. Browne, C.J. Phene and B.L. Sanden, 1996. Real-time irrigation scheduling of potatoes with sprinkler and subsurface drip systems. In *Proceedings International Conference on Evapotranspiration and Irrigation Scheduling*, eds. C.R. Camp, E.J. Sadler, and R.E. Yoder, 812–824. St. Joseph, MI: ASAE.

DeTar, W.R., C.J. Phene and D.A. Clark, 1994. Subsurface drip vs furrow irrigation: 4 years of continuous cotton on sandy soil. In *Beltwide Cotton Conference*, 542–545. Memphis, TN: National Cotton Council.

Devitt, D.A. and W.W. Miller, 1988. Subsurface drip irrigation of bermudagrass with saline water. Applied Agricultural Research, 3(3): 133–143.

Dhuyvetter, K.C., F.R. Lamm and D.H. Rogers, 1995. Subsurface drip irrigation (SDI) for field corn-An economic analysis. In *Proceedings of the 5th Internatonal Microirrigation Congress*, ed. F.R. Lamm, 395–401. St. Joseph, MI: ASAE.

Dirksen, C., 1978. Transient and steady flow from subsurface line sources at constant hydraulic head in anisotropic soil. *Transactions of the ASAE*, 21(5): 913–919.

Duncan, J., 1993. Buried drip gains more ground. *Grape Grower*, 25(4): 20–23.

Edwards, D.M., J.D. Eastin, R.A. Olson and R. German, 1970. Subsurface irrigation today and tomorrow in the Midwest. In *Proceedings of the National Irrigation Symposium*, Gl-Gll. St. Joseph, MI: ASAE.

El-Gindy, A.M. and A.M. El-Araby, 1996. Vegetable crop response to surface and subsurface drip under calcareous soil. In *Proceedings International Conference on Evapotranspiration and Irrigation Scheduling*, eds. C.R. Camp, E.J. Sadler, and R.E. Yoder, 1021–1028. St. Joseph, MI: ASAE.

Evett, S.R., T.A. Howell and A.D. Schneider, 1995. Energy and water balances for surface and subsurface drip irrigated corn. In *Proceedings of the 5th International Microirrigation Congress*, ed. F.R. Lamm, 135–140. St. Joseph, MI: ASAE.

Evett, S.R., T.A. Howell, A.D. Schneider, D.R. Upchurch and D.F. Wanjura, 1996. Canopy temperature based automatic irrigation control. In *Proceedings International Conference on Evapotranspiration and Irrigation Scheduling*, eds. C.R. Camp, E.J. Sadler, and R.E. Yoder, 207–213. St. Joseph, MI: ASAE.

Fangmeier, D.D., D.J. Garrot Jr., S.H. Husman and J. Perez, 1989. Cotton water stress under trickle irrigation. *Transactions of the ASAE*, 32(6): 1955–1959.

Ghali, G.S. and Z.J. Svehlik, 1988. Soil-water dynamics and optimum operating regime in trickle-irrigated fields. *Agricultural Water Management*, 13: 127–143.

Gibson, W., 1974. Hydraulics, mechanics and economics of subsurface and drip irrigation of Hawaiian sugarcane, 2: 639–648. In *Proceedings of the 15th Congress International Society Sugar*

Cane Technology, Honolulu, Hawaii: Hawaiian Sugar Planters' Assoc

Gilley, J.R. and E.R. Allred, 1974a. Infiltration and root extraction from subsurface irrigation laterals. *Transactions of ASAE*, 17(5):927–933.

Gilley, J.R. and E.R. Allred, 1974b. Optimum lateral placement for subsurface irrigation systems. In *Proceedings of the Second International Drip Irrigation Congress*, 234–239. Riverside, CA: University of California.

Goldberg, D., B. Gornat and D. Rimon, 1976. *Drip Irrigation Principles, Design and Agricultural Practices*. Kfar Shmaryahu, Israel: Drip Irrigation Scientific Publications.

Goldberg, D. and M. Shmueli, 1970. Drip irrigation-A method used under and and desert conditions of high water and soil salinity. *Transactions of the ASAE*, 13(1): 38–41.

Grattan, S.R., L.J. Schwankl, and W.T. Lanini. 1988. Weed control by subsurface drip irrigation. *CaliforniaAgricultural*, 42(3): 22–24.

Grattan, S.R., L.J. Schwankl and W.T. Lanini, 1990. Distribution of annual weeds in relation to irrigation method. In *Proceedings of the 3rd National Irrigation Symposium*, 148–153. St. Joseph, MI: ASAE.

Grimes, D.W., D.S. Munk and D.A. Goldhamer, 1990. Drip irrigation emitter depth placement in a slowly permeable soil. In *Proceedings of the 3rd National Irrigation Symposum*, 248–254. St. Joseph, MI: ASAE.

Gushiken, E.C., 1995. Irrigating with reclaimed water through permanent subsurface drip irrigation systems. In *Proceedings off the 5th International Microirrigation Congress*, ed. F.R. Lamm, 269–274. St.Joseph, MI:ASAE.

Hall, B.J., 1985. History of drip/trickle irrigation. In *Proceedings of the 3rd International Drip/Trickle Irrigation Congress*, 1–7. St. Joseph, MI: ASAE.

Howell, T.A., D.A. Bucks and J.L. Chesness, 1980. Advances in trickle irrigation. In *Proceedings of the 2nd National Irrigation Symposium*, 69–94. St. Joseph, MI: ASAE.

Hanson, B.R. and W.E. Bendixen, 1993. Salinity under drip irrigation of row crops. In *Proceedings of the International Exposition and Techni-*

cal Conference, 196–202. Arlington, VA: Irrigation Association.

Hanson, B.R., D.M. May, and W.E. Bendixen. 1997. *Wetting patterns under surface and subsurface drip irrigation*. ASAE Paper No. 97-2178. St. Joseph, MI: ASAE.

Hanson, E.G. and T.C. Patterson, 1974. Vegetable production and water-use efficiency as influenced by drip, sprinkler, subsurface, and furrow irrigation methods. In *Proceedings of the 2nd International Drip Irrigation Congress*, 97–102. Riverside, CA: University of California.

Hanson, E.G., B.C. Williams, D.D. Fangmeier and O.C. Wilke, 1970. Influence of subsurface irrigation on crop yields and water use. In *Proceedings of the National Irrigation Symposium*, D1-D13. St. Joseph, MI: ASAE.

Henggeler, J.C., 1995. A history of drip-irrigated cotton in Texas. In *Proceedings of the 5th International Microirrigation Congress*, ed. F.R. Lamm, 669–674. St. Joseph, MI: ASAE.

Henggeler, J., J. Kinnibrugh, W. Multer, D. Kight and R. Scott, 1996. *Economic impact resulting from the adoption of drip irrigation cotton*. Result Demonstration Report. College Station, TX: Texas Agricultural Extension Service, Texas A&M University.

Hiler, E.A. and T.A. Howell, 1973. Grain sorghum response to trickle and subsurface irrigation. *Transactions of the ASAE*, 16(4): 799–803.

Hills, D.J., F.M. Nawar and P.M. Waller, 1989a. Effects of chemical clogging on drip-tape uniformity. *Transactions of the ASAE*, 32(4): 1202–1206.

Hills, D.J., M.A.M. Tajrishy and Y. Gu., 1989b. Hydraulic considerations for compressed subsurface drip-tape. *Transactions of the ASAE*, 32(4): 1197–1201.

Howell, T.A., D.A. Bucks and J.L. Chesness, 1980. Advances in trickle irrigation. In *Proceedings of the 2nd National Irrigation Symposium*, 69–94. St. Joseph, MI: ASAE.

Howell, T.A., A.D. Schneider and S.R. Evett, 1997. Subsurface and surface microirrigation of corn-Southern High Plains. *Transactions of the ASAE*, 40(3): 635–641.

Huang, T.-J., Y.-H. Chang, P.-C. Yang and Y.-T. Fang, 1982. *Studies on subsurface drip irrigation for coarse-texture sug-*

arcane field in Taiwan. In Report of the Taiwan Sugar Research Institute, 13–32. Taiwan, Taiwan: Taiwan Sugar Reseach Institute.

Hutmacher, R.B., C.J. Phene, K.R. Davis and T.A. Kerby, 1993. Acala and pima cotton responses to subsurface drip irrigation: Water use, plant water relations, and yield. In *Proceedings of the Beltwide Cotton Conference*, 1221–1224. Memphis, TN: National Cotton Council.

Hutmacher, R.B., C.J. Phene, R.M. Mead, D. Clark, P. Shouse, S.S. Vail, R. Swain, M. van Genuchten, T. Donovan and J. Jobes, 1992. Subsurface drip irrigation of alfalfa in the Imperial Valley. In *Proceedings of the 22nd California/Arizona Alfalfa Symposium*, 22: 20–32, University of California and University of Arizona Cooperative Extension, Holtville, CA, December 9–10.

Henggeler, J.C., 1995. A history of drip-irrigated cotton in Texas. In *Proceedings of the 5th International Microirrigation Congress*, ed. F.R. Lamm, 669–674. St. Joseph, MI: ASAE.

Henggeler, J., J. Kinnibrugh, W. Multer, D. Kight and R. Scott, 1996. *Economic impact resulting from the adoption of drip irrigation cotton*. Result Demonstration Report. College Station, TX: Texas Agricultural Extension Service, Texas A&M University.

Hiler, E.A. and T.A. Howell, 1973. Grain sorghum response to trickle and subsurface irrigation. *Transactions of the ASAE*, 16 (4): 799–803.

Hills, D.J., F.M. Nawar and P.M. Waller, 1989a. Effects of chemical clogging on drip-tape uniformity. *Transactions of the ASAE*, 32(4): 1202–1206.

Hills, D.J., M.A.M. Tajrishy and Y. Gu, 1989b. Hydraulic considerations for compressed subsurface drip-tape. *Transactions of the ASAE*, 32(4): 1197–1201.

Howell, T.A., D.A. Bucks and J.L. Chesness, 1980. Advances in trickle irrigation. In *Proceedings of the 2nd National Irrigation Symposium*, 69–94. St. Joseph, MI: ASAE.

Howell, T.A., A.D. Schneider and S.R. Evett, 1997. Subsurface and surface microirrigation of corn-Southern High Plains. *Transactions of the ASAE*, 40(3): 635–641.

Huang, T.-J., Y.-H. Chang, P.-C. Yang and Y.-T. Fang, 1982. *Studies on subsurface drip irrigation for coarse-texture sugarcane field in Taiwan*. In Report of the Taiwan Sugar Research Institute, 13–32. Taiwan, Taiwan: Taiwan Sugar Research Institute.

Hutmacher, R.B., C.J. Phene, K.R. Davis and T.A. Kerby, 1993. Acala and pima cotton responses to subsurface drip irrigation: Water use, plant water relations, and yield. In *Proceedings of the Beltwide Cotton Conference*, 1221–1224. Memphis, TN: National Cotton Council.

Hutmacher, R.B., C.J. Phene, R.M. Mead, D. Clark, P. Shouse, S.S. Vail, R. Swain, M. van Genuchten, T. Donovan and J. Jobes, 1992. Subsurface drip irrigation of alfalfa in the Imperial Valley. *Proceedings of the 22nd California/Arizona Alfalfa Symposium*, 22:20–32, University of California and University of Arizona Cooperative Extension, Holtville, CA, December 9–10.

Hutmacher, R.B., C.J. Phene, K.R. Davis, S.S. Vail, T.A. Kerby, M. Peters, C.A. Hawk, M. Keeley, D. A. Clark, D. Ballard and N. Hudson, 1995. Evapotranspiration, fertility management for subsurface drip acala and pima cotton. In *Proceedings of the 5th International Microirrigation Congress*, ed. F.R. Lamm, 147–154. St. Joseph, MI: ASAE.

Hutmacher, R.B., S.S. Vail, J.G. Muthamia, V. Mwaja and R.C. Liu, 1985. Effect of trickle irrigation frequency and installation depth on tomato growth and water status. In *Proceedings of the 3rd International Drip/Trickle Irrigation Congress*, 2: 798–804. St. Joseph, MI: ASAE.

Isobe, M., 1972. Agronomic research in subsurface and drip irrigation. In *31st Report Hawaii Sugar Technology Annual Conference*, 23–26. Honolulu, Hawaii: Hawaiian Sugar Planters' Association.

Jorgenson, G.S. and K.N. Norum, 1992. Subsurface drip irrigation, B: Theory, practices and application. *Conference proceedings sponsored by California State University-Fresno and USDA ARS-Water Management Research Laboratory*. CATI Publication No. 92-1001, CSUF, Fresno, CA. Pages 212.

Knapp, K.C., 1993. *Economics of irrigation system investment.* In Subsurface Drip Irrigation-Theory, Practices and Application, 127–139. CATI Pub. No. 92 1001. Fresno, CA: California State University.

Kruse, E.G. and I. Israeli, 1987. *Evaluation of a subsurface dripirrigation system.* ASAE Paper No. 87-2034. St. Joseph, MI: ASAE.

Lamm, F.R., G.A. Clark, M. Yitayew, R.A. Schoneman, R.M. Mead and A.D. Schneider, 1995c. Installation issues for SDI systems. In *Proceedings of the International Exposition and Technical Conference*, 29–35. Arlington, VA.: Irrigation Association.

Lamm, F.R. and H.L. Manges. 1991. *Nitrogen fertilizer for drip-irrigated corn in northwest Kansas.* ASAE Paper No. 91-2596. St. Joseph, MI: ASAE.

Lamm, F.R., H.L. Manges, L.R. Stone, A.H. Khan and D.H. Rogers, 1995a. Water requirement of subsurface drip-irrigated corn in northwest Kansas. *Transactions of the ASAE*, 38(2): 441–448.

Lamm, F.R., A.J. Schlegel and G.A. Clark, 1997b. *Nitrogen fertigation for corn using SDI: A BMP.* ASAE Paper No. 97-2174. St. Joseph, MI: ASAE.

Lamm, F.R., W.E. Spurgeon, D.H. Rogers and H.L. Manges, 1995b. Corn production using subsurface drip irrigation. In *Proceedings 5th International Microirrigation Congress*, ed. F.R. Lamm, 388–394. St. Joseph, MI: ASAE.

Lamm, F.R., L.R. Stone, H.L. Manges and D.M. O'Brien, 1997a. Optimum lateral spacing for subsurface drip-irrigated corn. *Transactions of the ASAE*, 40(4): 1021–1027.

Lamm, F.R. and T.P. Trooien, 1998. SDI and the declining Ogallala. In *Proceedings of the 15th Annual Water and the Future of Kansas Conference*, Manhattan, KS, March 3, 1998. pp. 12–15.

Lamm, F.R. and T.P. Trooien, 1999. SDI research in Kansas after ten years. In *Proceedings of the Irrigation Association International Irrigation Show and Conference*, 1–8., Fairfax, VA: Irrigation Association.

Lanting, S., 1975. Subsurface irrigation-Engineering research. In *34th Report Hawaii Sugar Technology Annual Conference*, 57–62.

Honolulu, Hawaii: Hawaiian Sugar Planters' Association.

Lee, C.O., 1920. Irrigation tile. U.S. Patent No. 1,350, 229.

Lindsay, C.A., B.G. Sutton and N. Collis-George, 1989. Irrigation scheduling of subsurface drip-irrigated salad tomatoes. Acta Horticulturae, 247: 229–232.

Manges, H.L., W.E. Spurgeon, Z.M. Huang and D.J. Tomsicek, 1995. Subsurface dripline spacing and plant population for corn production. In *Proceedings of the 5th International Microirrigation Congress*, ed. F.R. Lamm, 382–387. St. Joseph, MI: ASAE.

Martin, E.C., D.C. Slack and E.J. Pegelow, 1996. Crop coefficients for vegetables in Central Arizona. In *Proceedings of the International Conference on Evapotranspiration and Irrigation Scheduling*, eds. C.R. Camp, E.J. Sadler, and R.E. Yoder, 381–386. St. Joseph, MI: ASAE.

McNamara, J.B., 1970. Subirrigation-The basis of tomorrow's agriculture. In *Proceedings of the National Irrigation Symposium*, C1-C13. St. Joseph, MI: ASAE.

Mead, R.M., R.B. Hutmacher and C.J. Phene, 1993. *Subsurface drip irrigation of alfalfa.* In Subsurface Drip Irrigation-Theory, Practices and Application, 145–146. CATI Pub. No. 92 1001. Fresno, CA: California State University.

Mikkelsen, R.L., 1989. Phosphorus fertilization through drip irrigation. *Journal of Production Agriculture*, 2(3): 279–286.

Mitchell, W.H., 1981. Subsurface irrigation and fertilization of field corn. *Agronomy Journal*, 73(6): 913–916.

Mitchell, W.H. and D.L. Sparks, 1982. Influence of subsurface irrigation and organic additions on top and root growth of field corn. *Agronomy Journal*, 74(6): 1084–1088..

Mitchell, W.H. and H.D. Tilmon, 1982. Underground trickle irrigation: The best system for small farms? *Crops Soils*, 34: 9–13.

Mizyed, N. and E.G. Kruse, 1989. Emitter discharge evaluation of subsurface trickle irrigation systems. *Transactions of the ASAE*, 32(4): 1223–1228.

Moore, R.C. and J.C. Fitschen, 1990. The drip irrigation revolution in the Hawaiian sugarcane industry. In *Proceedings of the 3rd National Irrigation Syruposium*, 223–227. St. Joseph, MI: ASAE.

Neibling, H. and R. Brooks, 1995. Potato production using subsurface drip irrigation-Water and nitrogen management. In *Proceedings of the 5th International Microirrigation Congress*, ed. F.R. Lamm, 656–663. St. Joseph, MI: ASAE.

Nightingale, H.I., C.J. Phene and S.H. Patton, 1985. Trickle irrigation effects on soil chemical properties. In *Proceedings of the 3rd International Drip/Trickle Irrigation Congress*, 2: 730–735. St. Joseph, MI: ASAE.

O'Brien, D.M., D.H. Rogers, E.R. Lamm and G.A. Clark, 1997. *Economics of SDI for corn in western Kansas*. ASAE Paper No. 97–2072. St. Joseph, MI: ASAE.

O'Brien, D.M., D.H. Rogers, F.R. Lamm and G.A. Clark, 1998. An economic comparison of subsurface drip and center pivot sprinkler irrigation systems. *Applied Engineering in Agriculture*, 14(4):391–398.

Onken, A.B., C.W. Wendt, O.C. Wilke, R.S. Hargrove, W. Bausch and L. Barnes, 1979. Irrigation system effects on applied fertilizer nitrogen movement in soil. *Soil Science Society of America Journal*, 43(2): 367–372.

Or, D., 1995. Soil water sensor placement and interpretation for drip irrigation management in heterogeneous soil. In *Proceedings of the 5th International Microirrigation Congress*, ed. E.R. Lamm, 214–221. St. Joseph, MI: ASAE.

Oron, G., Y. DeMalach, L. Gillerman and I. David, 1995. Pear response to saline water application under subsurface drip irrigation. In *Proceedings of the 5th Int'l. Microirrigation Congress*, ed. F.R. Lamm, 97–103. St. Joseph, MI: ASAE.

Oron, G., J. DeMalach, Z. Hoffman and R. Cibotaru, 1991. Subsurface microirrigation with effluent. *Journal of Irrigation and Drainage Engineering*, 117(1): 25–36.

Oron, G., Y. DeMalach, Z. Hoffman, Y. Keren, H. Hartman and N. Plazner, 1990. Wastewater disposal by sub-surface trickle irrigation. *Water Science and Technology*, 23:2149–2158.

Oron, G., M. Goemans, Y. Manor and J. Feyan, 1995. Polio virus distribution in the soil-plant system under reuse of secondary wastewater. *Water Research*, 29(4):1069–1078.

Payne, W.A., B. Gerard and M.C. Klaij, 1995. Subsurface drip irrigation to evaluate transpiration ratios of pearl millet. In *Proceedings of the 5th International Microirrigation Congress*, ed. E.R. Lamm, 923–931. St. Joseph, MI: ASAE.

Phene, C.J., 1974. High-frequency porous tube irrigation for water-nitrogen management in humid regions. In *Proceedings of the 2nd International Drip Irrigation Congress*, 166–171. Riverside, CA: University of California.

Phene, C.J., 1995a. The sustainability and potential of subsurface drip irrigation. In *Proceedings of the 5th International Microirrigation Congress*, ed. F.R. Lamm, 359–367. St. Joseph, MI: ASAE.

Phene, C.J., 1995b. Research trends in microirrigation. In *Proceedings 5th International Microirrigation Congress*, ed. F.R. Lamm, 6–24. St. Joseph, MI: ASAE.

Phene, C.J. and O.W. Beale, 1976. High-frequency irrigation for water nutrient management in humid regions. *Soil Science Society of America Journal*, 40(3): 430–436.

Phene, C.J. and O.W. Beale. 1979. Influence of twin-row spacing and nitrogen rates on high-frequency trickle-irrigated sweet corn. *Soil Science Society of America Journal*, 43(6): 1216–1221.

Phene, C.J. and R. Ruskin. 1995. Potential of subsurface drip irrigation for management of nitrate in wastewater. In *Proceedings of the 5th International Microirrigation Congress*, ed. F.R. Lamm, 155–167. St. Joseph, MI: ASAE.

Phene, C.J. and D.C. Sanders, 1976. High-frequency trickle irrigation and row spacing effects on yield and quality of potatoes. *Agronomy Journal*, 68(4): 602–607.

Phene, C.J., B. Bar-Yosef, R.B. Hutmacher, S.H. Patton, K.R. Davis and R.L. McCormick, 1986. Fertilization of high-yielding subsurface trickle irrigated tomatoes. In *Proceedings of the 34th Ann. Calif. Fertilizer Confernce*, 33–43, Fresno, CA. Sponsored by the Soil Improvement Committee of California Fertilizer Association.

Phene, C.J., K.R. Davis, R.B. Hutmacher and R.L. McCormick, 1987. Advantages of subsurface irrigation for processing tomatoes. *Acta Horticulturae*, 200: 101–114.

Phene, C.J., K.R. Davis, R.B. Hutmacher, B. Bar-Yosef, D.W. Meek and J. Misaki, 1991. Effect of high frequency surface and subsurface drip irrigation on root distribution of sweet corn. *Irrgaton Science*, 12(2): 135–140.

Phene, C.J., W.R. DeTar and D.A. Clark, 1992a. Real-time irrigation scheduling of cotton with an automated pan evaporations ystem. *Applied Engineering in Agriculture*, 8(6): 787–793.

Phene, C.J., J.L. Fouss and D.C. Sanders, 1979. Water-nutrient-herbicide management of potatoes with trickle irrigation. *American Potato Journal*, 56: 51–59.

Phene, C.J., R.B. Hutmacher, and K.R. Davis, 1992b. Two hundred tons per hectare of processing tomatoes-Can we reach it? *HortTechnology*, 2(1): 16–22.

Phene, C.J., R.B. Hutmacher, K.R. Davis and R.L. McCormick, 1990. Water- fertilizer management of processing tomatoes. Acta *Horticulturae*, 277: 137–143.

Phene, C.J., R.L. McCormick, K.R. Davis, J.D. Pierro, and D.W. Meek, 1989. A lysimeter feedback irrigation controller system for evapotranspiration measurements and real time irrigation scheduling. *Transactions of the ASAE*, 32(2): 477–484.

Phene, C.J., R. Yue, I.P. Wu, J.E. Ayars, R.A. Schoneman and B. Meso, 1992c. *Distribution uniformity of subsurface drip irrigation systems*. ASAE Paper No. 92-2569. St. Joseph, MI: ASAE.

Phene, R.C., 1996. Real-time irrigation scheduling with automated evaporation pan systems. In *Proceedings of the International Conf on Evapotrans-piration and Irrigation Scheduling*, eds. C.R. Camp, E.J. Sadler, and R.E. Yoder, 1093–1098. St. Joseph, MI: ASAE.

Philip, J.R, 1968. Steady infiltration from buried point sources and spherical cavities. *Water Resources Research*, 4(5): 1039–1047.

Philip, J.R., 1968. Steady infiltration from buried point sources and spherical cavities. *Water Resources Research*, 4(5):1039–1047.

Pier, J.W. and T.A. Doerge, 1995a. Nitrogen and water interactions in trickle-irrigated watermelon. *Soil Science Society of America Journal*, 59(1): 145–150.

Pier, J.W. and T.A. Doerge,1995b. Concurrent evaluation of agronomic, economic, and environmental aspects of trickle-irrigated watermelon production. *Journal of Environmental Quality*, 24(1): 79–86.

Pier, J.W. and T.A. Doerge, 1992. What happens near a quasi-linear point source? Water Resources. Res., 28(1): 47–52.

Pierzgalski, E., 1995. Application of subsurface irrigation on a hop plantation. In *Proceedings of the 5th International Microirrigation Congress*, ed. F.R. Lamm, 729–734. St. Joseph, MI: ASAE.

Plant, Z., A. Carmi and A. Grava, 1996. Cotton root and shoot responses to subsurface drip irrigation and partial wetting of the upper soil profile. *Irrigation Science*, 16(3): 107–113.

Plant, Z., M. Rom and A. Meiri., 1985. Cotton response to subsurface trickle irrigation. In *Proceedings of the 3rd International Drip/Trickle Irrigation Congress*, Volume II: 916–920. St. Joseph, MI: ASAE.

Powell, N.L. and E.S. Wright, 1993. Grain yield of subsurface microirrigated corn as affected by irrigation line spacing. *Agronomy Journal*, 85(6): 1164–1169.

Rolston, D.E., R.S. Rauschkolb, C.J. Phene, R.J. Miller, K. Uriu, R.M. Carlson and D.W. Henderson, 1979. *Applying nutrients and other chemicals to trickle-irrigated crops*. University of California Bulletin No. 1893. Davis, CA: University of California.

Rose, J.L., R.L. Chavez, C.J. Phene and M.M.S. Hile, 1982.Subsurface drip irrigation of processing tomatoes. In *Proceedings of the Specialty Conference on Environmentally Sound Water and Soil Management*, eds. E.G. Kruse, C.R. Burdick, and Y.A. Yousef, 369–376. New York, NY: ASCE.

Rubeiz, I.G., N.F. Oebker and J.L. Stroehlein, 1989. Subsurfacedrip irrigation and urea phosphate fertigation for vegetables oncalcareous soils. *Journal of. Plant Nutrition*, 12(12): 1457–1465.

Rubeiz, I.G., J.L. Stroehlein and N.F. Oebker. 1991. Effect ofirrigation methods on urea phosphate reactions in calcareoussoils. *Communications in Soil Science and Plant Analysis*, 22(5&6): 431–435.

Ruskin, R., 1992. *Reclaimed water and subsurface irrigation*. ASAE Paper No. 92-2578. St. Joseph, MI: ASAE.

Ruskin, R., P. Van Voris and D.A. Cataldo, 1990. Root intrusionprotection of buried drip irrigation devices with slow-releaseherbicides. In *Proceedings of the 3rd National Irrigation Symposium*, 211–216. St.Joseph, MI: ASAE.

Sadler, E.J., C.R. Camp and W.J. Busscher, 1995. Emitter flowrate changes caused by excavating subsurface microirrigation tubing. In *Proceedings of the 5th International Microirrigation Congress*, ed. F.R. Lamm, 763–768. St. Joseph, MI: ASAE.

Sammis, T.W., 1980. Comparison of sprinkler, trickle, subsurface, and furrow irrigation methods for row crops. *Agronomy Journal*, 72(5): 701–704.

Scherm, H. and A.H.C. van Bruggen, 1995. Comparative study of microclimate and downy mildew development in subsurface drip- and furrow-irrigated lettuce fields in California. *Plant Disease*, 79(6): 620–625.

Schwankel, L.J., S.R. Grattan and E.M. Miyao, 1990. Drip irrigation burial depth and seed planting depth effects on tomato germination. In *Proceedings of the 3rd Nat. Irrigation Symposium*, 682–687. St. Joseph, MI: ASAE.

Schwankel, L.J. and T.L. Prichard, 1990. Clogging of buried drip irrigation systems. *California Agriculture*, 44(1): 16–17.

Senock, R.S., J.M. Ham, T.M. Loughin, B.A. Kimball, D.J. Hunsaker, P.J. Pinter, G.W. Wall, R.L. Garcia, and R.L. LaMorte, 1996. Sap flow in wheat under free-air $CO_2$ enrichment. *Plant, Cell & Environment*, 19(2): 147–158.

Shani, U., S. Xue, R. Gordin-Katz and A.W. Warrick, 1996. Soil-limiting flow from subsurface emitters. I. Pressure measurements. *Journal of Irrigation and Drainage Engineering*, 122(5): 291–295.

Shrive, S.C., R.A. McBride and A.M. Gordon, 1994. Photosynthetic and growth responses of two broad-leaf tree species to irrigation with municipal landfill leachate. *Journal of Environmental Quality*, 23(3): 534–542.

Solomon, K.H. and G. Jorgensen, 1992. Subsurface drip irrigation. *Grounds Maintenance*, 27(10): 24, 26.

Sterrett, S.B., B.B. Ross and C.P. Savage Jr., 1990. Establishment and yield of asparagus as influenced by planting and irrigation method. *Journal of the American Society for Horticultural Science*, 115(1): 29–33.

Sutton, B.G., R.J. Stirzaker, C.J. Doney and S.D. English, 1985. Solar powered drip irrigation for vegetables. In *Proceedings of the 3rd International Drip/Trickle Irrigation Congress*, 2: 589–593. St. Joseph, MI: ASAE.

Thomas, A.W., H.R. Duke and E.G. Kruse, 1977. Capillary potential distributions in root zones using subsurface irrigation. *Transactions of the ASAE*, 20(1): 62–67, 75.

Thomas, A.W., E.G. Kruse and H.R. Duke, 1974. Steady infiltration from line sources buried in soil. *Transactions of the ASAE*, 17(1): 125–133.

Thompson, T.L. and T.A. Doerge, 1996a. Nitrogen and water interactions in subsurface trickle-irrigated leaf lettuce. I. Plant response. *Soil Science Society of America Journal*, 60(1): 163–168.

Thompson, T.L. and T.A. Doerge, 1996b. Nitrogen and water interactions in subsurface trickle-irrigated leaf lettuce. H. Agronomic, economic, and environmental outcomes. *Soil Science Society of America Journal*, 60(1): 168–173.

Van Bavel, C.H. M., J. Ahmed, S.I. Bhuiyan, E.A. Hiler and A.G. Smajstrla, 1973. Dynamic simulation of automated subsurface irrigation systems. *Transactions of the ASAE*, 16(6): 1095–1099.

Tollefson, S., 1985a. The Arizona system: Drip irrigation design for cotton. In *Proceedings of the 3rd International Drip/Trickle Irrigation Congress*, 1: 401–405. St. Joseph, MI: ASAE.

Tollefson, S., 1985b. Subsurface drip irrigation of cotton and small grains. In *Proceedings of the 3rd International Drip/Trickle Irrigation Congress*, 2: 887–895. St. Joseph, MI: ASAE.

Vaziri, C.M. and W. Gibson, 1972. Subsurface and drip irrigation for Hawaiian sugarcane. In *31st Report Hawaii Sugar Technology Annual*

*Conference*, 18–22. Honolulu, Hawaii: Hawaiian Sugar Planters' Association.

Warrick, A.W., D.O. Lomen and A. Amoozegar-Fard. 1980. Linearized moisture flow with root extraction for three dimensional, steady conditions. *Soil Science Society of America Journal* 44(5): 911–914.

Warrick, A.W. and U. Shari, 1996. Soil-limiting flow from subsurface emitters. II. Effect on uniformity. *Journal of Irrigation and Drainage Engineering*, 122(5): 296–300.

Welsh, D.F., U.P. Kreuter and J.D. Byles, 1995. Enhancing subsurface drip irrigation through vector flow. In *Proceedings of the 5th, International Microirrigation Congress*, ed. F.R. Lamm, 688–693. St. Joseph, MI: ASAE.

Wendt, C.W., A.B. Onken, O.C. Wilke, R. Hargrove, W. Bausch and L. Barnes, 1977. Effect of irrigation systems on the water requirement of sweet corn. *Soil Science Society of America Journal*, 41(4): 785–788.

Whitney, L.E, 1970. Review of subsurface irrigation in the Northeast. In *Proceedings of the National Irrigation Symposium*, Fl-F8. St. Joseph, MI: ASAE.

Whitney, L.F. and K.M. Lo, 1969. Plastic orifice inserts for subsurface irrigation. *Transactions of the ASAE*, 12(5): 602–607.

Yue, R., C.J. Phene, F. Dale, J.E. Ayars, R.A. Schoneman, I.-P. Wu and L. Kong, 1993. *Field uniformity of subsurface drip irrigation*. In Subsurface Drip Irrigation-Theory, Practices and Application, 181–183. CATI Pub. No. 92 1001. Fresno, CA: California State University.

Zachmann, D.W. and A.W. Thomas, 1973. A mathematical investigation of steady infiltration from line sources. *Soil Science Society of America Journal*, 37(4): 495–500.

Zetzsche, J.B. and J.S. Newman, 1966. Subirrigation with plastic pipe. *Agricultural Engineering*, 47(1): 74–75.

Zimmer, A.L., M.J. McFarland and J. Moore, 1988. *Upward free water movement from buried trickle emitters*. ASAE Paper No. 88-2063. St. Joseph, MI: ASAE.

Zoldoske, D.F. 1993. The future of irrigation is buried. In *Proceedings of the International Irrigation Expo and Technical Conference*, 86–88. Arlington, VA.: Irrigation Association.

Zoldoske, D.F., S. Genito and G.S. Jorgensen, 1995. Subsurface drip irrigation (SDI) on turfgrass: A university experience. In *Proceedings of the 5th International Microirrigation Congress*, ed. F.R. Lamm, 300–302. St. Joseph, MI: ASAE.

Zoldoske, D.F. and E.M. Norum, 1997. *Final Report. Progress report on the Lehman Farms Project: A case study in the conversion of an old vineyard from flood to surface drip (SDI) and subsurface drip (SSDI) irrigation*. Center for Irrig. Tech., Pub. No. 970702. Fresno, CA: California State University.

# 6

**Books/bulletins/journals and proceedings/reports**

Anonymous, 2000. Installation of Water Systems. In *Water reuse for irrigation*. CRC Press. Pages 87–97.

Avars, J., D. Bucks, F. Lamn and F. Nakayama, 2002. Microirrigation installation. In *Microirrigation for crop production*. Pages 22–43.

Banta, E.R. 2000. MODFLOW [Electronic Resource]: Modular Ground Water Model – Documentation of Packages for Stimulating Evotranspiration with a Segment Function (ETS 1) and Drains with Return Flow (DRT1). The U.S. Geological Survey.

Cason, K.J., 2002. Elevated drip irrigation systems works. *Countryside & Small Stock Journal*, 87(3): 1–20.

Dasberg, S., 2000. *Drip irrigation*. New York: Springer-Verlag Telos. Pages 67–77.

Experimental analysis of local pressure losses for microirrigation laterals. *Journal of Irrigation and Drainage Engineering*, 130(4): 318–324. July 2004.

New strategy for optimizing water application under trickle irrigation. *Journal of Irrigation and Drainage Engineering*, 128(5): 227–297. September/ October 2002.

Scherer, T.F., 2002. System installation: Design and control of sprinkler systems for crop disease research, *American Society of Agricultural and Biological Engineers*, St. Joseph, MI, 100(4): 45–56.

Shrock, C.C., 2001. Drip Irrigation: An Introduction. Oregon State University, *Extension Service*, Pages 67–89.

Water management of irrigated-drained fields in the Jordan Valley South of Lake Kinneret. *Journal of Irrigation and Drainage Engineering*, 131(4): 364–374. July/August 2005.

**Web page links**

Bertauski, Tony, 2001. How to install drip irrigation in new plantings: Maintenance and trouble shooting tutorials-The Drip Store. Pages 4–7. Trident Technical College. http://www.dripirrigation.com/drip_tutorial.php?page_view=head

Eartheasy Corporation, 2006. Simple methods of drip irrigation systems. Pages 5–8.

http://eartheasy.com/grow_drip-irrigation.htm

Institute of Food and Agricultural Sciences. 2000. Treating irrigation systems. Pages 2–4. http://www.eidis    .ifas.ufl.edu/scrips/htmlgen.exe?DOCUMENT_AE080

Luchsinger, Peter, 2006. Drip irrigation basic maintenance. Pages 3–7. http://www.dripirrigation.com/drip_tutorial.php

Northern Garden Supply, 2004. Planning your drip irrigation. Pages 4–9. http://www.dripirrigation.ca/HowTo_Plan.asp

Plaster irrigation and infrastructure systems, 2005. Irrigation systems maintenance and installation. Pages 8–10. http://topirrigation.co.uk/copyright.html

Shock, Clinton C., 2006. Drip irrigation an introduction. Pages 11–12. http://www.corpinfo.net/drip.html

Stryker, Jess, 2000. Landscape sprinkler and drip irrigation. Pages 1–4. http://www.irrigationtutorials.com/install.html

Sydney Water Corporation, 2006. Water system installation. Pages 5–7. http://www.sydneywater.com.au/SavingWater/InYourGarden/WateringSystems/

Wilson, C., 2006. Drip irrigation installation benefits. Pages 12–14. http://www.ext.colostate.edu/Pubs/garden/04702.html

# 7

**Literature cited**

1. Anónimo. 1983. *Insumos israelíes de riego.* Agricultura de las Américas. Febrero, paginas 6–17.

2. Anónimo. 1982. *La Automatización del riego.* Irrinews. ISSN 304-3606. Bet Degam-Israel. 25(1982): 12–13

3. Anonymous. *Computerized irrigation control system handout.* Motorola Inc. Agro-Control Department, Fresn,o CA, 93710

4. Anonymous. *Instruction manual for installation and operation of "Free Flow" media filters.* Water Management Products Division. P.O. Box 352 Corona, CA, 91720.

5. Anonymous. *Instruction manual – Irrigation Controler Model AG-7*, Rain Bird, CA.

6. Anonymous. *Irrigation water management using the neutron probe.* Buchanan Circle Corp., Pacheco, CA, 94553.

7. Bauder, J.W., L.D. King and G.L. Westesen, 1982. *Scheduling irrigation with evaporation pans.* Coop. Ext. Ser., Montana State Univ., Bozeman, Bulletin. 1262, Form Circular. 1211, Pages 1–23.

8. Goldberg, D., B. Gornat and D. Rimon, 1976. *Drip irrigation: principles, designing and agriculture practices.* Israel Drip Irrigation Scientific Publications, Pages 236–250.

9. Howell, T.A., J.L. Hatfield, H. Yamada and K.R. Davis, 1984. *Evaluation of cotton canopy temperature to detect crop water stress. Transactions of ASAE*, 27 (1): 84–88.

10. Jackson, R.D., 1982. Canopy temperature and crop water stress. In *Advances in irrigation*, ed. D.I. Hillel. I: 43–85. New York: Academic Press.

11. Nakayama, F.S. and D.A. Bucks, 1986. *Trickle irrigation for crop production designing, operation and management.* Amsterdam,The Netherlands: Elsevier Science Publishers B.V. Pages 188–210, 308 and 311.

12. Phene, C.J. and T.A. Howell, 1984. Soil sensor control of high frequency irrigation. *Transactions of ASAE*, 27(2): 386–391, 396.

**Books/bulletins/journals and proceedings/reports**

Abraham, N., Hema and Saritha Subramannian, 2000. Irrigation automation based on soil electrical conductivity and leaf temperature. *Agricultural Water Management*, 45(2, July): 145–157.

Ben-Gal, A., N. Lazorovitch and U. Shani. 2004, November. Subsurface drip irrigation in gravel-filled cavities. *Vadose Zone Journal*, 3: 1407–1413.

Anonymous, 2003. *Paddy and Water Environment*. 1(4): 157–214.

Assouline S., S. Cohen, D. Meerbach, T. Harodi and M. Rosner. 2002. Microdrip irrigation of field crops: Effect on yield, water uptake and drainage in sweet corn soil. *Soil Science Society of America Journal*, 66: 228–235

Broner, I., 2002. *Micro-irrigation for orchard and row crops*. Colorado State. University, Cooperative Extension. Colorado, USA.

Burt, C.M., 2000. *Selection of irrigation methods for agriculture: Drip micro irrigation*. Irrigation Training and Research Center (ITRC), CA.

Casey Francis, X.M. and E. Derby Nathan, 2002. Improved design for an automated tension infiltrometer. *Soil Science Society of America Journal*, 66: 64–67.

Clemmens, A.J., 2006. Canal automation. *Resource Magazine*. Pages 7–8. ASABE.

De Wit, C.T., 1999. The efficient use of labor, land and energy in agriculture. *Agricultural Systems*, 4(4, October): 279–228.

Dukes, M.D. and J.M. Scholberg, 2004. *SDI, Subsurface drip, automatic irrigation, arachis hypogaea, zea mays, irrigation water use efficiency, high frequency irrigation, sweet corn, peanut, TDR, (time domain reflectometry)*. Paper number 042188, ASAE Annual Meeting.

Giuseppina, Crescimanno and Paolo Garofalo, 2006, August. Management of irrigation with saline water in cracking clay soils. *Soil Science Society of America Journal*, 70: 1774–1787.

Hansonn, F. and Martin, 2002. Drip irrigation of row crops. What is the state of the art? *Proceedings of the 4th Decennial Irrigation Symposium*, ASAE.

Johnson D.O., F.J. Arriaga and B. Lowery. 2005. Automation of a falling head permeameter for rapid determination of hydraulic conductivity of multiple samples. *Soil Science Society of America Journal*, 69: 828–833.

K.F. Bronson, J.D. Booker, J.P. Bordovsky, J.W. Keeling, T.A. Wheeler, R.K. Boman, M.N. Parajulee, E. Segarra and R.L. Nichols, 2004. *Site-specific irrigation and nitrogen management for cotton production in the southern high plains*. *Agronomy Journal*, 98: 212–219.

Kang Yaohu, Hai-Jun Liu and Shi-Ping Liu, 2002. *Irrigation system in agriculture*. Paper number 022285, ASAE Annual Meeting.

Keller, J. and R.D. Bliesner, 2004. *Sprinkler and trickle irrigation*. New York (USA): Van Nostrand Rienhold.

Keller, J. and R.D. Bliesner, 2000. *Sprinkler and trickle irrigation*. Caldwell, NJ: Blackburn Press

Knox, J.W. and E.K. Weatherhead, 2003. *Trikcle irrigation in england and wales*. R&D Technical report W6-707/TR (also R&D Technical Summary W6_070/TS). Enviroment Agency, Bristol.

Luthra, S.K., M.J. Kaledhonkar, O.P. Singh and N.K. Tyagi, 2000. Design and development of an auto irrigation system. *Agricultural Water Management*, 53(2, June): 169–181.

Mehmet, Ozay and Hasan Ali Bicak, 2002. Modern and traditional irrigation technologies in the eastern Mediterranean.

Melby, P., 2003. *Simplified irrigation design*. New York (USA): Van Nostrand Reinhold.

Modern methods of irrigation. Earth and Environmental Science, *Geo Journal*, 35 (1): 59–63. 2004.

Morari, F. and L. Giardini, 2002. Irrigation automation with heterogeneous vegetation: the case of the Padova botanical garden. *Agricultural Water Management*, 55(3, June): 183–201.

Nadler, Arie, Eran Raveh, Uri Yermiyahu and S.R. Green, 2003. Evaluation of TDR use to monitor water content in stem of lemon trees and soil and their response to water stress. *Soil Science Society of America Journal*, 67: 437–448.

Natural Resource Conservation Service. National Engineering Handbook: Section 15. Chapter 1, Soil-Plant-Water Relationships; Chapter 2, Irrigation Water Requirements; Chapter 4, Border Irrigation; Chapter 5, Furrow Irrigation; Chapter 7, Trickle Irrigation; Chapter 11, Sprinkler Irrigation. United States Department of Agriculture, U.S.A.

Oztekin T., L.C. Brown, P.M. Holdsworth and A. Kurunc, 1999. DRAINMOD: Modeling, runoff, subsurface drainage, wastewater irrigation. *Applied Engineering in Agriculture*, 15(5): 449–455.

Pereira, L.S., T. Oweis and A. Zairi, 2002. Irrigation management under water scarcity. *Agricultural Water Management*, 57(3, December): 175–206.

Playan, E. and L. Mateos, 2006. Modernization and optimization of irrigation systems to increase water productivity. *Agricultural Water Management*, 80(1–3 February): 100–116.

R., Aragüés, E. Playán, R. Ortiz and A. Royo, 1999, September. A new drip-injection irrigation system for crop salt tolerance evaluation. *Soil Science Society of America Journal*, 63: 1397–1403.

Rodríguez de Miranda, Fabio, 2003. Automated irrigation control, micro irrigation, spatially-variable irrigation, distributed control. Paper number 031129, ASABE Annual Meeting.

Rodríguez, L. and M. Leal, 1997. Automation of the localized irrigation. Short communication. *Agrotecnia de Cuba*, 27(1).

Royo, R. Aragüés, E. Playán, and R. Ortiz, 2000. Salinity–grain yield response functions of barley cultivars assessed with a drip-injection irrigation system. *Soil Science Society of America Journal* 64: 359–365.

Sakai, K. 2001. *Nonlinear Dynamics and Chaos in Agricultural Systems.*

Schwankl, L. and T. Prichard, 2001. *Chemigation in tree and vine micro irrigation systems*. Agriculture and Natural Resources Publication 21599. University of California, Davism CA.

Shock, C.C., R.J. Flock, E.B.G. Feibert, C.A. Shock, A.B. Pereira and L.B. Jensen, 2005. *Irrigation monitoring using soil water tension*. Oregon State University Extension Service. EM 8900. Pages 1–6

Shock, Clinton C., Erik B.G. Feibert, Lamont D. Saunders and Eric P. Eldredge, 2002. Water Management: Allium Cepa, (onion), Solanum Tuberosum, (potato), watermark, granular matrix sensor, soil water potential, and SDI. *Proceedings of the World Congress of Computers in Agriculture and Natural Resources*, (13–15 March, at Iguacu Falls-Brazil) 701P0301, Pages 809–816.

Souza, C.F., D. Or and E.E. Matsura, 2004. A variable-volume TDR probe for measuring water content in large soil volumes. *Soil Science Society of America Journal*, 68: 25–31.

Wang D., M.C. Shannon, C.M. Grieve, P.J. Shouse and D.L. Suarez, 2002, September. Ion partitioning among soil and plant components under drip, furrow, and sprinkler irrigation regimes: field and modeling assessments. *Journal of Environmental Quality*, 31: 1684–1693.

Wesseling, J. 1998. Automated farm surface irrigation systems worldwide. International Commission on Irrigation. *Agricultural Water Management*, 51(1, April): 101.

Yamamoto Y., I. Ikai, M. Kume, Y. Sakai, A. Yamauchi, H. Shinohara, T. Morimoto, Y. Shimahara, M. Yamamoto and Y. Yamaoka. 2000. New simple technique for hepatic parenchymal resection using a cavitron ultrasonic surgical Aspirator® and Bipolar Cautery equipped with a channel for water dripping. *World Journal of Surgery*, 23(10): 1032–1037.

Yang, S., Pei Lu, L. Okushima and S. Sase, 2004. *Precision irrigation system based on detection of crop water stress with acoustic emission technique.* Pages: 444–447. School of Electron. Inf. & Autom., Tianjin University of Science & Technology, China.

Yunseop, K., R.G. Evans, W.M. Iversen, F.J. Pierce and J.L. Chavez, 2006, Wireless sensor network, site-specific, control, water management, *GPS. ASAE Annual Meeting.*

**Web page links**

Abdullah Ali, Mohammed, 2000. Modern irrigation systems. http://www.uae.gov.ae/uaeagricent/wateranddam/moderirrigation_e.stm

Ansari Ahsno, 2006, January. Irrigation Automation. http://www.control.com/1026151936/index

Basics of vegetable crop irrigation. Alabama Cooperative Extension System http://www.aces.edu/department/extcomm/publications/anr/anr-1169/anr-1169.html

Bob How, 2003, July. Irrigation Automation. http://www.almegcontrols.com/continuous.htm

Brouwer C. Irrigation water management. Chapter 6. http://www.fao.org/docrep/s8684e/s8684e07.htm

Christensen, S.A. Irrigation and its automation in glasshouses. Acta Hort. (ISHS) 2:76–79. http://www.actahort.org/books/2/2_12.htm

Hinduweb search. http://hinduwebsite.com/utilities/webDir/webDir.asp?/Science/Agriculture/Practices_and_Systems/Irrigation

Humphreys A.S. and T.J. Trout, 2003. Automation of Border and Basin Irrigation. Pages 293–300. www.pubs.asce.org/

Irrigation journals http://www.greenmediaonline.com/ij/2001/0110/

Lee, Eric, 2000, May. Irrigation Automation Forum. http://www.homeautomationforum.com/ubb/Forum1/HTML/000067.html

Leon, New and Roland E. Roberts. Drip irrigation for greenhouse vegetable production. http://aggie-horticulture.tamu.edu/greenhouse/hydroponics/drip.html

Leskovar, D. Integration of fertigation and soil moisture control for automation of subsurface drip irrigation system. http://twri.tamu.edu/water_resources_research/2004/leskovar_proposal.pdf#search='Automation%20in%20Drip%20Irrigation

Mandavia, Anil B. Drip Irrigation Management Information Systems. Sardar Saroar Narmada Nigam Limited, Gandhinagar, Gujarat-India. www.fao.org

Penn, Boyd W., 2000, May. Irrigation, Automation Forum. http://www.homeautomationforum.com/ubb/Forum1/HTML/000067.html

Reid, J.1997. The Evolution of automation in precision agriculture. http://www.frc.ri.cmu.edu/~mcm/seminar.html

Sanders, Douglas C. Drip or trickle irrigation systems: An operations and troubleshooting checklist. North Carolina State University. http://www.ces.ncsu.edu/depts/hort/hil/hil-33-b.html

Shock, Clinton C., B.G. Erik, Feibert, and Lamont D. Saunders. Automation of subsurface drip irrigation for onion production. http://www.cropinfo.net/AnnualReports/1996/ondrip96.htm

Springer, Leon. Drip irrigation. http://www.dripworksusa.com/info/DripIrrigation.html

Subsurface Drip Irrigation; Demonstration and Research Project; The University of Arizona, Department of Soil, Water and Environmental Science http://cals.arizona.edu/crops/irrigation/azdrip/azdripindex.html

Sydney water system http://www.sydneywater.com.au/SavingWater/InYourGarden/WateringSystems/

Tillet, Nick, 2005. Advance automation technologies for sustainable agricultural production. http://orgprints.org/6770/

Using soil moisture sensors for making irrigation management decisions in Virginia. Virginia Cooperative Extension, Virginia Tech. http://www.ext.vt.edu/pubs/rowcrop/442-024/442-024.html

William, J. Drip irrigation for vegetable production. http://agalternatives.aers.psu.edu

Yang, Shifeng, L. Pei, L.Okushima, S. Sase and G. Qiu. 2003. System based on estimation of water stress in tomato automatic and precise irrigation. St. Joseph, MI: American Society of Agricultural and Biological Engineers, www.asabe.org

# 8

**Literature cited**

1. *Aplique los plaguicidas correctamente: Guía para usuarios comerciales de plaguicidas,* 1967. Versión al español. EPA (Agencia de Protección Ambiental Federal)

2. BAR-RAM Irrigation Systems: Instructions for the operation and maintenance. Pages 3–5.

3. Goyal, M.R, J. Román, F. Gallardo Covas, L.E. Rivera and R. Otero Dávila, 1983. *Chemigation via trickle systems in vegetables and fruit orchards.* Paper presented at 1983 meeting of the American Society of Agricultural Engineers. Montana State University, Montana.

4. Harrison, D.S., 1974. *Injection of liquid fertilizer materials into irrigation systems.* Gainesville, FL: University of Florida, Pages 3–6, 10–11.

5. *Proceedings of the National Symposium on Chemigation.* Tifton-CA: Rural Development Center of University of Georgia, August 1982.

6. Rolston, D.E., R.S Rauschkolb, C.J. Phene, R.J. Miller, K. Uriu, R.M. Carlson and D.W. Henderson, 1979. Applying nutrients and other chemicals to trickle irrigated crops. University of California. Pages 3–12.

7. Smajstrla, A.G., D.S. Harrison, J.C. Good and W.J. Becker, 1982. *Chemigation Safety*. Agricultural Engineering Fact Sheet, Florida Cooperative Extension Service, Institute of Food and Agricultural Sciences, University of Florida.

**Books/bulletins/journals and proceedings/reports**

Adamsen, F.J., D.J. Hunsaker and H. Perea, 2005. Border strip fertigation: Effect of injection strategies on the distribution of bromide. *Transactions of the ASAE*, 48(2): 529–540.

Anbumozhi, V., K. Matsumoto and E. Yamaji, 2001. Sustaining agriculture through modernization of irrigation tanks: An Opportunity and a Challenge for Tamilnadu, India. *Agricultural Engineering International: the CIGR Journal of Scientific Research and Development*, III(10): 2–12.

Asadi, M.E., 2002. *Impacts of fertigation via sprinkler irrigation of nitrate leaching and corn yield in an acid sulfate soil in Thailand*. Doctoral Dissertation, Dissertation No: WM-00-02. Asian Institute of Technology, Bangkok, Thailand.

Blanco, F.F., M.V. Folegatti and E. Casarini, 2003. *Effects of nitrogen and potassium on dry matter production of tomato irrigated with saline water*. Paper number 032235, ASAE Annual Meeting

Cook, D.W.M., P. G. Long and S. Ganesh, 1999. The combined effect of delayed application of yeast biocontrol agents and fruit curing for the inhibition of the post harvest pathogen Botrytis cinerea in Kiwifruit. *Postharvest Biology and Technology*, 16:233–243.

Cullum, R.F., M.T. Moore and C.M. Cooper, 2006. *Assimilation of agrichemicals and sediments in runoff within drainage ditches and constructed wetlands*. Paper number 062004, ASAE Annual Meeting

De Neve, S. and G. Hoffman, 2000. Influence of soil compaction on C and N mineralization from soil organic matter and crop residues. *Biology and Fertility of Soils*, 30:544–549

Droby, S., 2004. *Non-chemical treatments in post harvest*. Lecture notes. International Research and Development Course on Post harvest Biology and Technology, The Volcanic Center, Israel.

Dukes, M.D., J.M. Scholberg and T.A. Hanselman, 2005. *Quantification of nitrogen movement under drip irrigated vegetable production*. Paper number 052240, ASAE Annual Meeting.

Farahani, H.J., D.L. Shaner, G.W. Buchleiter and G.A. Bartlett, 2006. Evaluation of a low volume agro-chemical application system for center pivot irrigation. Applied Engineering in Agriculture, 22(4): 517–528.

Irtwange, S. V., 2006. Application of biological control agents in pre- and post harvest operations. Agricultural Engineering International: the CIGR Ejournal. Invited Overview No. 3, VIII: 1–10.

King, B. A., R. W. Wall, D. C. Kincaid and D. T. Westermann, 2005. Field testing of a variable rate sprinkler and control system for site-specific water and nutrient application. Applied Engineering in Agriculture, 21(5): 847–853.

Laegreid, M., O. C. Bockman and O. Kaarstad, 1999. Agriculture fertilizers and environment, CABI Publ., Wallingford and Norsk Hydro ASA, Oslo. Pages: 5–12.

Lamm, F. R., A. J. Schlegel and G. A. Clark, 2004. Development of a best management practice for nitrogen fertigation of corn using SDI. Applied Engineering in Agriculture, 20(2): 211–220.

Leib, B. G., A. R. Jarrett, M. D. Orzolek and R. O. Mumma, 2000. Drip chemigation of Imidacloprid under plastic mulch increased yield and decreased leaching caused by rainfall. Transactions of the ASAE, 43(3): 615–622.

Leib, B.G., J.D. Jabro, and A.R. Jarrett, 2003. *Pesticide movement under drip chemigation: Model simulations and field measurements*. Paper number 032022, ASAE Annual Meeting.

Li, J., J. Zhang and M. Rao, 2005. Modeling of water flows and nitrate transport under surface drip fertigation. *Transactions of the ASAE*, 48(2): 627–637.

Marouelli, W.A. and P.M. Waller, 1999. Oil drop generator for foliar chemigation: Theory and laboratory evaluation. *Transactions of the ASAE*, 42(5): 1289–1302.

Montero, J., J.M. Tarjuelo and J.F.Ortega, 2000. Heterogeneity analysis of the irrigation in fields with medium size sprinklers. *Agricultural Engineering International: CIGR Journal of Scientific Research and Development*. Volume II.

Oztekin, S. and Y. Soysal, 2000. Comparison of adsorption and desorption isoteric heats for some grains. *Agricultural Engineering International: CIGR Journal of Scientific Research and Development*, 2(4): 2–16.

Pathak, B.K., F. Kazama and T. Iida, 2004. Monitoring of nitrogen leaching from a tropical paddy field in Thailand. *Agricultural Engineering International: CIGR Journal of Scientific Research and Development*, 6(4): 2–9.

Pimentel, D., M. Pimentel and M. Karpenstein-Machan, 1999. Energy use in agriculture: An Overview. *Agricultural Engineering International: CIGR Journal of Scientific Research and Development*. Volume I.

Pochi, D. and D. Vannucci, 2002. Prediction of pesticide distribution on the ground based on sprayer boom movements. *Agricultural Engineering International: CIGR Journal of Scientific Research and Development*, IV(19): 12–22.

Ray, C., 2001. Managing nitrate problems for domestic wells in irrigated alluvial, aquifers. *Journal of Irrigation and Drainage Engineering*, 127(1): 49–53.

Schwankl, L. and T. Prichard, 2001. *Chemigation in tree and vine micro irrigation systems.* Agriculture and Natural Resources Publication 21599. University of California, Davis, CA.

Sinai, G., A. Haramati and V. Gorbonos, 2005. Attenuation with soil depth of dynamic fluctuations in the ratio of chemicals to irrigation water in chemigation. *Applied Engineering in Agriculture*, 21(3): 357–370.

Stevenson, B.A., T. McLendon and E.F. Redente, 2000. Effects of soil fumigation and seeding regimes on secondary succession in a semiarid shrub land. *Arid Soil Reseach and Rehabilitation*, 14: 87–99.

Sumner, H.R., C.C. Dowler and P.M. Garvey, 2000. Application of agrochemicals by chemigation, pivot attached sprayer systems, and conventional sprayers. *Applied Engineering in Agriculture*, 16(2): 103–107.

Vaishnava, V.G., D.K. Shelke and P.R. Bharambe, 2002. *Drip irrigation and fertigation for sugarcane in deep black soils.* Paper number 022203, ASAE Annual Meeting.

Wilson, W.S., A.S. Ball and R.H. Hinton, 1999. *Managing risks of nitrates to humans and the environment.* Cambridge: The Royal Society of Chemistry.

Yuan, Z., C.Y. Choi, P.M. Waller and P. Colaizzi, 2000. Effects of liquid temperature and viscosity on venturi injectors. *Transactions of the ASAE*, 43(6): 1441–1447.

**Web page links**

Northwest Coalition for Alternatives to Pesticides, 1999. Does government registration mean pesticides are safe? *Journal of Pesticide Reform.* www.pesticide.org/BasicRegistration.pdf

Vineyard Water System, 2000. Chlorine injection guidelines. http://www.sonorapacific.com/files/Chlorine_Injection.pdf

Gelski, J., 2003. Avoid filter frustration. http://www.growermagazine.com/home/02-03filters.html

Shock, C.C., R.J. Flock, E.B.G. Feibert, A.B. Pereira and M. O'Neill, 2005. *Drip irrigation guide for growers of hybrid Poplar.* Oregon State University Extension Service. EM 8902, Pages 1–8. http://extension.oregonstate.edu/catalog/pdf/em/em8902.pdf

Sanders, C., 2001. *Drip or trickle irrigation systems: An operations and trouble shooting checklist.* College of Agriculture and Life Sciences North Carolina State University. http://www.ces.ncsu.edu/depts/hort/hil/hil-33-b.html

Anonymous, 1999. Annual Irrigation Survey. *Irrigation Journal.* http://www.greenindustry.com/ij/1999/0299/299

USDA Agricultural Research Service, 2001. *Pesticide properties database.* http://wizard.arsusda.gov/acsl/acslhome.html

USDA, 2005. *Pesticide Data Program.* Washington DC: US Department of Agriculture. http://www.ams.usda.gov/science/pdp

McCauley, L.A., 2006. Studying health outcomes in farm worker populations exposed to pesticides. *Environmental Health Perspec-*

*tives*, 114. http://www.ehponline.org/members/2006/8526/8526.html

USGS, 2005. *Irrigation methods*. http://ga.water.usgs.gov/edu/irmethods.html

# 9

**Literature cited**

1. Abbott, J.S., 1985. Emitter clogging: causes and prevention. *ICID Bulletin*, 34(2):17.

2. Boswell, M.J., 1985. *Micro-irrigation design manual.* In *Hardie Irrigation*, Chapter 3: 1–15. California, CA.

3. *Chlorine Kit.* Taylor Chemicals Inc., 7300 York Road, Baltimore-MD- 21204.

4. Ford, H.W., 1980. *The use of chlorine in low volume systems where bacterial slimes are a problem.* Lake Alfred AREC Research Report CS75-5, HWF100, Florida, FL.

5. Ford, H.W., 1980. *Water quality test for low volume irrigation.* Lake Alfred AREC-Florida Research Report CS79-6 HWF100, Florida, FL.

6. Ford, H.W., 1980. *Using a DPD test kit for measuring free available chlorine.* Lake Alfred AREC Research Florida Report CS79-1, Florida, FL.

7. Ford, H.W., 1980. *Estimating chlorine requirements.* Lake Alfred AREC Research Florida Report CS80-1, Florida, FL.

8. Ford, H.W., 1979. *A key for determining the use of sodium hyper chlorite to inhibit iron and slime clogging of low pressure irrigation systems in Florida.* Lake Alfred AREC Research Report CS79-3, Florida, FL.

9. Goyal, M.R. y L.E. Rivera, 1984. Riego por Goteo: Quimigación. Servicio de Extensión Agrícola, Recinto Universitario de Mayagüez, Río Piedras. IA61 Serie 4, Páginas 1–15.

10. Goyal, M.R. y L.E. Rivera, 1985. Riego por Goteo: Servicio y Mantenimiento. Servicio de Extensión Agrícola, Recinto Universitario de Mayagüez, Río Piedras. IA64 Serie 5, Páginas 1–18.

11. Nakayama, F.S. and D.A. Bucks, 1986. *Trickle irrigation for crop production.* Elsevier Science Publisher B.V., The Netherlands. Pages 142–157 and 179–183.

12. Rivera, L.E. y M.R. Goyal, 1986. Riego por Goteo: Sistemas de Filtración. Recinto Universitario de Mayagüez, Rio Piedras. IA66 Serie 7, Páginas 1–42.

**Books/bulletins/journals and proceedings/reports**

*American Conference of Governmental Industrial Hygienists* (ACGIH), 1999. TLVs and BEIs. Threshold limit values for chemical substances and physical agents. Biological Exposure Indices. Cincinnati, OH.

AWWA, 1999. *Water quality & treatment: A handbook of community water supplies.* 5th ed.

Barreras, J.T., 2001. *Evaluation of retrievable drip tape irrigation systems.* ITRC Paper No, 2001.

Environmental Protection Agency, 1999. *Alternative disinfectants and oxidants guidance manual.* Pages. 30–31

Environmental Protection Agency, 1999. *Trouble shooting guide for small ground water systems with hypochlorination.*

Golfinopoulos, S.K., 2002. Formation of organic by-products during chlorination of natural waters. *Journal of Environmental Monitoring*, 4(4): 910–916.

Hassan, Farouk A., 1999. Micro irrigation management and maintenance. *Agro Industrial Management*, Pages 120–124

Hills, D.J. and M.J. Brenes, 2001. Micro irrigation of wastewater using drip-tape. *Applied Engineering in Agriculture*. Pages 303–308.

Ivahnenko, T. and J. Bundash, 2004. Chloroform in the hydrologic system. *Journal of Engineering Mechanics*, Pages 20–24

Keller, J. 2000. *Sprinkler and trickle irrigation.* New York: Van Nostrand Reinhold, Pages 120–123

Matesi, S.W., 2003. Organic by-products of drinking water chlorination, *Global Nest*, 1: 143–156.

Miyamoto, K. and T.Yamamoto, 2005. The effect of chlorine on emitter clogging induced by algae and protozoa and the performance of drip irrigation. *Transactions of the ASAE*, 48(2): 519–527.

Nemery, B. P.H.M. Hoet and D. Nowak, 2002. Indoors swimming pools, water chlorination and respiratory health, Eur. Respir. J., 19 (May): 790–793.

Nieuwenhuijsen, M., M. Toledano and N. Eaton, 2000. Chlorination disinfection by-products in water and their association with adverse reproductive outcomes: A review. *Occupational and Environmental Medicine*, 57: 73–85.

Nikolaou, A.D. and S.K. Golfinopoulos, 2002. Investigation of the formation of organic by-products during chlorination of surface waters. *International Conference Protection and Restoration of the Environment VI*, Skiathos-Greece, 1: 115–122.

Nikolaou, A.D., S.K. Golfinopoulos, T.D. Lekkas and M.N. Kostopoulou, 2004. DBP levels in chlorinated drinking water: Effect of humic substances. *Journal of Environmental Monitoring and Assessment*, 93: 301–319.

Palacios, M., J.F. Pampillon and M.E. Rodriguez, 2002. Organohalogenated compounds levels in chlorinated drinking waters and current compliance with quality standards throughout the European Union. *Water Research*, 34(3): 1002–1016.

Park, H.S., T.M. Hwang, J.W. Kang, H., Choi and H.J. Oh 2001. Characterization of raw water for the ozone application measuring ozone consumption rate. *Water Research*, 35(11): 2607–2614.

Qassim, A., 2003. *Micro-irrigation: A situation analysis*. Pages 1–5.

Raleigh, M., 2005. Development of Micro irrigation Systems. *Irrigation Science*, 10: 47–50.

Reckhow, D.A. P.C. Singer and R.L. Malcolm, 1999. Chlorination of humic materials: byproduct formation and chemical interpretations. *Environmental Science & Technology*, 24(11): 1655–1664.

Richardson, S.D., A.D. Thruston, T.V. Caughran, Collette T.W., Patterson K.S. and B.W. Lykins, 1999. Chemical by-products of chlorine and alternative disinfectants. *Food Technology*, 52(4): 58–66.

Santana Vargas, Julio M., 2000. *Patrones de humedecimiento bajo el sistema de micro riego subterraneo*. Tesis de Maestía, Agronomía y suelos. Universidad de Puerto Rico Recinto Universitario de Mayagüez. Pages 5–8.

Santiago, L. 2006. Limpieza general en sistemas de riego. *Gaceta Sanitaria*, Pages 20–25.

Sung, W., 2002. Corrosion control and chlorination, discolored water and nitrification. *Proceedings Water Quality Technology Conference*, MWRA, Southborough, MA, USA, Pages 1683–1686.

Thokal, R., D. Mahale and A. Ganpat Powar. 2004. *Drip irrigation systems: Clogging and its prevention*. Pointer Publshers. Pages 125–126

Threadgill, E.D., D.E. Eisenhauer, J.R. Young and B. Bar-Yosef, 1999. *Chemigation*. St. Joseph, MI: American Society of Agricultural Engineers. Pages 749–780.

Tsai, L.S., R. Higby and J. Schade, (1999). Disinfection of poultry chiller water with chlorine dioxide: consumption and byproduct formation. *Journal of Agricultural and Food Chemistry*, 43: 2768–2773.

U.S. Environmental Protection Agency. 1999. *Integrated Risk Information System (IRIS) on Chlorine*. National Center for Environmental Assessment, Office of Research and Development, Washington, DC.

**Web page links**

Alam, M., T.P. Trooien, F.R. Lamm and D.H. Rogers. 1999. Chlorination for subsurface drip irrigation (SDI) systems, MF-2361. Kansas State University, Manhattan, K.S. www.ext.vt.edu/pubs/bse/442-757/442-757.html

Albrecht, Julie A. Water Resources and Irrigation Specialist. http://www.ianr.unl.edu

Alshammary, S.F., Y.L. Qian and S.J. Wallner, 2004. Growth response of four turfgrass species to salinity. *Agricultural Water Management*, 66(1): 97–111. www.c3.org/chlorine_issues/new_perspective.html

Clark, G.A. and A.G. Smajstrla, 1992. Treating irrigation systems with chlorine. Circular 1039 Florida Cooperative Extension Service, I.F.A.S, University of Florida, Gainesville, FL. www.microirrigationforum.com/new/archives

Clark, Gary and William Lamont. Maintaining drip irrigation systems. www.oznet.ksu.edu

D.J. Hills, MA Tajrishy, G. Tchobanoglous, 2000. *Transactions of the ASAE*. asae.frymulti.com

Dukes, Michael and Johanes Scholberg. 2004. Automated subsurface drip irrigation based on soil moisture. http://asae.frymulti.com/abstract.asp

Foley, J.P. and, S.R. Raine 2000. Drip irrigatiom in the Australian cotton industry: A scoping study. www.ncea.org.au

Granberry, Darby and William Terry Kelley. 2005. Drip Chemigation: Injecting fertilizer, acid and chlorine. http://pubs.caes.uga.edu/caebspubs/pubcd/B1130.htm

Gross, M.A. and S.W. Jones, 1999. Stratified sand filter and ozonation for wastewater reuse. *Proceedings NOWRA Conference, 8th Annual Conference and Exhibit*, Jekyll Island, GA. www.swrcb.ca.gov/ab885/docs/techonsite/chapter10.pdf

Hochmuch, George and Allen Smajsrtla. 1999. Fertilizer application and management for micro (drip)- irrigated vegetables in Florida. http://edis.ifas.ufl.edu

Jonas, Dan, 2000. Drip and trickle irrigation. www.plt.org/ irrigation

Karen, M. Emergency disinfection of water supplies. Ohio State University, Extension Factsheet. http://www.ag.ohio-state.edu/

Kovach, S. P., 1984. Injection of fertilizers into drip irrigation systems for vegetables. Circular 606. Florida Coop. Ext. Ser., Univ. Fla., Gainesville. edis.ifas.ufl.edu/AE080

Lamm F.R., 2002. Advantages and disadvantages of subsurface drip irrigation. http://www.oznet.ksu.edu/sdi/REPORTS/2002

Laosheng Wu, 2002. Drip Irrigation using low quality water. http://www.greenmediaonline.com

Qassim, A., 2003. Micro-irrigation: A situation analysis. Pages 1–5. wca-infonet.org

Rogerio, V. and P. Cleide, 2003. Chemigation with benomyl and fluazinam and their fungicidal effects in soil for white mold control on dry beans. http://www.scielo.br/scielo.php

Texas Agricultural Extension Service, 2000. Tablet Chlorination. http://texaserc.tamu.edu

Valiantzas, J.D., 2006. Hydraulic analysis and optimum design of multidiameter irrigation laterals. *Journal of Irrigation and Drain Engineering*, 2006 ink.aip.org

Vander Hoek, W., F. Konradsen, J.H.J. Ensink, M. Mudasser and P.K. Jensen, 2001. Irrigation water as a source of drinking water: is safe use possible? *Tropical Medicine & International Health*, 6: 46–54. www.c3.org/chlorine_knowledge

# 10

**Literature cited**

1. Design, Installation and Performance of Trickle Irrigation Systems. *ASAE Engineering Practice*, ASAE EP. 405, 2005.

2. Drip Irrigation Management. Cooperative Extension Service, Division of Agricultural Sciences, University of California, Berkeley-CA. 1981.

3. Goldberg D.B., B. Gornat and D. Rimon, 1976. *Drip Irrigation: Principles, Design and Agricultural Practices*. Kfar Shmaryahu, Israel: Drip Irrigation Scientific Publications.

4. Goyal, M.R. y L.E. Rivera, 1985. Riego por goteo: servicio y mantenimiento. Colegio de Ciencias Agrícolas, Servicio de Extensión Agrícola, Recinto Universitario de Mayagüez, Río Piedras. Serie 5, IA 64

5. Goyal, M.R., L.E. Rivera, M. Martínez y N. Rojas, 1984. Principios de riego por goteo. Colegio de Ciencias Agrícolas, Servicio de Extensión Agrícola, Recinto Universitario de Mayagüez, Río Piedras. Serie 3, IA 62.

6. Jensen E., 1980. *Design and operations of farm irrigation systems*. ASAE Monograph No. 3. American Society of Agricultural Engineers, St. Joseph, MI.

7. Drip/Trickle Irrigation in Action. Volumes I and II. *Proceedings of the 3rd International Drip/Trickle Irrigation Congress*. Fresno, CA, November 18–21, 1985. American Society of Agricultural Engineers, St. Joseph - MI. Publication No. 10–85.

**Books/ bulletins/ journals and proceedings / reports**

Ahuja, Lajpat R., Liwang Ma and Terry A. Howell, 2002. *Agricultural system models in field research and technology transfer*. Boca Raton, Fl: John Wiley

Ascough, G.W. and G.A. Kiker, 2002. The effect of irrigation uniformity on irrigation water requirements. *Journal of the Water South Africa*,

28(2): 235–242. http://www.doaj.org/ http://www.wrc.org.za/publications/watersa_.htm

Bergmann, L. and G.E. Weismantel, 2000. *Filtration technology handbook*. North Carolina, NC: John Wiley.

Binnie, C., K. Martin and S. George, 2002. *Basic water treatment*. London: IPC.

Bomo, A.M., A. Husby, T.K. Stevik and J.F. Hanssen, 2003. Removal of fish pathogenic bacteria in biological sand filters. *Water Resources*, 37: 2618–2626.

Bouwer, E.J., H.H.M. Rijnaarts, A.B. Cunningham and R. Gerlach, 2000. Biofilms in porous media. In *Bio-films II: Process Analysis and Applications*, ed. J.D. Bryers, 123–158. New York: Wiley-Liss. Pages 123–158.

Burt, Charles M. and Stuart W. Styles, 2000. *Drip and micro irrigation for trees, vines, and row crops*. Irrigation Training and Research Center (ITRC), CA.

Christopherson, S.H., J.L. Anderson and D.M. Gustafson, 2001. Evaluation of recirculating sand filters in Minnesota. American Society of Agricultural Engineers, St. Joseph, MI. Pages 207–214.

Healy, M.G., M. Rodgers and J. Mulqueen, 2004. Recirculating sand filters for the treatment of synthetic dairy parlour washings. *Journal of Environmental Quality*, 33:713–718.

Horwatich, J.A., S.R. Corsi and R.T. Bannerman, 2004. *Effectiveness of a pressurized stormwater, filtration system in green bay*. U.S. Geological Survey Reston, VA.

Julich, W. and J. Schubert, 2000. *Proceedings of the International Riverbank Filtration*. Dusseldorf-Germany.

Keller, J. and R.D. Bliesner, 2002. *Sprinkler and trickle irrigation*. Blackburn Press.

Oosthuizen, L.K., P.W. Botha, B. Groové and J.A. Meiring, 2005. Cost-estimating procedures for drip-micro and furrow irrigation systems. J. Water South Africa, 31: 3: 403–406.

Phansalkar, S.J., 2000. *Appropriate drip irrigation technologies: A socio-economic assessment*. New Delhi, India: International Development Enterprises.

Postel, Sandra and Paul Polak, 2002. *Drip irrigation for small farmers: a new initiative to alleviate hunger and poverty*. New Delhi: International Development Enterprises.

Ray, C. and Dordrecht, 2002. *Riverbank filtration: Understanding contaminant biogeochemistry and pathogen removal*. Boston: Kluwer Academic Publishers.

Rodgers, M., J. Mulqueen and M.G. Healy, 2004. Surface clogging in an intermittent stratified sand. *Soil Science Society of America Journal*, 68: 1827–1832

Sanewe, A., M. Plessis and G. Backeberg, 2003. Water Utilization in Agriculture. *Water research Comission*, Pages 156–187

Schwankl, L. and T. Prichard, 2001. *Chemigation in tree and vine micro irrigation systems*. Agriculture and Natural Resources Publication 21599. University of California, Davis, CA.

**Web page links**

Aung, K.H. and F.S. Scherer, 2003. Introduction to micro-irrigation. http://www.ext.nodak.edu/extpubs/ageng/irrigate/ae1243w.htm

Dorota Z. Haman, Allen G. Smajstrla and Fedro S. Zazueta, 2002. Screen filters for irrigation systems. http://edis.ifas.ufl.edu/WI009

Doulto USA Co., 2006. Stainless steel commercial filters. http://doultonusa.com/commercial_industrial_filters/commercial_and_industrial_water_filters.htm

Gelski, J., 2003. Avoid filter frustration. http://www.growermagazine.com/home/02-03filters.html

Hanson, F. and M. Drip, 2000. Irrigation of row crops: What is the state of the art? *Proceedings of the 4th Decennial Irrigation Symposium*. ASAE. http://www.oznet.k-state.edu/sdi/Abstracts/Drip%20Irrigation%20of%20Row%20Crops.htm

Hoitink and Matthew S. Krause, 2001. New approaches to control of plant pathogens in irrigation water. http://ohioline.osu.edu/sc173/sc173_13.html

Jains Co., 2006. All types of filter systems. www.jains.com

Orival, 2006. Water filtration solutions. http://www.orival.com

SABI Design Norms. South African Irrigation Institute, Strand - RSA. http://www.sabi.co.za/html/designnorms.html

Sanders, Douglas, 2001. Drip or trickle irrigation systems: An outline of components. http://www.ces.ncsu.edu/depts/hort/hil/hil-33-a.html

Scanaway Filters Co., 2006. A large amount of filter selection. http://news.thomasnet.com/fullstory/464484/954

Schwankl, Lawrence J. The orchard: Micro irrigation systems. http://fruitsandnuts.ucdavis.edu/crops/papers/Chapter_06.pdf#search='drip%20irrigation%20filtering'

Shock, C.C., 2006. Drip irrigation: An introduction. Malheur Experiment Station, Oregon State University. EM 8782. http://www.cropinfo.net/drip.htm

Shock, C.C., R.J. Flock, E.B.G. Feibert., A.B. Pereira and M. O'Neill, 2005. *Drip irrigation guide for growers of Hybrid Poplar*. Oregon State University Extension Service. EM 8902, Pages 1–6. http://extension.oregonstate.edu/catalog/pdf/em/em8902.pdf

Shock, C.C., R.J. Flock, E.B.G. Feibert, C.A. Shock, L.B. Jensen and J. Klauzer, 2005. *Drip irrigation guide for onion growers in the Treasure Valley*. Oregon State University Extension Service. EM 8901, Pages 1–8. <http://extension.oregonstate.edu/catalog/pdf/em/em8901.pdf

Shock, C.C., R.J. Flock, E.B.G. Feibert, C.A. Shock, A.B. Pereira and L.B. Jensen, 2005. *Irrigation monitoring using soil water tension*. Oregon State University Extension Service. EM 8900, Pages 1–6. http://extension.oregonstate.edu/catalog/pdf/em/em8900.pdf

Shuster, Christopher, 2001. Irrigation-Filtration Systems. http://grounds-mag.com/mag/grounds_maintenance_irrigationfiltration_systems/index.html

Smith Bryan, 2004. *Irrigation systems*. http://hgic.clemson.edu/factsheets/HGIC1705.htm

U.S. Filters Co., 2006. *Filtration in general*. www.usfilters.com

## 11

### Literature cited

1. Bar-Ram. *Instructions for the Maintenance of BAR-RAM Drip Irrigation Systems*. Israel.

2. *Design, Installation, and Performance of Trickle Irrigation Systems*, 2004. ASAE Engineering Standard No. ASAE-EP-405

3. *Drip Irrigation Management*. 1981. Division of Agricultural Sciences at University of California. Berkley, CA.

4. *Trickle irrigation in the Eastern United States*, 1980. Agricultural Cooperative Extension Service, Northeast Regional Agricultural Engineering.

### Books/bulletins/journals and proceedings/reports

Anonymous, 2000. *Water reuse for irrigation*. CRC Press

Ayars, J.E., R.A. Schoneman, F. Dale, B. Meso and P. Shouse, 2001. Managing subsurface drip irrigation in the presence of shallow ground water. *Agricultural Water Management*, 47(3): 243–264.

Bacarello, V., V. Ferro, G. Provenzano and D. Pumo, 1999. Evaluating pressure losses in drip-irrigation lines. *Journal of Irrigation and Drainage Engineering*, 123(1): 1–7.

Burt, C.M. and S.W. Styles, 1999. *Drip and micro irrigation for trees, vines, and row crops*.

Carpena, R.M., M.D. Dukes, Y.C. Li and W. Klassen, 2005. Field comparison of tensiometer and granular matrix sensor automatic drip irrigation on tomato. *Tropical Research and Education Center, IFAS, University of Florida*, 15(3): 1–12.

Chidenga, E.E., 2003. *Irrigation technology choices and operations and maintenance in smallholder systems in Zimbabw*. Wageningen: Wageningen Academic Publishers. Pages 1–205.

El-Hafedh, A.V. O.M., H. Daghari and M. Maalej, 2001. Analysis of several discharge rate-spacing-duration combinations in drip irrigation system. *Agricultural Water Management*, 52(1): 33–52.

Escoe, A. Keith, 2006. *Piping and pipelines assessment guide*. Elsevier Gulf Professional Publishing.

Howell, T.A., 2001. Enhancing water use efficiency in irrigated agriculture. *Agronomy Journal*, 93(4): 281–289.

Keller, J. 2000. *Sprinkler and Trickle Irrigation*.

Laosheng, Wu, 2000. Drip irrigation using low-quality water. *Irrigation Journal*, 50(3):18–20.

Mehmet, O., 2002. *Modern and traditional irrigation technologies in the eastern mediterranean*. International Development Research Center, Pages 130–132.

Meshkat, M., R.C., Warner and S.R. Warner, 1999. Comparison of water and temperature dristribution profiles under sand tube irrigation. *Transactions of ASAE*, 4:1657–1663.

Millera, S.D., G.F. Vanceb and R. Zhangb, 2001. Agroeconomic analyses of drip irrigation for sugarbeet production. *Agronomy Journal*, 93(3): 517–523.

Nakayama, F.S. and D.A. Bucks, 2002. *Trickle irrigation for crop production*. Elsevier Publishing Company. Pages 1–383.

National Research Council, 2002. A New Era For Irrigation.

Provenzano, G. and D. Pumo, 2006. Experimental analysis of local pressure losses for microirrigation laterals. *Journal of Irrigation and Drainage Engineering*, 132(2): 193–194.

Santana Vargas, Julio M., 2000. *Patrones de humedecimiento bajo el sistema de micro riego subterráneo*. UPR – RUM, Mayagüez, Puerto Rico.

Scherer, T.F., 2005. *Design and control of sprinkler systems for crop disease research*. St. Joseph, MI: The American Society of Agricultural and Biological Engineers. Paper number 052182, ASAE Annual Meeting.

Selker, J., 2004. *Irrigation system maintenance, groundwater quality and improved production*. Oregon State University Extension Service. Pages 1–56.

Snellen, W. Bart, 1999. *Irrigation scheme operation and maintenance*. Training Manual, No. 10. Food & Agriculture Organization of the United, Pages 1–48.

Thokal, R.T., D.M. Mahale and A.G. Powar, 2004. *Drip Irrigation System: Clogging and its Prevention*. Jodhpur-India: Pointer Publisher. Pages 1–107.

Tiwari, K.N., A. Singh and P.K. Mal, 2002. Effect of drip irrigation on yield of cabbage (*Brassica oleracea* L. var. capitata) under mulch and non-mulch conditions. *Agricultural Water Management*, 58(1):19–28.

**Web page links**

Alam, M., T.P Trooien and F.R. Lamm, 1999. Filtration and maintenance considerations for subsurface drip irrigation systems. Kansas State University. Pages 1–4. http://www.oznet.ksu.edu/library/ageng2/mf2361.pdf

Benham, Brian, 2002. Filtration, treatment and maintenance considerations for micro irrigation systems. Virginia State University. Page 1. http://www.ext.vt.edu/pubs/bse/442-757/figure1.html

Bertauski, Tony, 2001. HOW TO: Install drip irrigation in new plantings: Maintenance and trouble shooting tutorials- The Drip Store. Trident Technical College. http://www.dripirrigation.com/drip_tutorial.php?page_view=head

Bill Ashcroft, Tatura, 1994. Maintenance of Trickle Irrigation Systems. http://www.dpi.vic.gov.au/dpi/nreninf.nsf/childdocs/-2BAF4D73531CD1544A2568B3000505AF-FFB44D93F7BB6C37CA256C4A0083CDB7-0123B04778E33BD84A256DE-A0027B8C3-2FE8144CEA3D-02E0CA256C1A0020FAA8?open

Burt, C.M. and S.W., Styles. 2004. *Conceptualizing irrigation modernization through benchmarking and the rapid appraisal process*. In Irrigation and Drainage. ITRC Paper No. P 03-002. http://www.interscience.Wiley.com

Carrigan, E. and S. Knights, 2000. Sub-surface drip irrigation (SDI) – system maintenance. Department of Primary Industries and Fisheries. Page 1. http://www2.dpi.qld.gov.au/fieldcrops/17653.html

Components and design of a drip irrigation system. http://www.cropinfo.net/drip.htm#Components

Do-it-yourself drip irrigation. http://www.vintagerosery.com/dripirrigation.htm

Drip-irrigation. http://www.irrigationtutorials.com/dripguide.htm

Eartheasy.com (c) 2000–2006. http://www.eartheasy.com/grow_drip-irrigation.htm#d

Experimental analysis of local pressure losses for microirrigation laterals. *Journal of Irrigation and Drainage Engineering*, 130(4): 318–324. 2004. http://ascelibrary.aip.org/vsearch/servlet

How to use our technical support and design services. http://www.dripworksusa.com/tech.html

Irma, S. and R.B. Ferguson, 2005. *Subsurface drip irrigation research- I.* University of Nebraska-Lincoln Agricultural Research Division. Page 1. http://bse.unl.edu/Research/subsurface.htm

Irrigation equipment/ irrigation controllers, timers or clocks. http://www.waterboysprinkler.com/ieq-ctrl.html

Johnson, T.A., 2003. Water wisely with drip irrigation. Edited by Fine Gardening. Pages 52–55. http://www.taunton.com/finegardening/pages/g00005.asp

Kietzman, Shannon, 2006. What is drip irrigation? Pages 1–2. http://www.wisegeek.com/what-is-drip-irrigation.htm

Kizer, Michael A., 2002. Drip (trickle) irrigation system. Oklahoma Cooperative Extension Service. Pages 1–4. http://osuextra.okstate.edu/pdfs/F-1511web.pdf#search=%22trickle%20irrigation%20maintenance%2

New strategy for optimizing water application under trickle irrigation. *Grounds Magazine*, 128(5):287–297. *Journal of Irrigation and Drainage Engineering.* September/October 2002. http://grounds-mag.com/mag/grounds_maintenance_install_drip_irrigation/

Northern Garden Supply, 2004. Drip irrigation maintenance. How To Tutorials about Drip Irrigation. Page 1. http://www.dripirrigation.ca/HowTo_Maintain.asp

Sanders, D.C., 2001. An outline of components: In drip or trickle irrigation systems. Page 1. http://www.ces.ncsu.edu/depts/hort/hil/hil-33-a.html

Stone, J., 2003. Irrigation installation and maintenance time Savers, Part II. Pages 1–3. http://www.progardenbiz.com/issues/v1issue3/Irrigation-Installation.html

Water management of irrigated-drained fields in the Jordan Valley South of Lake Kinneret. *Journal of Irrigation and Drainage Engineering*, 2005, 131(4, July/August): 363–374. http://ascelibrary.aip.org/vsearch/servlet/

# 12

**Literature cited**

1. Howell, T.A. and E.A. Hiler, 1974. Designing trickle irrigation laterals for uniformity. *Journal of the Irrigation and Drainage Division*, ASCE Vol. 100, No. IR4, Proc. Paper 10983, Pages 443–454.

2. Howell, T.A. and E.A. Hiler, 1974. Trickle irrigation lateral design. *Transactions of America Society of Agricultural Engineers*, 17(5): 902–908.

3. Williams, G.S. and A. Hazen, 1960. *Hydraulic Tables*, 3rd ed. New York, NY: John Willey and Sons.

4. Wu, I.P., 1975. Design of drip irrigation main lines. *Journal of the irrigation and Drainage Division*, ASCE. Vol. 101: No. IR4, Proceeding Paper 11803, Pages 265–278.

5. Wu, I.P. and H.M. Gitlin, 1973. Hydraulics and uniformity for drip irrigation. *Journal of the Irrigation and Drainage Division*, ASCE, Vol. 99, No. IR3, Proceeding Paper 9786, Page 157–168.

6. Wu, I.P. and H.M. Gitlin, 1973. Hydraulics irrigation design based on uniformity. *Transactions of America Society Agricultural Engineers*, 17(3): 157–168.

7. Wu, I.P. and H.M. Gitlin, 1974, June. *Design of irrigation lines.* Technical Bulletin No.96, Hawaii Agricultural Experimental Station, University of Hawaii.

8. Wu, I.P. and H.M. Gitlin, 1975. Energy Gradient line for drip Irrigation and Drainage Division, ASCE Vol. 101, No. IR4, Proc. Paper 11750, Pages 323–326.

9. Wu, I.P. and H.M. Gitlin, 1975. *Drip irrigation designs on non uniform slopes.* Paper presented at the 1975 Winter Meeting of America Society of Agricultural Engineers, Chicago, IL.

10. Wu, I.P. and H.M. Gitlin, 1977. Design drip irrigation lines varying pipes sizes. *Journal of the Irrigation and Drainage Division*, ASCE, Vol. 103, No. IR4, Proc. Paper 13384.

11. Wu, I.P. and H.M. Gitlin. *Drip irrigation systems design.* Bulletin No. 144 and 156 of the Cooperative Extension Service, University of Hawaii.

**Books/bulletins/journals and proceedings/reports**

Aragués, R., E. Playán, R. Ortiz and A. Royo, 1999. A new drip-injection irrigation system for crop salt tolerance evaluation. *Soil Science Society of America Journal*, 63: 1397–1404.

Assouline, S. 2002. The effects of microdrip and conventional drip irrigation on water distribution and uptake. *Soil Science Society of America Journal*, 66: 1630–1636.

Assouline, S., S. Cohen, D. Meerbach, T. Harodi and M. Rosner, 2002. Microdrip irrigation of field crops: Effect on yield, water uptake and drainage in sweet corn. *Soil Science Society of America Journal*, 66, (January): 228–235

Christians, N.E., 2000. *The mathematics of turfgrass maintenance.*, Ann Arbor Press. Page 1–176

Czemerda, K. and T. McConnell, 2003. *Drip irrigation: Extension service.* West Virginia University.

EP 405 Design, installation and performance of trickle irrigation systems, 1995. St. Joseph, MI: Standards of American Society of Agricultural Engineering, 49085–49659.

Evett, S.R., T.A. Howell, A.D. Schneider, D.R. Upchurch and D.F. Wanjura, 2000. Automatic drip irrigation of corn and soybean. *Proceedings of the 4th Decennial National Irrigation Symposium*, , November 14–16. Pages 401–408.

Fanning, J.L., 2001. A field and statistical modeling study to estimate irrigation water use at benchmark farms study sites in Southwestern Georgia. U.S. Dept. of the Interior, U.S. Geological Survey.

K. Jack, 2000. *Sprinkler and trickle irrigation*, 12th ed. Blackburn Press, Pages 1–652.

Postel, S., 1999. *Pillars of sand: Can the irrigation miracle last?* W.W. Norton & Co, Pages 1–313

Postel, S., P. Polak, F. Gonzales and J. Keller, 2001. Drip irrigation for small farmers: A new initiative to alleviate hunger and poverty. *Water International*, 26(1): 3–13.

Royo, A., R. Aragués, E. Playán and R. Ortiz, 2000. Salinity—grain yield response functions of barley cultivars assessed with a drip-injection irrigation system. *Soil Science Society of America Journal*, 64: 359–365.

Sando, S.K., 2001. Irrigation of the angostura reclamation unit. 4th ed. U.S. Dept. of the Interior, U.S. Geological Survey, Pages 1–65

Swihart, J., 2002. Canal-lining demonstration project year 10 final report. 1st version. U.S. Dept. of the Interior.

Thompson, T.L., S.A. White, J. Walworth and G.J. Sower, 2003. Fertigation frequency for subsurface drip-irrigated broccoli. *Soil Science Society of America Journal*, 67: 910–918.

Thompson, T.L., T.A. Doerge and R.E. Godin, 2000. Nitrogen and water interactions in subsurface drip-irrigated cauliflower: I. Plant Response. *Soil Science Society of America Journal*, 64: 406–411.

Thompson, T.L., T.A. Doerge and R.E. Godin, 2002. Subsurface drip irrigation and fertigation of broccoli: I. Yield, quality, and nitrogen uptake. *Soil Science Society of America Journal*, 66(January): 186–192.

Thompson, T.L., T.A. Doerge and R.E. Godin, 2000. Nitrogen and water interactions in subsurface drip-irrigated cauliflower: II. Agronomic, economic and environmental outcomes. *Soil Science Society of America Journal*, 64: 412–418.

Thompson, T.L., T.A. Doerge and R.E. Godin, 2002. Subsurface drip irrigation and fertigation of broccoli: II. Agronomic, economic, and environmental outcomes. *Soil Science Society of America Journal*, 66(January): 178–185.

United States, Agricultural Research Service. 2001. Irrigation & Drainage: A national research plan to meet competing demands and protect the environment. Pages 1–27.

United States. Bureau of Reclamation. Denver Office. Engineering Division. 2003. Water management workshop session notes [microform].

United States. Congress. Senate. Committee on Indian Affairs, 2004. The Oglala Sioux Tribe Angostura Irrigation Project Rehabilitation and Development Act [microform]: iii, 41 p. U.S. G.P.O.

**Web page links**

Beaulieu, David. What are drip irrigation systems and how do they promote water conservation?

http://landscaping.about.com/cs/cheaplandscaping1/f/drip_irrigation.htm

Clark, G.A, C.D. Stanley, F.Z. Zazueta and E.E. Albregts, 2002. Farm ponds in Florida irrigation systems. University of Florida. http://edis.ifas.ufl.edu/AE143

Drip irrigation leach field. http://www.toolbase.org/Technology-Inventory/Sitework/drip-irrigation-leach-field

Drip irrigation systems. http://www.jains.com/irrigation/drip%20irrigation%20system.htm

Drip irrigation systems. http://www.1-hydroponics.co.uk/top-tips/drip-irrigation.htm

Drip irrigation. http://www.eartheasy.com/grow_drip-irrigation.htm#a

Drip irrigation. http://www.swfwmd.state.fl.us/waterres/drought/articles/drip.htm

Drip irrigation. http://en.wikipedia.org/wiki/Drip_irrigation

Drip watering tips. http://www.snwa.com/html/land_irrig.html

Easy drip irrigation. http://www.laspilitas.com/drip.htm

Haman, D.Z. and A.G. Smajstrla, 2003. Design tips for drip irrigation of vegetables. University of Florida. http://edis.ifas.ufl.edu/AE093/

Haman, D.Z. and I.Forrest, 2003. Principles of micro irrigation. University of Florida. http://edis.ifas.ufl.edu/WI007

Harrison, C. and O'Ney Susan, 2002. Design and modification of an installation method to stabilize small trapezoidal flumes in drinage ditches. http://www.srs.fs.usda.gov/pubs/rn/rn_srs011.pdf

Irrigation water management: Irrigation methods. http://www.fao.org/docrep/S8684E/s8684e07.htm#6.1%20when%20to%20use%20drip%20irrigation

Irrigation: Drip/microirrigation. http://ga.water.usgs.gov/edu/irdrip.html

Jess Stryker's. Drip irrigation design guidelines. http://www.irrigationtutorials.com/dripguide.htm

NETAP, 2006. Engineering & technical support division. http://www.netafim.com/Business_Divisions/Engineering_and_Technical Support/

O'Connell Landscape, 2006. Irrigation design process outline. http://www.oclandscape.com/articles/irrigationprocess_article.htm

Oztekin T, L.C. Brown, P.M. Holdsworth, A. Kurunc and D. Rector, 2000. Evaluating drainage design for wastewater irrigation applications to minimize impact on surface waters. http://asae.frymulti.com/abstract.asp?aid=5803&t=1

Pepsi drip/Easy drip irrigation. http://www.practicafoundation.nl/smartwater/EN/pepsidrip.htm

Relf, D. Irrigating the home garden http://www.ext.vt.edu/pubs/envirohort/426-322/426-322.html

Schattenberg, P. Drip tape irrigation useful, convenient for small acreage forage production. http://agnews.tamu.edu/dailynews/stories/SOIL/Oct2505a.htm

Shock, C.C., 2001. Drip Irrigation: An Introduction. http://www.cropinfo.net/drip.htm

Statewide IPM Program, 2004. Irrigation design. Agriculture and natural resources, University of California. http://www.ipm.ucdavis.edu/TOOLS/TURF/SITEPREP/irrdes.html

Stecks Nursery and Landscaping. 2006. Landscape design. http://www.atstecks.com/Design.htm

Stryker, Jess, 2005. Landscape sprinkler irrigation design tutorial. http://www.irrigationtutorials.com/sprinkler00.htm

Valenzuela, H. Crop production guidelines: Drip irrigation. HITAHR, University of Hawaii.

http://www.extento.hawaii.edu/kbase/reports/dripirrigation.htm

# 13

## Literature cited

1. Gillespie, V.A., A.L. Phillips and I-Pai Wu, 1979. Drip irrigation design equations. *Journal Irrigation and Drainage*, ASCE, 105 (IR3): 247–257 Paper # 14819.

2. Howell, T.A. and E. A. Hiler, 1974. Trickle irrigation lateral design. *Transactions American Society of Agricultural Engineers*, 17(5): 902–908.

3. Howell, T.A. and E.A. Hiler, 1974. Designing trickle irrigation laterals for uniformity. *Journal of the Irrigation and Drainage Division*, ASCE, 100 (IR4): 443–454, Paper 10983.

4. Williams, G.S. and A. Hazen, 1960. *Hydraulic Tables*, 3rd ed. New York: John Wiley and Sons.

5. Wu, I.P. and D.C. Fangmeir, 1974, December. Hydraulic design of twin-chamber trickle irrigation laterals. Technical Bulletin No. 216, The Agricultural experiment station, Tucson, Ariz.

6. Wu, I.P. and H.M. Gitlin, 1973, June. Hydraulics and uniformity for drip irrigation. *Journal of the Irrigation and Drainage Division*, ASCE, 99 (IR3): 157–168. Paper 9786.

7. Wu, I.P. and H.M. Gitlin, 1974, June. Design of drip irrigation lines. Technical Bulletin No. 96, Hawaii Agricultural Experiment Station, University of Hawaii, Honolulu, Hawaii.

8. Wu, I.P. and H.M. Gitlin, 1975, December. Energy gradient line for drip irrigation laterals. *Journal of the Irrigation and Drainage Division*, ASCE, 101 (IR4): 323–326. Paper 11750.

**Books/bulletins/journal and proceedings/reports**

Abreu, V.M. and Luís S.P., 2002. Sprinkler irrigation systems design using ISADim. Paper number 022254, ASAE Annual Meeting.

Assouline, S., 2002. The effects of microdrip and conventional drip Irrigation on water distribution and uptake. *Soil Science Society of America Journal*, 66: 1630–1636.

Brandt R.A. and W. B.William, 1988. Reliability in design. ASAE Distinguished Lecture No. 13, pages 1–27. Winter Meeting of the American Society of Agricultural Engineers, December, Chicago-Illinois 913C0888.

Camp, C.R., E.J. Sadler and W.J. Busscher, 1997. A Comparison of uniformity measures for drip irrigation systems. *Transactions of the ASAE*, 40(4): 1013–1020.

Camp, C.R., P.J. Bauer and P.G. Hunt, 1997. Subsurface drip irrigation lateral spacing and management for cotton in the southeastern coastal plain. Transactions of the ASAE, 40(4):993–999.

Chieng, S. and A. Ghaemi, 2003. Uniformity in a microirrigation with partially clogged emitters. Paper number 032097, 2003 ASAE Annual Meeting.

Manoliadis, O.G., 2002. Analysis of irrigation systems using sustainability-related criteria. *Journal of Environmental Quality*, 30: 1150–1153.

Moore, S., Y.J. Han, A.Khalilian, T.O. Owino and B.Niyazi, 2005. Instrumentation for variable-rate lateral irrigation system. Paper number 052184, ASAE Annual Meeting.

Zhu, H., C.L. Butt, M.C. Lamb and P.D. Blankenship, 2004. An implement to install and retrieve surface drip irrigation laterals. *Applied Engineering in Agriculture*, 20(1): 17–23.

**Web page links**

Evans, R. and R.E. Sneed, 1996. Selection and management of efficient low volume irrigations systems. North Carolina Cooperative Extension Service. EBAE-91-153. http://www.bae.ncsu.edu/programs/extension/evans/ebae-91-153.html

Fisher, G.W. and S. Turf, 2001. Innovations in Irrigation. http://grounds-mag.com/irrigation/grounds_maintenance_innovations_irrigation/

Gallion, G.B., 2005. Irrigating difficult spaces. http://grounds-mag.com/irrigation/grounds_maintenance_irrigating_difficult_spaces/

Haman, D.Z. and A.G. Smajstrla. Design tips for drip irrigation of vegetables. University of Florida.

http://edis.ifas.ufl.edu/AE093

Hill, R.W., 2000. Management of sprinkler irrigation systems. http://ucanr.org/alf_symp/2000/00-119.pdf

Kizer, M.A., 1990. Drip (trickle) irrigation systems. Oklahoma Cooperative Extension Fact Sheet.

<http://pods.dasnr.okstate.edu/docushare/dsweb/Get/Document-1443/f-1511%20web.pdf>

McGuirk, S., 2001, January. Irrigating steep-sloped landscapes. Grounds Maintenance. http://grounds-mag.com/irrigation/grounds_maintenance_irrigating_steepsloped_landscapes/

Solomon, K.H., 1988. Irrigation system selection. Center for Irrigation Technology Irrigation Notes. California State University, Fresno, California 93740-0018. http://www.wateright.org/site2/publications/880105.asp

Stryker, J. Drip irrigation design guidelines. http://www.irrigationtutorials.com/dripguide. htm

Subsurface Drip Irrigation Systesms, 2000. Geoflow surface drip systems. http://www.geoflow. com/landscape%20(inclu%20golf)/design_general.htm

Szolosi, 2005. *Drip/Trickle.* Irrig8right. Pratt Water Pty Ltd. http://www.irrig8right.com.au/Irrigation_Methods/Micro_Irrigation/Drip/Drip_Trickle/Details_DR.htm

# 14

## Literature cited

1. ASAE Standards, 1996. Field evaluation of microirrigation systems. ASAE EP405.1 and EP458. St. Joseph, MI: American Society of Agricultural Engineers,Pages 1–7.

2. Smajstrla A.G., B.J. Boman, D.Z. Haman, D.J. Pitts and F.S. Zazueta, 2002. Field evaluation of microirrigation water application uniformity. Cooperative Extension Service, Department of Agricultural Engineering, Gainesville, FL: University of florida.

3. Bralts, V.G. and C.D. Keesme, 1982. Drip irrigation field Uniformity estimation. Paper No. 82-2062 Presented at Summer meeting of American Society of Agricultural Engineers.

## Books/bulletins/journal and proceedings/reports

Ben-Gal, A., N. Lazorovitch and U. Shani, 2004. Drip irrigation in gravel-filled cavities. *Vadose Zone Journal,* 3(4), 1407–1413.

Assouline, S., S. Cohen, D. Meerbach, T. Harodi and M. Rosner, 2002. Micro drip irrigation of field crops: Effect on yield, water uptake, and drainage in sweet corn. Soil Science Society Of American Jornal, 66: 228–235.

Burt, C.M., A. Clemmens, R. Bliesner and L. Hardy, 2000. Selection of irrigation methods for agriculture. American Society of Civil Engineers (ASCE), USA. Chapter 1: 1–26.

Wang, D., M.C. Shannon, C.M. Grieve, P.J. Shouse and D.L. Suarez. Ion Partitioning among Soil and Plant Components under Drip, Furrow, and Sprinkler Irrigation Regimes: Field and Modeling Assessments. *Journal of Environmental Quality,* 31(5), 1684–1693.

Heldman R.H., 2003. *Encyclopedia of agricultural, food and biological engineering: Drip irrigation.* New York: Marcel Dekker. Pages 206–211

Heuvelink, E., 2005. *Tomatoes: Irrigation.* The Netherlands: Wageningen University. Page 171.

Ozay, M. and H.A. Bicak, 2002. Modern and traditional irrigation technologies in the eastern mediterranean. Chapter 9, Pages 90–97.

Punmia, B.C. and B.B.L. Pande, 2005. *Irrigation and water power enginering.* New Delhi: Laxmin Publications (P) LTD. Page 89.

Rossi, G., A. Cancelliere and L.S. Pereira, 2003. *Tools for drought mitigation in mediterranean regions.* Netherlands: Kluwer Academic Publisher. Part III: 162–259.

Shock, C.C., E. Feibert, L. Saunders and E.P. Eldredge, 2002. Water management, allium cepa, onion, solanum tuberosum, potato, watermark, granular matrix sensor, soil water potential, sdi. World Congress of Computers in Agriculture and Natural Resources. Iguacu Falls Brazil, Pages 13–15.

## Web page links

Boman, B. and S.Shukla, 2003. Hydraulic considerations for citrus micro irrigation system. University of Florida, Page 1. http://edis.ifas.ufl. edu/CH156

Hla, A.K. and T.F. Scherer, 2003. Introduction to micro-irrigation. North Dakota State University. NDSU Extension Service. Page 1. http://www. ext.nodak.edu/extpubs/ageng/irrigate/ae1243w. htm

Kizer, M.A. Drip (trickle) irrigation systems. Oklahoma Cooperative Extension Services. Pages 1–4.

http://osuextra.okstate.edu/pdfs/F-1511web.pdf

Smajstrla, A.G., B.J. Boman, D.Z. Haman, D.J. Pitts and F.S. Zazueta, 2006. Field evaluation of microirrigation water application uniformity. University of Florida, Page 1. http://edis.ifas.ufl. edu/AE094

Rain Bird, 2006. Distribution uniformity for sprinkler irrigation. Rain Bird Corporation, Page 1. http://www.rainbird.com/ag/du.htm

Anonymous, 2001. Irrigation system evaluation. Westland Water District. Pages 7–8. http://www.westlandswater.org/wtrcon/handbook/eval6.htm

Emerson, M., 2000. Irrigation techniques: Micro systems save water and labor. Coachella Valley Water District. Page 1. http://www.cvwd.org/lush&eff/lsh&ef7.htm

Qassim, A., 2003. Sprinkler irrigation. Institute of Sustainable Irrigated Agriculture (ISIA). Pages 1–4. http://www.wca-infonet.org

Stryker, J., 2003. Landscape sprinkler design tutorial: Sprinkler spacing. Jess Stryker's Irrigation Tutorials. Pages 15. http://www.irrigationtutorials.com/sprinkler15.htm

Stryker, J., 2003. Landscape sprinkler design tutorial: Place sprinkler heads. Jess Strykers Irrigation Tutorials. Page 17. http://www.irrigationtutorials.com/sprinkler15.htm

Bloomer, D., 2005. Evaluating drip irrigation systems. Web Page Bloomer Associates. Pages 1–2. http://www.nzwine.com/assets//evaluating_irrigation_systems.pdf

# 15

## Literature cited

1. Crespo Ruiz, M., M.R. Goyal, C. Chao de Baez and L.E. Rivera, 1988. Nutrient uptake and growth characteristics of nitrogen fertigated sweet peppers under drip irrigation and plastic mulch. *Journal of Agriculture of the University of Puerto Rico*, 72(4): 575–584.

2. Davies, J.W., 1975. *Mulching effects on plant climate and yield.* Technical Note no. 136. Secretariat of the World Meteorological Organization, Geneva-Switzerland. Pages 1–92.

3. Goyal, M.R., 1983. Labor input requirements for experimental production of summer peppers under drip irrigation. *Journal of Agriculture of the University of Puerto Rico*, 67(1): 22–27.

4. Goyal, M.R., 1987. Title and abstracts, "*Trickle irrigation in humid regions: Puerto Rico*". Agriculture Experiment Station, Río Piedras. Pages 1–20.

5. Goyal, M.R., R.G. Luna, E. Recio de Hernández and L.E. Rivera, 1984. Effects of

plastic mulch types on sizing and crop performance of drip irrigation winter and summer peppers. *Journal of Agriculture of the University of Puerto Rico*, 68(3): 297–305.

6. Goyal, M.R., T. Persaud and L.E. Rivera, 1988. Labor input requirements for experimental production of drip irrigated vegetables. *Journal of Agriculture of the University of Puerto Rico*, 71(4): 349–358.

7. Goyal, M.R. and L.E. Rivera, 1985. *Trickle irrigation scheduling in vegetables.* In: Drip/Trickle Irrigation in Action. Third Internacional Drip/Trickle Congreso by American Society of Agricultural Engineers at Fresno-CA. Vol. II. Paper No. M-7.

8. Goyal, M.R., C.L. Santiago and C. Chao, 1984. How plastic mulch types affect growth paramenters of drip irrigated summer peppers. *Journal of Agriculture of the University of Puerto Rico*, 68(4): 365–373.

9. Hopen, H.J. and N.F. Oebkar, 1976. *Vegetable crop responses to synthetic mulches: An annotated bibliography.* National Agricultural Plastics Association Tech. Bull. #1, U.S.A.

10. Liu, L.C., M.A. Padilla, M.R. Goyal and J.G. Ibanez, 1987. Integrated weed management in transplanted tomatoes and peppers under drip irrigation. *Journal of Agriculture of the University of Puerto Rico*, 71(4): 349–58.

11. Rivera, L.E. and M.R. Goyal, 1985. Mulch types on soil temperature at varying depths of drip irrigated summer and winter peppers. *Journal of Agriculture of the University of Puerto Rico*, 69(1): 121–123.

12. Rivera L.E. and M.R. Goyal, 1986. Mulch types for soil moisture retention in drip irrigated summer and winter peppers. *Journal of Agriculture of the University of Puerto Rico*, 70(4): 63–68.

13. Santiago, C.L. and M.R. Goyal, 1985. Nutrient uptake and solute movement in drip irrigated summer peppers. *Journal of Agriculture of the University of Puerto Rico*, 69(1): 63–68.

14. Waggoner, P.E., P.M. Miller and H.C. Rao, 1960. *Plastic mulching: Principles and benefits.* Connecticut Agriculture Experiment Station, New Haven, CT. Bulletin 643.

**Books/bulletins/journal and proceedings/reports**

Blanco-Canqui, H., R. Lal, W.M. Post and L.B. Owens, 2006. Changes in long-term no-till corn growth and yield under different rates of stover mulch. *Agronomy Journal*, 98(1): 1128–1136.

Books, S., 2000. *Mulch it! A practical guide to using mulch in the garden and landscape.* Storey Publishing. Pages 1–128.

Bresbin Publications, Inc. 1962. The fantastic business of plastic plants and flowers. *Modern Plastics*, 39: 94–97, 205–206.

Buclon, F., 1966. Comparisons of agricultural uses of plastics in France, Italy, Japan, Russia and the United States. *Proceedings of National Agriciulture Plastics Congress*, 7: 21–33.

Caamal-Maldonado, J.A., J.J. Jiménez-Osornio, A. Torres-Barragán and A.L. Anaya, 2001. The use of allelopathic legume cover and mulch species for weed control in cropping system. *Agronomy Journal*, 93(1): 27–36.

Cornell, J.T., 1989. The recycling of plastic in agriculture. *Proceedings of National Agriciulture Plastics Congress*, 21: 60–64.

Dalrymple, D.G, 1973. *A global review of greenhouse food production.* USDA Report 89.

Decoteau, D.R., M.J. Kasperbaur and P.G. Hunt, 1989. Mulch surface color affects yield of fresh-market tomatoes. *Jounal of American Society for Horticultural Science*, 114: 216–219.

Duppong, L.M., K. Delate, M. Liebman, R. Horton, F. Romero, G. Kraus, J. Petrich and P.K. Chowdbury, 2004. The effect of natural mulches on crop performance, weed suppression and biochemical constituents of catnip and St. John,s Wort. *Crop Science Journal*, 44(3): 861–869.

Hall, B.J., 1971. Comparisson of drip and furrow irrigation for market tomatoes. *Proceedings of National Agriciulture Plastics Congress*, 10: 19–27.

Hair, M., L. Coit and T. Boland, 2001. *Michigan gardener's guide,* Revised ed. Cool Springs Press. Pages 1–272.

Hall, B.J., 1963. Continous polyethylene tube covers for cucumbers. *Proceedings of National Agriciulture Plastics Congress* 4: 112–132.

Harper-Lore B. and M. Wilson, 2000. *Roadside use of native plants.* Island Press. Pages 1–665.

Hillel, D., 2003. *Introduction to environmental soil physics.* Academic Press. Pages 1–494.

Hodges, L. and J.R. Brandle, 1996. Windbreak: An important component in a plasticulture system. *Horticulture Technology*, 6: 177–181.

Hopen, J.H. and N.F. Oebker, 1976. *Vegetable crop responses to synthetic mulches.* Special publication # 42. University of Illinois.

Huang, Z, 1989. *The research and application of plastic films in China.* Chinese Plastics Mulch Research Association.

Jensen, M.H. and A.J. Malter, 1994. *Protected agriculture—A global review.* World Bank technical paper no. 253. The World Bank, Washington, DC.

Jensen, M.H., 1965. Concluding results of air-supported row covers for early vegetable production. *Proceedings of National Agriciulture Plastics Congress*, 6: 100–112.

Jensen, M.H., 1967. A new approach to high yields. *American Vegetable Grower*, 15(2): 16, 23–24.

Jensen, M.H., M. Valenzuela and D.D. Fangmeier, 1998. Using non-woven floating covers on summer squash for exclusion of whitefly—transmited gemini viruses. *Proceedings of National Agriciulture Plastics Congress*, 27: 159–164.

Ji, S. and P.W. Unger, 2001. Soil water accumulation under different precipitation, potential evaporation and straw mulch conditions. *Soil Science Society of America Journal*, 65(2): 442–448.

Jiménez, C.C., M. Tejedor, F. Diaz and C.M. Rodríguez, 2005. Effectiveness of sand mulch in soil and water conservation in an arid region, Lanzarote, Canary Islands, Spain. *Journal of Soil and Water Conservation*, 60(1): 63–67.

Kasperbauer, M.J, 2000. Strawberry yield over red vs. black plastic mulch. *Crop Science Journal*, 40(1): 171–174.

Laflen, J.M., J. Tian and C.H. Huang, 2000. *Soil erosion and dryland farming.* CRC Press. Pages 1–744.

Lamont, W.J., K.A. Sorensen and C.W. Averre, 1990. Painting aluminum strips on black plastic mulch reduces mosaic symptoms on summer squash. *Horticulture Science*, 25: 1305.

Larson, R.A., 1993. Impact of plastics in the floriculture industry *Horticulture Technology*, 3: 28–34.

Loy, B., J. Lindstrom, S. Gordon, D. Rudd and O. Wells, 1989. Theory and development of wavelength selective mulches. *Proceedings of National Agriulture Plastics Congress*, 21: 193–197.

Loy, J.B. and. O.S. Wells, 1982. A comparison of slitted polyethylene and spunbonded polyester for plant row covers. *Horticulture Science* 17: 405–407.

Ma, L. and H.M. Selim, 2005. Pesticide transport in mulch amended soils: A two-compartment model. *Soil Science Society of America Journal*, 69(2): 318–327.

Mansour, N.S., 1991. *The use of field covers in vegetable production. Proceedings of International Workshops on Implied Vegetable Production Through the use of fertertilizers*, Mulching and Irrigation, Chaing Mai University, Thailand.

Martius, C., H. Tiessen and P.L.G. Vlek, 2001. *Managing organic matter in tropical soils: Scope and (developments in plant and soil sciences)*. Springer Press. Pages 1–248.

Morrison, C.T., W.T. Green and P. Hadley, 1989. Energy exchange by plastic row covers. *Proceedings of National Agriulture Plastics Congress*, 21: 269–275.

Orzolek, M.D. and J.H. Murphy, 1993. The effect of colored polyethylene mulch on the yield of squash and pepper. *Proceedings of National Agriulture Plastics Congress*, 24: 157–161.

Reich, L., 2001. *Weedless gardening.* Workman Publishing Company. Pages 1–176.

Rice, P.J., L.L. McConnell, L.P. Heighton, A.M. Sadeghi, A.R. Isensee, J.R. Teasdale, A.A. Abdul-Baki, J.A. Harman-Fetcho and C.J. Hapeman, 2001. Runoff loss of pesticides and soil: A comparison between vegetative mulch and plastic mulch in vegetable production **systems.** *Journal of Environmental Quality*, 30(5): 1808–1821.

Robert, W.J.,C. Simpkins and P. Kendall, 1976. Using solar energy to heat plastic film greenhouses. *Proceedings of Solar Energy Fuel-Food Workshop.* University of Arizona, Tucson. Pages 142–159.

Robert, W.J. and D.R. Mears, 1969. Double covering a film greenhouse using air to separate film layers. *Transactions of ASAE*, 12: 32–33, 38.

Roose, E.J., R. Lal, C. Feller, B. Barthes and B.A. Stewart, 2005. *Soil erosion and carbon dynamics (advances in soil science)*. CRC Press. Pages 1–376.

Schabenberger, O. and F.J. Pierce, 2001. *Contemporary statistical models for the plant and soil sciences*. CRC Press. Pages 1–738.

Selim, H.M., L. Zhou and H. Zhu, 2003. Herbicide retention in soil as affected by sugarcane mulch residue. *Journal of Environmental Quality*, 32(4): 1445–1454.

Sheldrake, R. and R. Langhans, 1961. Heating study with plastic greenhouses. *Proceedings of National Horticultural Plastics Congress*, 2: 16–17.

Takakura, T., 1988. Protected cultivation in Japan. *Symposium on High Technology in Protected Cultural Acta Horticultural*, 230: 29–37.

Thompson, H.C. and W.C. Kelly, 1957. *Vegetable corps*. 5th ed. New York: McGraw-Hill Book Co, Inc. Chapter 7: Pages 86–106.

Tejedor M., C.C. Jiménez and F. Díaz, 2002. Soil moisture regime changes in tephra-mulched soils: Implications for soil taxonomy. *Soil Science Society of America Journal*, 66(1): 202–206.

Van der Meulen, E.S., L. Nol and L.H. Cammeraat, 2006. Effects of irrigation and plastic mulch on soil properties on semiarid abandoned fields. *Soil Science Society of America Journal*, 70(2): 930–939.

Warrick, A.W., 2001. *Soil physics companion.* CRC Press. Pages 1–400.

Witter, S.H., 1993. World-wide use of plastic in horticultural production. *Horticulture Technology*, 3: 6–27.

Wittwer, S.H. and N. Castilla, 1995. Protected cultivation of horticultural crops worldwide. *Horticulture Technology*, 5: 6–23.

**Web page links**

Advantages and disadvantages of mulch. http://www.virtualsciencefair.org/2004/kwol4s0/public_html/adv.dis.htm

American society for plasticulture. www.plasticulture.com

Anonymous, 2003. About plastics. American Plastics Council. http://www.americanplasticscouncil.org/benefits/about_plastics/history.html

Anonymous, 2006. Plastic mulch. Wikimedia foundation, Inc. http://en.wikipedia.org/wiki/Plastic_mulch

Bramlage, Georgene A., 2006. Advantages and disadvanteges of plastic nulch. http://landscaping.suite101.com/article.cfm/landscape_fabric_pros_co

Durham, S., 2003. Plastic mulch: Harmful or helpful? Agricultural Research Service. http://www.highbeam.com/library/docFree.asp?DOCID=1G1:105480610

Garthe, J.W., 2002. Used agricultural plastic mulch as a supplemental boiler fuel: An overview of combustion test results for public dissemination. Energy Institute. Penn State University. Pages 1–7.

http://environmentalrisk.cornell.edu/C&ER/PlasticsDisposal/AgPlasticsRecycling/References/Garthe2002b.pdf

Hawaii Cooperative Extension Service, 1996. Vegetable crops update. Vol. 6 No. 4. http://www2.hawaii.edu/~hector/vegcropupdate/1996/August96.html

Jensen, M.H. Plasticulture in the global community—View of the past and future. http://www.plasticulture.org/history_global_community.htm

LaLiberte, K., 2006. The right mulch makes a difference. http://www.gardeners.com/The-Right-Mulch-Makes-a-Difference/default/5013.page?SC=

Lamont, W.J., 1991. The use of plastic mulches for vegetable production. Food and Fertilizer Technology Center. Kansas State University. www.agnet.org/library/article/eb333.html

Marr, C.W., 2006. Plastic mulches for vegetables. Kansas State University Agricultural Experiment Station and Cooperative Extension Service. Pages 1–4. http://www.hfrr.kstate.edu/DesktopModules/ViewDocument.aspx?DocumentID=1093#search=%22plastic%20mulch%22

Marvel, R., 2006. Plastic mulch. http://www.robertmarvel.com/Plastic_Mulch.html

Masiunas, J.B., 2003. Weed management in fruit and vegetable crops. University of Illinois. http://www.nres.uiuc.edu/research/r-masiunas.html

Miles, C., G. Becker, K. Kolker, C. Adams, J. Nickel and M. Nicholson, 2004. Alternatives to plastic mulch for organic vegetable production. Washington States University. Pages 1–14. http://agsyst.wsu.edu/MulchReport04.pdf

Miles, C., L. Garth, M. Sonde and M. Nicholson, 2003. Searching for alternatives to plastic mulch. Washington States University. Pages 1–7. http://agsyst.wsu.edu/MulchReport03.pdf

Mulching benefits/organic and inorganic mulch types. http:/www.gardenstew.com/blog/e3-15-mulching-benefits—organic-and-inorganic-mulch-types.html

Pons, L., 2003. More than meets the eye: New findings on how mulch color can affect food plants. Agricultural Research Service. http://www.highbeam.com/library/docFree.asp?DOCID=1G1:108550679

Sanders, D.C., 2001. Using plastic mulches and drip irrigation for vegetable production. North Carolina State University Horticulture Information Leaflets. Pages 1–6. http://www.ces.ncsu.edu/depts/hort/hil/hil-33.html

Sanders, D.C., 2001. Using plastic mulches and drip irrigation for vegetables production. www.ces.ncsu.edu/depts/hort/hil/hil-8033.html

Steinegger, D. H. and A. Greving, 2000. Mulches. University of Rhode Island Landscape Horticulture Program. Pages 1–6. http://www.uri.edu/ce/factsheets/sheets/mulch.html

Sweat, M.S. Plasticulture technology for vegetable production. http://nfrec-sv.ifas.ufl.edu/mulch.htm

Wall, T.E. and E., Maynard, 2006. Plastic mulch and weed control. Piketon Research and Extension Center, Ohio State University. http://southcenters.osu.edu/hort/plastic.htm

Waterer, D. and J. Bantle, 2005. Microclimate modification with plasticulture. University of Saskatchewan, Dept. of Horticulture Science. http://www1.agric.gov.ab.ca/$department/deptdocs.nsf/all/faq8266?opendocument

# 16

## Literature cited

1. Doorenbos, J. and W.O. Pruitt, 1999. *Crop water requirements*. FAO Irrigation and Drainage Paper 24, Rome-Italy.

2. Jensen, M.N., 1980. *Design and operation of farm irrigation systems*. Monograph # 3. St. Joseph, MI: American Society of Agricultural Engineers.

3. Withers, B. and S. Vipond, 1980. *Irrigation design and practice*. New York: Cornell University. Pages 21–33.

## Books/bulletins/journal and proceedings/reports

2001. Agroeconomic analyses of drip Irrigation for sugarbeet production. *Agronomy Journal*, 93(3) 517–523.

Assouline, S., 2002. The effects of microdrip and conventional drip irrigation on water distribution and uptake. *Soil Science Society of America Journal*, 66: 1630–1636.

Broner, I., 2002. *Micro-irrigation for orchard and row crops*. University Cooperative Extension. Colorado, USA. Pages 3–7.

Burt, C.M., 2000. *Selection of irrigation methods for agriculture: Drip/micro irrigation*. California, CA: Irrigation Training and Research Center (ITRC). Pages 1–6.

Cassel, F., 2001. Agroeconomic analyses of drip Irrigation for sugarbeet production. *Agronomy Journal*, 93: 517–523.

Coelho, E.F., 1999. Root distribution and water uptake patterns of corn under surface and subsurface drip irrigation. *Plant Soil Journal*, 26: 123–136.

Environ, J., 2006. Analysis of irrigation systems using sustainability-related criteria. *Plant Soil Journal*, 30: 1150–1153.

Escoe, K., 2006. *Piping and pipelines assessment guide*. Elsevier Gulf Professional Publishing.

Experimental analysis of local pressure losses for microirrigation laterals. *Journal of Irrigation and Drainage Engineering*, 130(4): 318–324. July 2.

Gilley, J.R. and R.J. Supalla, 1999. Economic analysis of energy saving practices in irrigation. *Transactions of ASAE*, 26: 1784–1792.

Hansonn, F. and J. Martin, 2002 Drip irrigation of row crops: What is the state of the art? *Proceedings of the 4th Decennial Irrigation Symposium*, ASAE. Pages 4–20.

Heldman, R., 2003. *Encyclopedia of agricultural, food and biological engineering*. San Luis Obispo-CA: California Polytechnic State University. Pages 4–20.

Keller, J. and R.D. Bliesner, 2004. *Sprinkler and trickle irrigation*. New York, U.S.A: Van Nostrand Reinhold Publishers. Pages 28–40.

Knox, J.W. and E.K. Weatherhead, 2003. *Trickle irrigation in England and Wales*. Bristol-England: R&D Technical Report W6-070/TR Environment Agency.

Lascano, R.J., 2000. A general system to measure and calculate daily crop water use. *Agronomy Journal*, 92: 821–832.

Manoliadis, O., 2006. Analysis of irrigation systems using sustainability-related criteria. *Environment Journal*, 30: 1150–1153.

Meiri, A. and J. Letey, 2003. Evaluation of a model for irrigation management under saline conditions: I. Effects on plant growth. *Soil Science Society of America Journal*, 67: 71–76.

Melby, P., 2003. *Simplified irrigation design*. Van Nostrand Reinhold Publishers. Pages 14–30

Mergos, G.J., 2000. Sustainability issues and technology choice in irrigation investment. *Water Resource Management Journal*, 5: 244–251.

Moynihan, M.J., and D.Z. Haman, 2001. Microirrigation systems for small-scale farms in the Rio Cobre basin area of Jamaica. *Applied Engineering in Agriculture Journal*, 8: 617–623.

2002 (September/October). *New strategy for optimizing water application under trickle irrigation*. Grounds Magazine, 128(5): 287–297.

Pair, C.H., 2003. *Irrigation*. 5th ed. Irrigation Association. Pages 10–45.

Rosner, M., S. Assouline, S. Cohen, D. Meerbach and T. Harodi, 2002. Microdrip irrigation of field crops: Effect on yield, water uptake, and drainage in sweet corn. *Soil Science Society of America Journal*, 66: 228–235.

Santana, J. M., 2003. *Patrones de Humedecimiento bajo el sistema de micro riego subterráneo*. Report. Caribbean Atmospheric Research Center, University of Puerto Rico at Mayagüez.

Schwankl, L. and T. Prichard, 2001. *Chemigation in tree and vine micro irrigation systems.* Agriculture and Natural Resources Publication #21599. Oregon State University.

Shock, C.C., E.B. Flock and R.J. Feibert, 2005. *Irrigation monitoring using soil water tension.* University Extension Service. EM 8900. Oregon State. Pages 1–6.

Thomas, F. 2005. *Design and control of sprinkler systems for crop disease research.* American Society of Agricultural and Biological Engineers, St. Joseph, Michigan. Vol. 3.

Thomas, L., T.A. Doerge and R.E. Godin, 2002. Subsurface drip irrigation and fertigation of broccoli: II Agronomic, economic, and environmental outcomes. *Soil Science Society of America Journal*, 66: 178–185.

Wopereis, M., 2004. Agro-Economic characterization of rice production in a typical irrigation scheme in Burkina Faso. *Agronomy Journal*, 96: 1314–1322.

**Web page links**

Bellows, B., 2004. Irrigation techniques. http://www.attra.org/downloads/water_quality/irrigation.pdf

Bertauski, T., 2001. How to: Install drip irrigation in new plantings. Maintenance and trouble shooting tutorials—The drip store. Trident Technical College. http://www.dripirrigation.com/drip_tutorial.php?page_view=head

Brouwer, C. Drip irrigation. Chapter VI In: *Surface Irrigation Management.* http://www.fao.org/docrep/S8684E/s8684e07.htm

Burt, C.M and S.W. Styles, 2004. Conceptualizing irrigation modernization through benchmarking and the rapid appraisal process In: "*Irrigation and Drainage*". ITRC Paper No. P 03-002. http://www.interscience.Wiley.com

Drip micro-sprays information and drip irrigation education, 2004. www.plumbingsupply.com/ed-micro.html

Dugan, K., 2005. Small farms: Crops. www.smallfarms.wsu.edu/crops/dripIrrigation.html

FRN., 2004. Supply water directly to plant roosts with pitcher and drip irrigation. http://www.farmradio.org/english/radio-scripts/71-10script_en.asp

Lamm, F., 2003. Drip irrigation laterals on center pivot irrigation. http://www.oznet.k-state.edu/irrigate/MDI.htm

LearnAbout: Garden irrigation, 2006. www.irrigation.learnabout.info/

Reaves, E., 1999. Latest farm and ranch survey reveals pulse of irrigation. http://www.irrigation.org/ibt/_9911/p57.html

Rogers, D., 1999. Maintaining drip irrigation systems. http://www.wcainfonet.org/servlet/CDSServlet?status=ND0xMjY1LjEyNjE1JjY9ZW4mMzM9ZG9jdW1lbnRzJjM3PWluZm8~

Sanders, D.C., 2006. Drip or trickle irrigation systems: An outline of components. http://www.ces.ncsu.edu/depts/hort/hil/hil-33-a.html

Scherer, T.F. and A.K. Hla, 2003. Introduction to micro-irrigation. http://www.ext.nodak.edu/extpubs/ageng/irrigate/ae1243w.htm

Shock, C.C, 2006. Drip irrigation: An introduction. http://www.cropinfo.net/drip.htm

Shortt, R., 2005. Drip irrigation—How much is enough? www.omafra.gov.on.ca/english/crops/hort/news/vegnews/2005/vg0805a3.htm

Sideman, E., 2005. Drip, drip, drip-is it better than a downpour. http://www.mofga.org/mofga/other/mofgj05es.html

Smajstrla, A.G. and D.Z. Haman, 2003. Design tips for drip irrigation of vegetables. http://edis.ifas.ufl.edu/AE093

Strauss, A., F. Zee, K. Hummer, W. Nishijima, R. Kai, M. Yamasaki and R. T. Ramasaki, 2006. Preliminary yields of southern highbush blueberry in Waimea, Hawaii. http://www.ctahr.hawaii.edu/oc/freepubs/pdf/F_N-12.pdf

Stryker, J., 2005. Drip irrigation design guidelines. http://www.irrigationtutorials.com/dripguide.htm

Stryker, J., 2005. How to install a landscape sprinkler system or drip irrigation systems. www.irrigationtutorials.com/install.htm

USGS., 2005. Irrigation techniques. http://ga.water.usgs.gov/edu/irmethods.html

WWD., 2001. Irrigation system evaluation. http://www.westlandswater.org/wtrcon/handbook/eval6.htm

# Index